Nitrogen, phosphorus and sulphur utilization by fungi

Nitrogen, phosphorus and sulphur utilization by fungi

Symposium of the British Mycological Society held at the University of Birmingham April 1988

Edited by

Lynne Boddy, R. Marchant & D. J. Read

Published for The British Mycological Society by
CAMBRIDGE UNIVERSITY PRESS
Cambridge
New York New Rochelle Melbourne Sydney

CAMBRIDGE UNIVERSITY PRESS
Cambridge, New York, Melbourne, Madrid, Cape Town,
Singapore, São Paulo, Delhi, Tokyo, Mexico City

Cambridge University Press
The Edinburgh Building, Cambridge CB2 8RU, UK

Published in the United States of America by Cambridge University Press, New York

www.cambridge.org
Information on this title: www.cambridge.org/9780521106245

First published 1989
First paperback edition 2011

A catalogue record for this publication is available from the British Library

ISBN 978-0-521-37405-7 Hardback
ISBN 978-0-521-10624-5 Paperback

Contents

Contributors

Lynne Boddy, *School of Pure and Applied Biology, University of Wales, College of Cardiff, PO Box 917, Cardiff CF2 1XH, UK.*

M. D. Coffey, *Department of Plant Pathology, University of California at Riverside, CA 92521-0122, USA.*

P. D. Crittenden, *Department of Botany, University of Nottingham, Nottingham NG7 2RD, UK.*

J. Dighton, *Institute of Terrestrial Ecology, Merlewood Research Station, Grange-over-Sands, Cumbria LA11 6JU, UK.*

S. Gianinazzi, *Laboratoire de Phytoparasitologie, Station d'Amélioration des Plantes, INRA, BV 1540, 21034 DIJON CEDEX, France.*

V. Gianinazzi-Pearson, *Laboratoire de Phytoparasitologie, Station d'Amélioration des Plantes, INRA, BV 1540, 21034 DIJON CEDEX, France.*

B. A. D. Hetrick, *Department of Plant Pathology, Kansas State University, Throckmorton Hall, Manhatten, Kansas 665016, USA.*

D. H. Jennings, *Department of Genetics and Microbiology, University of Liverpool, P.O. Box 147, Liverpool L69 3BX, UK.*

A. R. Langdale, *Department of Plant Sciences, University of Sheffield, Sheffield S10 2TN, UK.*

J. R. Leake, *Department of Plant Sciences, University of Sheffield, Sheffield S10 2TN, UK.*

R. Marchant, *Department of Biological and Biomedical Sciences, University of Ulster, Coleraine BT52 1SA, Northern Ireland, UK.*

D. G. Ouimette, *Department of Plant Pathology, University of California at Riverside, CA 92521-0122, USA.*

N. D. Paul, *Institute of Environmental and Biological Sciences, University of Lancaster, Baillrigg, Lancaster LA1 4YQ, UK.*

D. J. Read, *Department of Plant Sciences, University of Sheffield, Sheffield S10 2TN, UK.*

Anthony H. Rose, *Zymology Laboratory, School of Biological Sciences, University of Bath, Bath BA2 7AY, Avon, UK.*

J. C. Slaughter, *Department of Brewing and Biological Sciences, Heriot Watt University, Edinburgh EH1 1HX, UK.*

A. B. Tomsett, *Department of Genetics and Microbiology, The University of Liverpool, P.0. Box 147, Liverpool L69 3BX, UK.*

M. Wainwright, *Department of Microbiology, University of Sheffield, Sheffield S10 2TN, UK.*

D. R. Walters, *Department of Plant Sciences, West of Scotland College, Auchincruive, Nr Ayr KA6 5HW, UK.*

Preface

Nitrogen, phosphorus and sulphur are the major mineral nutrient elements in the biosphere and are essential for growth of both fungi and autotrophic plants. In terrestrial ecosystems fungi play a central role in the release of these elements from organic residues and therefore they indirectly influence not only plant nutrition but also primary productivity. Also, fungi which interact with autotrophs as mutualists or parasites directly influence the availability of these nutrients for the plant. Thus, understanding nutrient relationships of fungi and of fungus-plant interactions is both important and rewarding and these subjects have attracted a great deal of attention in recent years. This volume contains key papers from the 6th General Meeting of the British Mycological Society on the subject of *Nitrogen, phosphorus and sulphur metabolism*. The book is broadly based, covering four main areas: physiology and metabolism of nitrogen, phosphorus and sulphur in fungi (Chapters 1 to 6); the role of these minerals in pathogenic relationships with plants (Chapters 7 & 8); their role in mutualistic relationships with plants (Chapters 9 to 12); and the role of fungi in cycling of these elements within ecosystems (Chapter 13).

The literature concerning the physiology and biochemistry of nitrogen and phosphorus metabolism in fungi is extensive and in the first chapter Jennings has synthesized some of the major features while presenting linking hypotheses and suggesting areas for future research. The theme of nitrogen metabolism is continued in the following chapter in which Tomsett provides an account of the genetics and biochemistry of nitrate assimilation, particular emphasis being placed on ascomycetes. The utilization of inorganic sulphur is considered in Chapters 3 to 5. In the chapters by Rose and Slaughter it is apparent that much past work has been concentrated on hemiascomycetous yeasts, presumably because of their importance in the food and brewing industry, and the need to extend such studies to mycelial fungi is stressed. In contrast, inorganic sulphur oxidation has been investigated almost exclusively in mycelial fungi. Again, however, much further work is needed before such processes can be fully understood. As Wainwright points out in his chapter (4) on the utilization of reduced sulphur compounds, it is still unclear what 'benefits' the fungus obtains from the ability to metabolise such compounds - the processes do not appear to be favourable in energetic terms and it is not clear whether they are of quantitative significance in sulphur cycling within ecosystems. These three chapters therefore highlight the opportunities for further research on fungal sulphur metabolism.

The sixth chapter deals with the metabolism of phosphorus compounds, but covers the fascinating and much neglected subject of the role of phosphorus compounds as antifungal agents. Here Coffey & Ouimette examine the effect of phosphonates on Oomycetes and give some indication of how knowledge of fungal metabolism may be put to practical use.

To complement the first six chapters, which deal with the way in which fungi metabolise nitrogen, phosphorus and sulphur compounds, the following six chapters examine the processes involved in capture and transport of these nutrients by fungi growing in association with living plants, and consider the consequences of these processes for the plant. Walters (Chapter 7) and Paul (Chapter 8), in complementary chapters, consider pathogenic associations, and since little is known about necrotrophic and hemibiotrophic infections emphasis is placed on biotrophic infections which are better understood. Taken together, they provide a picture of nutrient fluxes between host plant and fungus and changes in location, content and concentrations of nitrogen, phosphorus and sulphur within whole plants and plant tissues when infected by biotrophic fungi.

The next three papers concern mutualistic relationships. It is now recognised that the roots of the majority of land plants grow in mutualistic association with fungi to form mycorrhizas. Because they receive a regular supply of carbon from their hosts the fungi involved are amongst the most active components of the soil microbial community. Their role in the nitrogen and phosphorus nutrition of the autotroph has been extensively studied but the contributions presented here (Chapters 9 to 11) emphasize recent developments in our understanding of the biology of mycorrhizas, some of which challenge conventional notions of their function. Plant ecologists increasingly recognize the overwhelming importance of nitrogen as the factor determining productivity of ecosystems, and new information presented by Read and his colleagues (Chapter 9) suggests that ericoid and some ectomycorrhizal fungi may be directly involved in the mobilization, assimilation and transport of organic nitrogen compounds. The role of mycorrhizas in the phosphorus nutrition of plants is well documented but the contributions of Hetrick (Chapter 10) and Gianinazzi-Pearson & Gianinazzi (Chapter 11) present new insights and syntheses, the latter relating in particular to biochemical and ultrastructural features of mycorrhizas in general and the former to whole plant aspects of the vesicular-arbuscular mycorrhizal symbiosis.

The other type of mutualistic association involving fungi and plants is the lichen symbiosis. As Crittenden explains in Chapter 12, knowledge of nitrogen, phosphorus and sulphur relationships in lichens is comparatively rudimentary, partly because lichenology has not benefitted from the same economic impetus as have studies of other fungal systems, but also because of the experimental difficulties imposed by lichens. Crittenden's review of current knowledge of the nitrogen relations of mat-forming lichens indicates the boundless opportunities for future research. Nonetheless, he emphasizes that a number of recent advances have been made and illustrates some experimental approaches which may help further to elucidate these relations.

In addition to pathogenic and mutualistic associations, fungi are also important as decomposers and as such play a crucial role in nutrient cycling within ecosystems. Unfortunately, although detailed nutrient budgets have been obtained for a number of ecosystems, the specific contribution of fungi to these budgets has received relatively little attention. In the final chapter Dighton & Boddy attempt to illustrate the importance of fungi in nitrogen, phosphorus and sulphur cycling in temperate forests, and to point to the many gaps which still exist in our knowledge of their role at the ecosystem level.

We are grateful to all those who contributed to the symposium, and especially to Dr Chris Caten whose skillful organization of the domestic arrangements was a major contribution to the success of the meeting. The Society wishes to thank the British Council and the Royal Society of London for donations in support of speakers visiting from outside the UK.

Lynne Boddy, University of Wales, College of Cardiff
Roger Marchant, University of Ulster
David Read, University of Sheffield

Chapter 1

Some perspectives on nitrogen and phosphorus metabolism in fungi

D. H. Jennings

Department of Genetics and Microbiology, University of Liverpool, P.O. Box 147, Liverpool L69 3BX, UK

Introduction

Nitrogen and phosphorus are two of the major elements in fungal nutrition. It is tempting to consider initially the ecological role of these two elements because they can exist in many chemical forms in natural habitats. Thus, the availability of the various forms might be expected to have very considerable selective influence on the presence of particular species within a habitat. However, I have argued (Jennings, 1987a) that minor elements may be equally if not more important in determining fungal niche. For that reason I will initially consider the relationship between fungi and the two elements from the metabolic and physiological standpoint. This consideration does then engender some thoughts which are relevant to fungal ecology.

There is a large body of literature concerning the physiology and biochemistry of nitrogen and to a lesser extent phosphorus. Beever & Burns (1980) have provided an excellent review of the uptake, storage and utilization of phosphorus in fungi, hence I have placed much more emphasis on nitrogen. When we consider the molecular details of nitrogen metabolism in fungi, the information from three species – *Aspergillus nidulans*, *Neurospora crassa* and *Saccharomyces cerevisiae* – far outweighs what we know for all other fungi. Combined biochemical and genetic analysis now allows an understanding of how the various processes of catabolism and anabolism of nitrogen in these three fungi are regulated, but extrapolation to other species must be undertaken cautiously. Nevertheless, there is sufficient information on other species to point to other aspects which need to be taken into account in any overview of the role of nitrogen in the physiology and ecology of fungi

Nitrogen metabolism in *Aspergillus nidulans*, *Neurospora crassa* and *Saccharomyces cerevisiae*

These fungi can use diverse sources of nitrogen including ammonium, nitrite, nitrate, numerous amino acids, acetamide, purines and proteins. Ammonium, glutamate and glutamine tend to be the favoured sources; use of the others requires the synthesis of the necessary enzymes or activation of previously existing enzymes. Maximal growth rates are obtained with these three nitrogen sources and growth on them leads to the most intense depression of a whole range of activities, including membrane transport linked to utilization of other nitrogen compounds. This depression of activities – termed nitrogen catabolite control, regulation or repression – must be lifted if *de novo* synthesis of enzymes catabolizing other nitrogen compounds is to take place. Nitrogen catabolite control might be considered to be an adaptive mechanism which gives priority to the utilization of nitrogen sources. Thus an amino acid such as arginine is only utilised effectively when ammonium, glutamate or glutamine are absent from the medium.

Nitrogen catabolite control is a transcriptional control system that governs many nitrogen-catabolic enzymes and all the evidence is that glutamine acts as the metabolic signal leading to metabolic control (Wiame, Grenson & Arst, 1985; Davis, 1986). In *S. cerevisiae* control appears to be a negative system in which glutamine represses the target enzymes, though it is possible that ammonium also has a role (Wiame *et al.*, 1985). The effect of glutamine is presumed to be mediated by an as yet unidentified protein (GDHCR) which is the product of the *gdhCR* locus (which has been named for the constitutive state of the NAD-linked glutamate dehydrogenase activity but which is not a structural gene for *either* glutamate dehydrogenase). Mutants of this locus are derepressed for many glutamine-controlled enzymes. Indeed in the presence of a *gdhCR* mutation, synthesis of all the proteins (enzymes, permeases and probably regulatory proteins) except one have been found to be derepressed under nitrogen-repress-ing conditions. The exception is glutamine synthetase, synthesis of which is repressed even in a *gdhCR* mutant.

In the two filamentous fungi it appears that glutamine opposes a positive regulatory element encoded by *areA*, in the case of *A. nidulans*, and by *nit2* in the case of *N. crassa* (Wiame *et al.*, 1985; Davis, 1986; and see Tomsett, Chapter 2). The regulator produced is required for expression of nitrogen-catabolizing enzymes including nitrate reductase (Cove, 1979; Marzluf, 1981).

Nitrogen catabolite control can be seen as one aspect of the ability of these three fungi to adapt to a changing environment with respect to the form and amount of nitrogen present. In *S. cerevisiae* there is rapid adaptation to new nitrogen sources. Thus, only 3 - 5 min elapses between the addition of arginine or urea to the growth medium and the appearance of the enzymes degrading arginine or allantoin (which is broken down to ammonia via urea); incidentally, the time required for induction is the same as that needed for the induction of β-galactosidase in *Escherichia coli* (Cooper, 1982a). It must be realised that not only is there induction of the necessary catabolic enzymes within the cytosol but also the necessary permease(s). Indeed a prominent part of nitrogen catabolite control is via inducer exclusion, i.e. the inducer for a particular metabolic pathway is unable to enter the protoplasm of the fungus. Of particular interest in this respect is the effect of ammonium (probably through the *gdhCR* gene product GDHCR) on the general amino acid permease (Wiame *et al.*, 1985). This transport system has more than 20 natural substrates including all the L-amino acids which are constituents of proteins. No activity of the permease is detectable in wild type cells grown in the presence of ammonium. The available data indicate that the permease can be activated and inactivated and that ammonium promotes the inactivation process, thus excluding inducers taken up by this particular transport system.

In growing green plants, nitrogen metabolism is envisaged as a flux of the element from inorganic form via amino acid to protein. In fungi, the ability, indeed the necessity, to use a range of nitrogen sources means that both catabolic and anabolic activity must take place simultaneously. For ordered growth to occur, there must be sophisticated regulation of metabolism particularly in those situ-

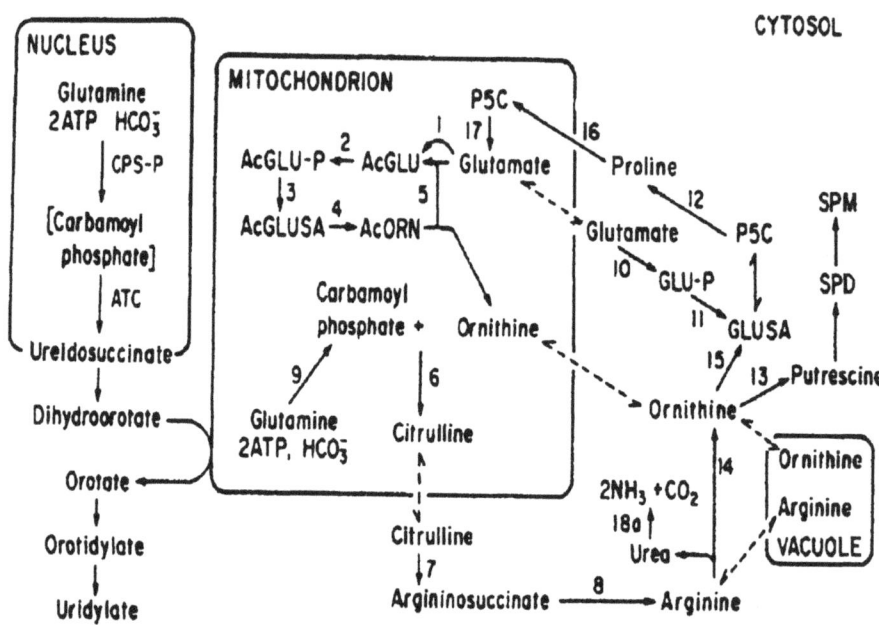

Fig. 1.1A (above): Organisation of arginine, pyrimidine, proline and polyamine metabolism in *Neurospora crassa*. The locations of reactions 16 and 17 have not been proved but are shown as in *Saccharomyces cerevisiae*. Fig. 1.1B (facing page): Organisation of selected enzymes of arginine metabolism in *S. cerevisiae*. Abbreviations: ATC, aspartate carbamoyltransferase; CPS-P, pyrimidine-specific carbamoyl-P synthetase; Ac-, acetyl; GLU, glutamate; GLUSA, glutamate semialdehyde; ORN, ornithine; P5C, pyrroline 5-carboxylate; SPD, spermidine; SPM, spermine. Identities of the enzymes: 1, acetylglutamate synthase; 2, acetylglutamate kinase; 3, acetylglutamyl-P reductase; 4, acetyl-ornithine transaminase; 5, acetylornithine-glutamate acetyltransferase; 6, ornithine carbamoyltransferase; 7, argininosuccinate synthetase; 8, argininosuccinate lyase; 9, carbamoyl-P synthetase A; 10, glutamate kinase; 11, glutamyl-P reductase; 12, pyrroline 5-carboxylate reductase; 13, ornithine decarboxylase; 14, arginase; 15, ornithine transaminase; 16, proline oxidase; 17, pyrroline 5-carboxylate dehydrogenase; I8a, urease; 18b, urea amidohydrolase. From Davis (1986).

ations when common intermediate(s) exist. Perhaps the best studied carbon source has been arginine in which the catabolic pathway consists of the hydrolysis of arginine to ornithine and urea, the breakdown of urea to ammonia and carbon dioxide and the conversion of ornithine to glutamate. The anabolic pathway involves the synthesis of ornithine and of carbamoyl phosphate, with the conversion of these two compounds to arginine. Questions are posed as to how catabolic and anabolic processes can both proceed with ornithine as a common intermediate and how both activities take place given that ornithine and carbamoyl phosphate might be diverted into other pathways. Arginine metabolism has been studied most extensively in *N. crassa* and *S. cerevisiae* and we now know that similar metabolic problems are solved in different ways in these two species (Davis, 1986).

A major difference between *N. crassa* and *S. cerevisiae* is the location of carbamoyl phosphate synthetase A (responsible for the conversion of glutamine to carbamoyl phosphate) and ornithine carbamoyl-transferase (responsible for the conversion of carbamoyl phosphate and ornithine to citrulline). In *N. crassa* these two enzymes are in the mitochondrion; in *S. cerevisiae* they are in the cytosol. Thus both these enzymes differ in kinetic characteristics as befits their different locations (Fig. 1.1). Nevertheless, while there are these very distinct differences between the two organ-

Fig. 1.1B. The organization of arginine metabolism in *S. cerevisiae*

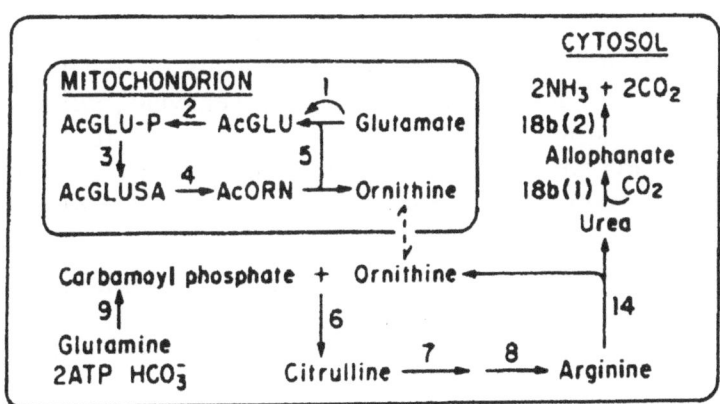

isms, this should not distract us from seeing a similar role of cell compartments in helping to modulate metabolic activity, either by storage, as is the case of the vacuole, or by separating one pathway from another, as is the case when enzymes are located in the mitochondrion. Thus for instance, when arginine is added to a culture of *N. crassa* in minimal medium, the vacuole allows the rapid onset of catabolism of arginine by discharging ornithine into the cytosol, while arginine minimises a futile ornithine cycle by preventing the ornithine in the cytosol from reaching the ornithine carbamoyl transferase in the mitochondrion (Bowman & Davis, 1977). The vacuole acts as a store for most of the amino acids which are basic; 90% of them are vacuolar (with acidic amino acids being predominantly in the cytosol) (Messenguy, 1987).

It is clear that, since ammonium, glutamate and glutamine bring about nitrogen catabolite control, the enzyme glutamate dehydrogenase must play a key role in nitrogen metabolism as well as also being a branch point between carbon and nitrogen metabolism. This enzyme activity can catalyze either the reductive amination of 2-oxoglutarate to produce glutamate or the oxidative deamination of glutamate to produce ammonia. In the three fungi considered here, there are two enzymes, one requiring NADPH and apparently serving to synthesize glutamate (NADP-GDH), the other requiring NAD^+ and leading to the production of 2-oxoglutarate and ammonium (NAD-GDH). A great deal is known about NADP-GDH from *N. crassa*. It is a hexamer of six identical subunits, each of which is a polypeptide of 452 amino acid residues; the polypeptide sequence is known (Holder *et al.*, 1975) and there is much information about allosteric behaviour (Kinsey *et al.*, 1980); the gene for the enzyme (*am*) has been sequenced (Kinnaird & Fincham, 1983). More recently, considerable attention has been given to the effects of specific amino acid replacements on enzyme properties (Fincham, Kinnaird & Burns, 1985).

However, despite the considerable molecular knowledge, little is known about the regulation of NADP-GDH or of NAD-GDH (Marzluf, 1981; Cooper, 1982a). One difficulty may be that both enzymes appear to be controlled by both nitrogen and carbon circuits. Another is that glutamine can be synthesised via the activity of glutamine synthetase (GS) and glutamate synthase (glutamine (amide) : 2-oxoglutarate amino transferase, with acronym GOGAT). Thus:

$$\text{Glutamate} + \text{ATP} + \text{NH}_3 \xrightarrow{\ \text{GS}\ } \text{Glutamine} + \text{ADP} + \text{P}_i$$

$$\text{Glutamine} + 2\text{-oxoglutarate} + \text{NADPH} + \text{H}^+$$

$$\xrightarrow[\ ]{\text{GOGAT}} 2\text{Glutamate} + \text{NADP}^+$$

The foregoing is a somewhat superficial overview of the regulation of the metabolism of relatively simple nitrogen compounds which ensure the coordinated production of the specific amino acids necessary for protein synthesis. Much is known about these pathways; those in *A. nidulans* and *N. crassa* have been reviewed by Pateman & Kinghorn (1976) and those in *S. cerevisiae* by Jones & Fink (1982). It is not appropriate to consider here how enzyme activity of a pathway is regulated by controls specific to that pathway but an example is provided in Fig. 1.2 which gives control patterns observed for transport and catabolism of arginine, proline and the allantoin pathway intermediates in *S. cerevisiae*. However, attention needs to be drawn to the fact that many enzymes of amino acid biosynthesis are under complex general or cross-pathway control. Thus, starvation for any one of a number of amino acids will lead to derepression of the same spectrum of enzymes of several pathways. Those pathways which are now known to have an enzyme or enzymes under general control include arginine, histidine, tryptophan, lysine, leucine, isoleucine and valine. As the number of enzymes in a pathway under general control can differ, the pattern of regulation occurring when the general control mechanism is triggered can be complex. This general control is very relevant to amino acid metabolism during starvation which will be considered below. I will not, however, discuss how the relatively simple nitrogen compounds enter the three fungi as this is treated in detail elsewhere (Pateman & Kinghorn, 1976; Eddy, 1980; Iaccarino, Guardiola & De Felice, 1980; Wolfinbarger, 1980a; Cooper, 1982b; Eddy, 1982; Wiame *et al.*, 1985).

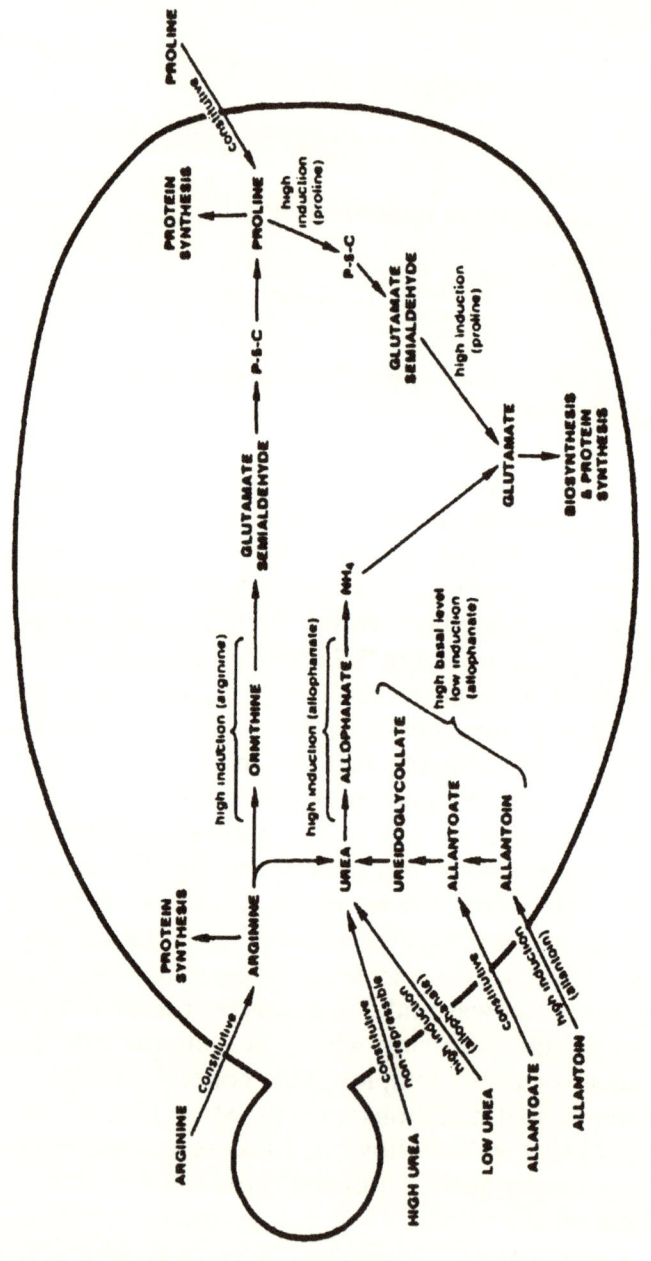

Fig. 1.2. Summary of control patterns observed from the transport and catabolism of arginine, proline and allantoin pathway intermediates. From Cooper (1982a).

If cells of a wild-type culture of *S. cerevisiae* undergoing exponential growth are transferred to distilled water, cell division continues for one or two generations, although the time interval between divisions increases. The cells are drawing upon their stores of nitrogen compounds to synthesise the necessary proteins. In particular the vacuolar pools of nitrogen are mobilised. Under the conditions in which amino acids are released from the vacuole, their metabolism is such as to be no longer under general or cross-pathway control. While information is available about the molecular basis of the general control system (Cooper, 1982a; Henry, Klig & Loewy, 1984; Messenguy, 1987), there are many uncertain aspects (Cooper, 1982a). In particular we do not know the relationship between the control system and flux of amino acids into and out of the vacuole or the relation between it and the cell cycle.

Proteins in yeast cells are constantly degraded. It is this process of degradation which allows the cell to adapt its protein composition to changes in the external environment. The rate of protein degradation increases in cells growing normally from around 0.5 - 1.0 to 2 - 3% h^{-1} under starvation conditions (Lopez & Gancedo, 1979) and proteinases are clearly involved. There is now considerable information about proteinases in *S. cerevisiae* and *N. crassa* (Wolf, 1980, 1986; Jones, 1984). The proteinases of the vacuole are very unspecific and it is this nonspecificity which, together with the properties of other vacuolar enzymes, have led to the vacuole being equated with a lysosome. Studies with mutants leave little doubt that the proteinases found in the vacuole are involved in rather nonspecific protein degradation induced by nitrogen starvation, and make both exogenous and endogenous peptides accessible for primary metabolism. Jones (1984) points out that the proteinases of the vacuoles are brought into play when the only source of amino acids for adaptation and restructuring is within the cell itself. The vacuole can therefore be viewed as the resting place of machinery to be mobilized in 'emergencies'.

It is becoming increasingly clear that proteinases play a highly significant specific role in the metabolism and development of cells of *S. cerevisiae*. Thus proteinases have been implicated in septum formation, sporulation, catabolite inactivation, carbon starvation-induced degradation of NADP-dependent glutamate dehydrogenase, nitrogen starvation-induced degradation of

NAD-dependent glutamate dehydrogenase and of glutamate synthase, enzyme secretion, localization of mitochondrial enzymes, processing of enzyme precursors, production of the killer toxin and of the pheromone α-factor, destruction of the α-factor, degradation of nonsense proteins and nonsense fragments and cell cycle-regulated protein degradation (Jones, 1984).

Both *Aspergillus* species and *N. crassa* can secrete extracellular proteinases, the production of which is regulated by carbon-, nitrogen-, sulphur- and even phosphorus-catabolite repression (Cohen, 1980, 1981; North, 1982). Starvation for any of the these elements results in the production of proteinase. In *Aspergillus*, relief of catabolite repression is sufficient for production; in *N. crassa*, protein must also be present. It should be noted that for growth to occur using the products of protein breakdown, it is not necessary for there to be the release of free amino acids. In *N. crassa*, there is an oligopeptide transport system capable of transporting peptides with up to five amino acid residues (Wolfinbarger, 1980b).

Nitrogen metabolism in other fungi

A number of themes could be followed when considering nitrogen metabolism in the less studied fungi. For example, there are the two pathways of lysine biosynthesis in fungi (Vogel, 1964): one involves α,ε-diaminopimelic acid and is associated with the Oomycetes *sensu stricto*; the other involves α-aminoadipic acid and is found in the Chytridiales, Mucorales and the so-called higher fungi. The presence of these pathways is associated with other biochemical features (Fig. 1.3; LéJohn, 1971). In particular, NADP-linked glutamate dehydrogenase is confined to the Ascomycotina, Basidiomycotina and Deuteromycotina, and within the lower fungi the NAD-linked glutamate dehydrogenases can be divided into three classes on the basis of their regulatory properties (LéJohn, 1971). This argues the need for concerted further study of the nitrogen metabolism of lower fungi.

There are differences in the specificity of nitrogen sources for growth. All fungi seem able to utilise some kind of organic nitrogen source. The great majority of fungi can use ammonium as the sole source of nitrogen. A number of fungi can use nitrate in addition to ammonium but few Basidiomycotina and Saprolegnia-

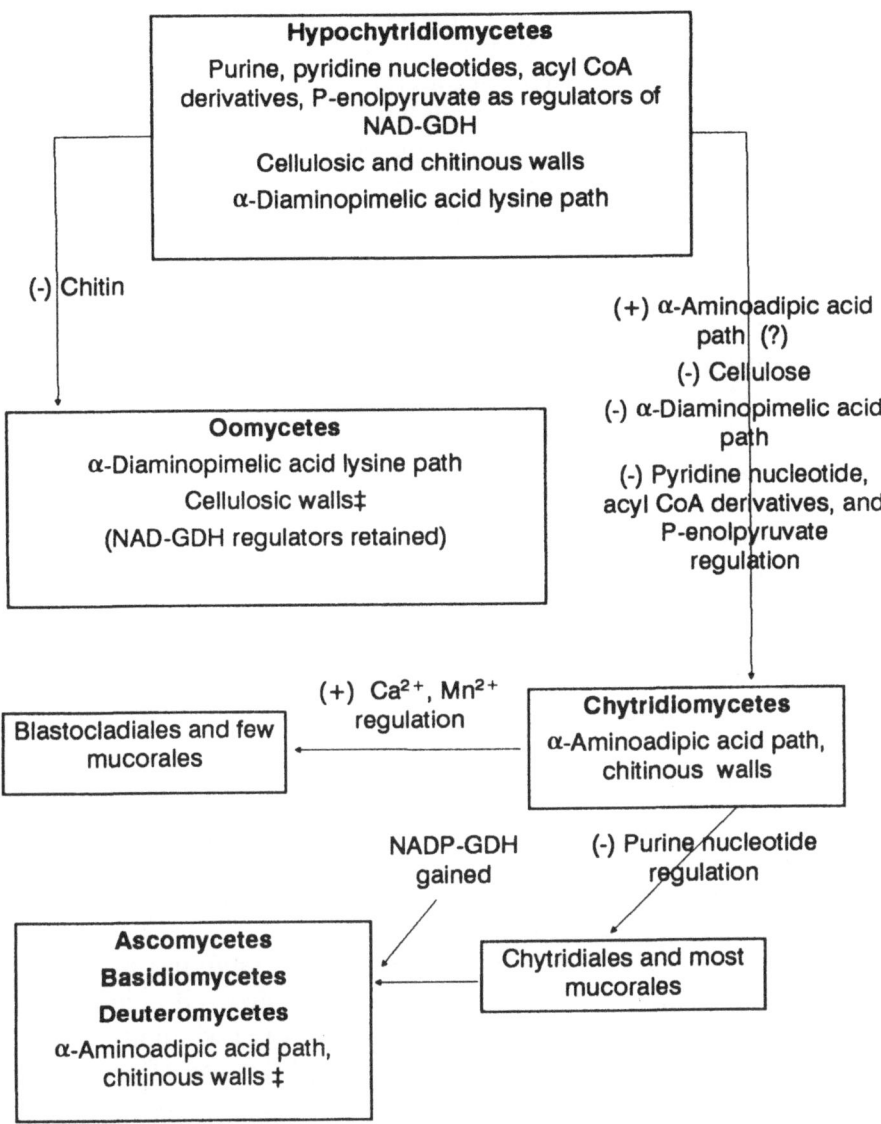

Fig. 1.3. Suggested path in the evolution of NAD-linked glutamate dehydrogenases in fungi and the relationship to cell wall structure, lysine biosynthesis and phylogeny; (+) signifies acquisition of ability; (-) loss of capacity. The suffix, ‡, means that exceptions are known. After LéJohn (1971).

ceae have this ability (Cochrane, 1963). There are indications that certain plant pathogens are apparently unable to use nitrate or ammonium as a source of nitrogen, and the situation is probably complex. Thus, in the two rust fungi whose nutrition has been critically assessed in axenic culture, it has been found that *Gymnosporangium juniperi-virginianae* can use either nitrate or ammonium as well as a wide range of amino acids and other organic nitrogen sources, whereas *Puccinia graminis* cannot use nitrate (Maclean, 1982). Further, in the presence of 8 - 80 mM ammonium, *P. graminis* and many rust fungi require a sulphur amino acid, e.g. 4 mM cysteine or 0·3 mM methionine. Likewise, while it is generally considered that species of *Phytophthora* prefer organic nitrogen sources, only very few species – and then only certain strains – have an absolute requirement (Hohl, 1983). In considering the ability of a fungus to grow on an inorganic nitrogen source, it must be remembered that absence of growth may not mean that the fungus cannot convert the nitrogen from that source into amino acids. It may simply be that the fungus has a specific additional requirement for one or a few amino acids. This appears to be the case for *Thraustochytrium* with respect to lysine; apparently the fungus cannot synthesise the amino acid (Paton & Jennings, 1988) though it seems to be able to metabolise inorganic nitrogen sources (Bahnweg, 1970).

With regard to ammonia assimilation in the higher fungi, it is clear, as indicated above, that both NAD-linked and NADP-linked glutamate dehydrogenases are present in vegetative mycelium (Fawole & Casselton, 1972). In *Coprinus cinereus*, NAD-GDH is subject to catabolite repression and urea derepression, while NADP-GDH is catabolite derepressed and repressed by urea, with the former enzyme seemingly involved in ammonia assimilation (Stewart & Moore, 1974). However when the carpophore of this fungus develops there is amplification of activity in the tricarboxylic acid and urea cycles in the cap (Moore, 1984). NADP-GDH is one of the enzymes which also increases in activity and the regulation appears to involve accumulation of acetyl CoA under conditions in which ammonia is limiting (Moore, 1981). NAD-GDH activity also increases in the cap; this enzyme appears to be required for amination of 2-oxoglutarate, the glutamate produced being converted to succinate via the glutamate oxidation loop involving glutamate decarboxylase and 4-aminobutyrate aminotransferase. These steps allow a modified tricarboxylic acid cycle to proceed in spite of the absence of detectable activity of 2-oxoglutarate dehydrogenase (Moore & Ewaze, 1976).

The increase in activity of NADP-GDH in the cap is accompanied by an increase in the activity of glutamine synthetase (Ewaze, Moore & Stewart, 1978). Moore, Horner & Liu (1987) have shown that the two enzymes are probably co-ordinately controlled and together they act as an ammonia scavenging system; ammonium apparently inhibits basidiospore formation in a manner similarto its inhibition of sporulation in *S. cerevisiae*.

Finally, with respect to primary nitrogen metabolism in *C. cinereus*, the stimulation of the urea cycle in the carpophore leads to accumulation of urea, which becomes part of the battery of low molecular weight compounds (including alanine, arginine, glutamate and trehalose) which contribute to the generation of the solute potential required for carpophore growth (Rao & Niederpruem, 1969; Moore, 1984).

It appears from nitrogen-15 nuclear magnetic resonance (^{15}N-NMR) studies that ammonia assimilation by beech (*Fagus sylvatica*) mycorrhizas occurs mainly via the glutamine synthetase/glutamate synthase (GS/GOGAT) pathway and that glutamate dehydrogenase has no significant role (Martin *et al.*, 1986). Higher plants, of course, use this pathway (Lea & Miflin, 1974) and likewise GDH plays no major role in ammonia assimilation (Miflin & Lea, 1980). On the other hand, when the ectomycorrhizal fungus *Cenococcum graniforme* is grown in a nitrogen-rich medium it has been found, again using ^{15}N-NMR, that here too ammonia assimilation occurs via glutamine synthetase (Martin, 1985). Interestingly, the same studies indicated that, as in *S. cerevisiae*, arginine may well be associated with polyphosphate. In contrast to the results of these ^{15}N-NMR studies, Genetet, Martin & Stewart (1984) found that with nitrogen-starved mycelium of *C. graniforme* labelling of amino acids from externally supplied ^{15}N-ammonium could only be explained by assimilation being mediated by NADP-GDH. This implies that when the fungus is part of the ectomycorrhizal association there is both carbon and nitrogen repression, the former probably as a consequence of the significant flux of carbohydrate from the host root into the fungal tissue (Lewis & Harley, 1965).

Finally, it is important to note that *Agaricus bisporus*, *Coprinus cinereus* and *Volvariella volvacea* can all use protein as the sole source of carbon, nitrogen and sulphur (Kalisz, Moore & Wood, 1986; Kalisz, Wood & Moore, 1987). Unlike *Aspergillus* and *N. crassa*, proteinase activity is not repressed by the presence of glucose, ammonium and sulphate in the medium. The presence of protein is required for induction of the proteinases, though in the case of *C. cinereus* proteinase activity is derepressed in protein-free media in the absence of sulphate. It is particularly interesting to find that, when these three fungi were grown with protein as sole source of carbon, very significant levels of ammonium were found in the medium. Since the amount of protein consumed contained far more nitrogen than required for the amount of biomass produced, it must be assumed that a proportion of the protein was catabolized leading to the production of non-nitrogen containing polymers or oxidised to provide the necessary energy for growth. As a result of such processes free ammonia would be produced. In the experiments referred to here one-third to one-half of the supplied protein appears to have been assimilated and metabolised in this manner (Kalisz *et al.*, 1986).

Some thoughts on nitrogen metabolism in fungi

The many studies on *A. nidulans*, *N. crassa* and *S. cerevisiae* have produced relatively clear pictures of how nitrogen metabolism in these fungi might be regulated. From these studies we appreciate the need for compartmentation of certain of the enzymes such that catabolic and anabolic processes can occur in the same cell or hyphal compartment when it is supplied with a particular organic nitrogen compound. The complexity of regulation of nitrogen metabolism is now being revealed and, even within seemingly closely related species, there are significant differences, in spite of the pathways of nitrogen metabolism being generally very similar. Nevertheless, it would seem that most catabolic enzymes are inducible with one of the degradative intermediates acting as an inducer. Overarching this induction process is nitrogen catabolite repression.

Cooper & Sumadra (1983) have pointed out that nitrogen catabolite repression tends to give a certain primacy to such compounds as ammonium, glutamate and glutamine as nitrogen sources. They referred to the relatively long standing presumption that the combination of induction and catabolite repression enables fungi to utilise good nitrogen sources, i.e. those which give high growth rates, in preference to ones which give low growth rates. A corollary would be that the provision of a good nitrogen source to cells growing on a poor one should shift metabolism such that the good source is used. These authors drew attention to the fact that this view does not address the fate of previously formed enzymes that were degrading the poor nitrogen source, or the effects of these enzymes on the metabolism of the good source. They went on to show that in functional terms nitrogen catabolite repression is long-term. Other processes such as transport of the nitrogen compounds into the cell are much more likely to be the sites of short-term regulation such that good nitrogen sources are utilised in preference to poor ones.

On the whole, *A. nidulans, N. crassa* and *S. cerevisiae* make poor use of organic compounds as sources of combined carbon but it is becoming increasingly clear that this is by no means true for all fungi. As indicated above, organic nitrogen compounds as complex as proteins are readily used as carbon sources by *A. bisporus, C. cinereus* and *V. volvacea*, and Read, Leake & Langdale (Chapter 9) have demonstrated that mycorrhizal fungi can do the same. It is possible that this ability is even more widespread. Loll & Bollag (1983), in their review of protein transformation in soil list 40 genera of fungi said to be proteolytic, though this does not necessarily mean that they can utilise protein as sole sources of carbon. Nevertheless, the ability to produce proteolytic enzymes is a necessary prerequisite of protein utilization.

The use of protein as a source of combined carbon, however, has certain consequences for the physiology of the fungus. Kalisz *et al.* (1986) point out that protein provides far more nitrogen than is necessary for the biomass of a fungus when it utilises protein as the sole source of carbon. It is not surprising therefore that ammonia is released into the medium. But when this occurs it raises the question as to how nitrogen metabolism is regulated particularly since ammonium is metabolically closely associated with nitrogen catabolite repression. While excess nitrogen can be lost as ammonia, the wall could be a sink for excess nitrogen through the polymerization of N-acetylglucosamine. This may be one reason why the wall of *A. bisporus* contains something like 45% (w/w) glucosamine, the amount of chitin and protein being 7-8 times greater than the relatively closely related *Schizophyllum commune* (Michalenko, Hohl & Rast, 1976).

The hypothesis that the wall might act as a sink for nitrogen in this way can only be tested by examining total wall composition of a range of fungi and relating the results to the physiology and ecology of the individual species concerned. Thus, I have pointed out (Jennings, 1982) that walls might be the sink for excess carbon in the face of a high C:N ratio of the substrate on which they are growing. Be that as it may, if fungi release ammonia when metabolising proteins, there must be some mechanism for preventing the compound from exerting its known inhibitory effects on developmental processes, such as sporulation, referred to above. Of course, once non-nitrogenous organic compounds as well as proteins are available for metabolism, it is likely that production of ammonia will decline or indeed cease altogether. In this respect, it is conceivable that a selective pressure towards mycorrhiza formation might be the provision of carbohydrate by the higher plant such that the fungus forming the association might be better able to cope with the utilization of protein available in the external environment.

It is pertinent here to refer to the contrasting situation where nitrogen is in short supply relative to degradable non-nitrogen organic compounds, Paustian (1985) and Paustian & Schnürer (1987a & b) have considered this matter with respect to soil fungi growing under carbon and nitrogen limitation and have demonstrated the importance of nitrogen recycling within the mycelium such that the element is preferentially allocated to hyphal extension. The importance of nitrogen recycling by fungi, especially wood decomposers, has been suggested by Merrill & Cowling (1966) and Levi & Cowling (1969), while Watkinson (1979, 1984) has emphasised the probability that in *Serpula lacrimans* mycelial development, including formation of strands, there is transport of nitrogen within the mycelium and reutilization of the element. This implies that, in view of what we know of regulation of nitrogen metabolism in fungi, there must be fairly rigid compartmentation such that breakdown of cell components can take place in one compartment producing a flux of nitrogen and other compounds to support normal metabolic activity in the next compartment. For recycling to take place with growth of mycelium over non-nutrient surfaces, there must be translocation of nitrogen. Currently the form(s) translocated is not known, though ^{14}C from exogenously supplied ^{14}C-aspartic acid is translocated very readily in mycelium of *S. lacrimans* (Brownlee & Jennings, 1982). However, Cárdenas & Hansberg (1982a & b) have presented good evidence that glutamine may act as the nitrogen carrier from the mycelial mat to the growing aerial mycelium of *Neurospora crassa*.

Watkinson (1984) has pointed out that a wood-decay fungus is able not only to extract nitrogen from wood but concentrate it, by which is meant that the C:N ratio in mycelium is much lower than in wood. It is not clear what is meant mechanistically by concentration (but see Dighton & Boddy, Chapter 13). What is certain is that a fungus must often degrade a considerable quantity of organic matter in wood not containing nitrogen in order to obtain the necessary supply of this element. The release into the surrounding medium of soluble residues from the organic matter so degraded would provide energy sources for competing micro-organisms. Thus, there is a need to remove such compounds. One way in which it might occur would be by metabolism of the compounds to carbon dioxide. The most effective ways in which this could occur would be by fermentation or by respiration in which electrons flow in the respiratory chain to a terminal oxidase which is cyanide-resistant, thus generating less ATP and being energetically less efficient than when electrons flow to cytochrome aa3 (cytochrome oxidase). This is essentially the proposition of Lambers (1980) with respect to the role of the cyanide-resistant pathway in higher plant cells. Such a pathway has been shown to be present in fungi (Lloyd, 1974) and, interestingly, also in beech mycorrhizal roots (Harley, 1978).

An interesting situation is found in those fungi which decompose wood on the forest floor and which can spread from one resource to another by mycelial cords or rhizomorphs. We know that the structures can absorb phosphate (Clipson, Cairney & Jennings, 1987; Cairney *et al.*, 1988) and can break down organic phosphate (Jennings, 1988). If these structures were able to break down protein, the supply of carbohydrate from the wood (Jennings, 1987b) would assist in any problems there might be due to the relatively unfavourable C:N ratio of the protein. In any case there is a need to examine nitrogen metabolism of these fungi, particularly the extent to which it might be different in mycelium in the wood, which is degrading lignocellulose, and in cords or rhizomorphs growing through the soil. There are clear differences in the biochemistry and physiology of the two types of mycelium (Jennings, 1984b). In terms of what is discussed here, it is of great interest that ligninolytic activity of *Phanerochaete chrysosporium* becomes apparent when the mycelium becomes starved of carbon, nitrogen or sulphur (Jefferies, Choi & Kirk, 1981). These are conditions in which one would expect certain enzymes of nitrogen metabolism repressed in growing mycelium to become derepressed. Nitrogen metabolism of saprotrophic basidiomycetes is clearly a fertile field for investigation.

Finally, we need to be alert to the possibility that in fungi proteins may act as stores of nitrogen. Candidates for proteins which may act in this way are: (i) Woronin bodies, now known not to be made up of ergosterol (Markham & Collinge, 1987); (ii) crystalloids often associated with microbodies (Maxwell *et al.*, 1975) and which are almost certainly enzymic, an outstanding example being the identification of crystalloids in peroxisomes of *Hansenula polymorpha* growing on methanol as being alcohol oxidase (Veenhuis, van Dijken & Harder, 1983); and (iii) the gamma particles of *Blastocladiella emersonii* zoospores (Hohn, Lovett & Bracker, 1984). The concept of enzymes being stores of nitrogen is becoming familiar to green-plant physiologists with ribulose 1,5-bisphosphate carboxylase-oxygenase (RUBISCO) being the prime example of a protein with both an enzymic and storage function (Millard, 1988). In respect to the above, Blayney & Marchant (1977) showed that protein inclusions, with a crystalline substructure, are formed during stipe elongation in *Coprinus cinereus* and degraded during the later stages of development. These investigations suggested that nitrogen in the chitin synthesised during elongation may be derived from the degraded protein inclusions.

Phosphorus physiology

As indicated in the Introduction, Beever & Burns (1980) provide a detailed consideration of uptake, storage and utilization of phosphorus transformations in fungi. Here I wish to focus on the central position of orthophosphate in the physiology and biochemistry of phosphorus in fungi. In contrast to my considerations of nitrogen, it is appropriate first to make an ecological comment.

There are many ecological situations where phosphorus is not readily available to higher plants. The marked ability of most soils to sequester phosphorus is such that the amount of available phosphorus is very low. There is no effect of bulk water on phosphorus movement and the lack of availability is associated with the diffusive movement of phosphorus (as phosphate) to plant roots. Consequently, the more competitive plants appear to be those which can extend the length of their root systems, since this is the most effective way of increasing uptake (Jennings, 1986). Vesicular-arbuscular mycorrhizas are seen as another means of increasing the length of the root system (Tinker, 1975; Hetrick, Chapter 10; Gianinazzi-Pearson & Gianinazzi, Chapter 11).

With regard to fungal mycelium spreading through soil there is, however, little evidence to suggest that phosphorus is not other than accessible. First, mycelium can grow towards sources of phosphorus. Second, we know that fungi can utilise sources of phosphorus other than soluble phosphate, namely condensed phosphates, insoluble phosphates and organic phosphorus compounds, including nucleic acids (Beever & Burns, 1980; Jennings, 1988). Further, some fungi, along with some bacteria and actinomycetes, are able to solubilize phosphate through secretion of organic acids (Khan & Bhatnager, 1977; Kucey, 1987). Finally, apart from the large fruit bodies produced by many basidiomycetes and some ascomycetes, fungi unlike higher plants do not have aerial structures to supply with the element.

Thus, fungal growth in nature may not be limited by phosphorus availability in spite of the low levels of phosphorus found in many environments (Beever & Burns, 1980). A study on the marine species *Rhodotorula rubra* grown in continuous culture supports this view (Button, Dunker & Morse, 1973). The yeast was found to be capable of growth at very low phosphate concentrations; the external concentration giving half maximal growth rate was 10·8 nM. Cells grown for many days in continuous culture under phosphorus limitation showed a higher maximum specific growth rate as a function of phosphate concentration than did cells grown for less than three days. Initial phosphate uptake rates that were in agreement with the steady-state flux in continuous culture were obtained by using cells and medium directly from such cultures. With this procedure, uptake rates were found to be about five hundred times greater than for cells grown in batch culture. Thus it appears that under conditions of phosphate limitation the organism synthesises more of the relevant transport system (Robertson & Button, 1979).

Similar studies of the kind just described need to be made using other fungi. If it is assumed that what has been found for *R. rubra* is generally applicable, and given the known flexibility of phosphate transport systems in fungi with respect to the internal phosphorus status, and the capability of fungi to produce extracellular phosphatases (Beever & Burns, 1980), it would seem that fungi can react very flexibly to differing availability of phosphorus in the external medium.

The above raises the query about the role of polyphosphates, which are ubiquitous in fungi (Kulaev, 1979). There has been a tendency to assume that they could act as stores of phosphorus, possibly because polyphosphates are found in both VA and ecto-mycorrhizas – those fungal associations which are known to benefit their higher plant partners in terms of the phosphate nutrition of these partners, allowing them to grow better in the face of a limited phosphate supply. Yet, if the role of polyphosphate was solely as a store for phosphorus, it would seem unlikely that it would be in a polymerised form which is energetically demanding, requiring the conversion of ATP to ADP. In any case, it is frequently found that, as in beech mycorrhizal roots (Jennings, 1964) and rhizomorphs of *Armillaria mellea* (Cairney *et al.*, 1988) growing in low orthophosphate concentrations, orthophosphate levels within the fungus are as high if not higher than those of polyphosphate. Finally, if polyphosphates functioned solely as stores of phosphorus, it would be surprising to find them in locales other than the vacuole. Yet as much as 30-35% of the total polyphosphate can be found in the wall of *S. cerevisiae* and substantial amounts have been found to be associated with nuclei (Kulaev & Vagabov, 1983). In this respect it is important to realise that, though polyphosphate is certainly located in the vacuole, how much is there is uncertain since it is not clear how much contamination occurs from vesicles from the endoplasmic reticulum during isolation of vacuoles (Kulaev & Vagabov, 1983).

Whatever view is held of the role of polyphosphates in fungi, account must be taken of the wealth of information now available on the metabolism of these compounds, which is well reviewed by Kulaev & Vagabov (1983). These authors believe that polyphosphates should be viewed as part of a micro-organism's armoury of 'metabolic traps' allowing it to function in the varying conditions of its external environment. Polyphosphates can be viewed as components of the homeostatic system equivalent to the hormonal and nervous regulation of multicellular organisms. Kulaev & Vagabov (1983) point out that 'availability of sufficient amounts of such valuable endogenous pools as high molecular weight polyphosphates makes micro-organisms less dependent on external conditions and, on the other hand, capable, at any suitable moment, of initiating growth and reproduction without any considerable lag period'. Polyphosphates may thus play a role equivalent to that ascribed to polyols, namely that these compounds are part of a system of 'physiological buffering' within fungal protoplasm (Jennings, 1984a).

Such a role of polyphosphates could be played out via a number of metabolic activities. Polyphosphate may act as a high-energy phosphate alternative to ATP (Kulaev & Vagabov, 1983). There is now good evidence that growing micro-organisms, including those fungi studied, control their adenine nucleotide content and also the proportion of the total pool present as ATP, ADP and AMP (Knowles, 1977). Furthermore, micro-organisms have evolved mechanisms for protecting the adenine nucleotide pools against fluctuations in the nutrient content of the environment. Polyphosphates may therefore play a similar protective role. There is evidence from ^{31}P-NMR studies on *Candida utilis* and *Brettanomyces intermedius* that under conditions of lowered ATP concentration within the cells, polyphosphate can provide an alternative source of energy (Nicolay *et al.*, 1983).

Polyphosphates are also involved in the maintenance of the cation/anion balance within fungi. It has been pointed out already that considerable amounts of basic amino acids may be linked to the negative charges of polyphosphate, but, under certain conditions, it has been shown that bivalent cations, particularly the magnesium and manganous ions, can also be so linked (Okorokov, Lichko & Kulaev, 1980; Lichko, Okorokov & Kulaev, 1982). The fact that magnesium can be so associated with polyphosphate may be part of those processes required to regulate the concentration of this bivalent cation within the cytoplasm, where it plays a significant role in phosphorus metabolism. With respect to the more general matter of cation/anion balance within a fungus, it needs to be remembered that orthophosphate will also contribute to the total negative charge within the cell. There is value in a high proportion of that charge being inorganic, such as ortho- and polyphosphate, since there is less need to produce organic acid anions to balance incoming inorganic cations. In this way, carbon is conserved for other functions within the organism.

Finally, the breakdown and synthesis of polyphosphate are two of those several processes which help to regulate cytoplasmic orthophosphate concentration. It has been shown (den Hollander *et al.*, 1981) that the concentration of orthophosphate could be an important determinant of the rate of glycolysis in *Saccharomyces cerevisiae*, depending upon the cytoplasmic pH. From these data and from other published information about the effects the interaction of orthophosphate and pH on the *in vitro* activity of hexokinase and phosphofructokinase, it could be surmised that the regulatory action of the cytoplasmic concentration on glycolysis is via the activity of these two enzymes.

Control of phosphorus concentration in the cytoplasm must also occur at the plasmalemma and the tonoplast. With respect to the former membrane, the best studied fungi are *S. cerevisiae* and *Neurospora crassa*. In the former species, there appear to be three transport systems for the monovalent anion: (i) a constitutive system which has low affinity for the anion; (ii) an inducible proton symport; and (iii) an inducible sodium symport, both of the latter having a high affinity for phosphate. Derepression of the latter two systems occurs when cells are incubated in the presence of a suitable substrate, such as glucose, in a medium lacking phosphate

(Borst-Pauwels, 1981; Nieuwenhuis & Borst-Pauwels, 1984). The protein synthesis inhibitor cycloheximide prevents derepression. At the same time that there is derepression of the high-affinity system, the low-affinity system disappears. The sum, at any one time, of the maximal rates of uptake by the two systems (constitutive and inducible) is constant, suggesting that the constitutive system is involved in both the high-affinity and low-affinity transport of the anion. More recent evidence (Jeanjean *et al.*, 1986) indicates that the transport system is composed of a cell-wall protein and another in the plasmalemma, that in the wall being involved in the high-affinity system.

A similar situation appears to hold for phosphate transport across the plasmalemma in *N. crassa* (Lowendorf, Slayman & Slayman, 1974; Lowendorf, Bazinet & Slayman, 1975; Lowendorf & Slayman, 1975; Burns & Beever, 1977; Beever & Burns, 1977, 1978). As with *S. cerevisiae*, there is a high-affinity system which is derepressed during phosphorus starvation. However, cycloheximide appears to have no effect on derepression. While there is no direct evidence that elements of the phosphate transport system in *N. crassa* involve cotransport of protons, the kinetic characteristics of the system, particularly with respect to pH, strongly suggest that this is likely to be the case (Sanders, 1986).

There is circumstantial evidence that phosphate transport into many fungi is likely to be very similar to what has been found in *S. cerevisiae* and *N. crassa*, namely that some elements of the system involve proton symport and the kinetics of transport being dependent on the internal phosphorus status of the organism (Beever & Burns, 1980; Clipson, Cairney & Jennings, 1987; Straker & Mitchell, 1987; Cairney *et al.*, 1988).

These various transport systems form part of what Metzenberg (1979) has termed the 'phosphorus acquisition system' of a fungus. The other parts are those enzymes which are concerned with 'garnering' phosphorus for the cell when there is insufficient orthophosphate in the external medium. *N. crassa* produces, extracellularly, an alkaline and an acid phosphatase, a 5'-nucleotidase and three nucleases all of which are repressible by phosphate (Metzenberg, 1979). Presumably, these enzymes release orthophosphate from their respective substrates which is then absorbed by the high affinity phosphate transport system. Further consideration of the complex field of extracellular phosphatase production by fungi is beyond the scope of this article, but the review by Beever & Burns (1980) provides some guidance. However, there is no doubt that many fungi are able to release phosphates from organic form. Thus, as is the case for nitrogen, these organisms are likely to have within their environment a much greater source of available phosphorus than the green vascular plant.

Clearly then, as postulated at the beginning of this section, it seems likely that phosphorus does not limit growth in many environments and the best perspective from which to view the relationship between phosphorus and fungi is probably from the standpoint of the regulation of phosphate concentration within the cytoplasm. This has to be achieved in the face of (i) the various ways that the fungus can acquire phosphate and the rate of transport into the cytoplasm, and (ii) the rates of processes within the cytoplasm leading to the net utilization of phosphate. There is clearly control of transport into the cytoplasm and presumably metabolic regulation of those processes utilising phosphate. In this context the vacuole, where phosphate and polyphosphate might be stored (along with those other sites for the storage of polyphosphate mentioned above), might be visualised as providing a reservoir to cushion large changes in cytoplasmic phosphate concentration which otherwise could have untoward effects on metabolism. These ideas can now be investigated experimentally since it is now possible to determine directly cytoplasmic phosphate concentration by ^{31}P-NMR, and a suitable organism would be one which could be grown in continuous culture, yet use a variety of phosphorus sources.

References

Bahnweg, G. (1970), Studies on the physiology of Thraustochytriales. 1. Growth requirements and nitrogen nutrition of *Thraustochytrium* spp., *Schizochytrium* sp., *Japonochytrium* sp., *Ulkenia* spp. and *Labyrinthuloides* spp. *Verofentlichungen der Instituts für Meeresforschung Bremerhaven*, 17, 245-268.

Beever, R. E. & Burns, D. J. W. (1977). Adaptive changes in orthophosphate uptake by the fungus *Neurospora crassa* in response to phosphorus supply. *Journal of Bacteriology*, **132**, 520-525.

Beever, R. E. & Burns, D. J. W. (1978). Does cycloheximide-induced loss of phosphate uptake activity in *Neurospora crassa* reflect rapid turnover? *Journal of Bacteriology*, **134**, 1176-1178.

Beever, R. E., & Burns, D. J. W. (1980). Phosphorus uptake, storage and utilization by fungi. *Advances in Botanical Research*, 8, 127-219.

Blayney, G. P. & Marchant, R. (1977). Glycogen and protein inclusions in elongating stipes of *Coprinus cinereus*. *Journal of General Microbiology*, 98, 467-476.

Borst-Pauwels, G. W. F. H. (1981). Ion transport in yeast. *Biochimica et Biophysica Acta*, 650, 88-127.

Bowman, B. J. & Davis, R. H. (1977). Arginine catabolism in *Neurospora*: cycling of ornithine. *Journal of Bacteriology*, 130, 285-311.

Brownlee, C. & Jennings, D. H. (1982). Long distance translocation in *Serpula lacrimans*. Velocity estimates and continuous monitoring of induced perturbations. *Transactions of the British Mycological Society*, 79, 143-148.

Burns, D. J. W. & Beever, R. E. (1977). Kinetic characterisation of the two phosphate uptake systems in the fungus *Neurospora crassa*. *Journal of Bacteriology*, 132, 511-519.

Button, D. K., Dunker, S. S. & Morse, M. L. (1973). Continuous culture of *Rhodotorula rubra*: kinetics of phosphate-arsenate uptake, inhibition, and phosphate-limited growth. *Journal of Bacteriology*, 113, 599-611.

Cairney, J. W. G., Jennings, D.H., Ratcliffe, R. G. & Southon, T. E. (1988). The physiology of basidiomycete linear organs. II. Phosphate uptake by rhizomorphs of *Armillaria mellea*. *New Phytologist*, 109, 327-334.

Cádenas, M. E. & Hansberg, W. (1984a). Glutamine requirement for aerial mycelium growth in *Neurospora crassa*. *Journal of General Microbiology*, 130, 1723-1732.

Cádenas, M. E. & Hansberg, W. (1984b). Glutamine metabolism during aerial mycelium growth of *Neurospora crassa*. *Journal of General Microbiology*, 130, 1733-1741.

Clipson, N. J. W., Cairney, J. W. G. & Jennings, D. H. (1987). The physiology of basidiomycete linear organs. I. Phosphate uptake by cords and mycelium in the laboratory and in the field. *New Phytologist*, 105, 449-457.

Cochrane, V. W. (1963). *Physiology of Fungi*. John Wiley: New York.

Cohen, B. L. (1980). Transport and utilization of proteins. In *Microorganisms and Nitrogen Sources*, ed. J. W. Payne, pp. 411-430. John Wiley: Chichester.

Cohen, B. L. (1981). Regulation of protease production in *Aspergillus*. *Transactions of the British Mycological Society*, 76, 447-450.

Cooper, T. G. (1982a). Nitrogen metabolism in *Saccharomyces cerevisiae*. In *The Molecular Biology of the Yeast Saccharomyces: Metabolism and Gene Expression*, eds. J. N. Strathern, E. W. Jones & J. R. Broach, pp. 399-461. Cold Spring Harbor Laboratory: New York.

Cooper, T. G. (1982b). Transport in *Saccharomyces cerevisiae*. In *The Molecular Biology of the Yeast Saccharomyces: Metabolism and Gene Expression*, eds. J. N. Strathern, E. W. Jones & J. R. Broach, pp. 39-99. Cold Spring Harbor Laboratory: New York.

Cooper, T. G. & Sumadra, R. A. (1983). What is the function of nitrogen catabolite repression in *Saccharomyces cerevisiae*? *Journal of Bacteriology*, 155, 623-627.

Cove, D. J. (1979). Genetic studies of nitrate assimilation in *Aspergillus nidulans*. *Biological Reviews*, 54, 291-327.

Davis, R. H. (1986). Compartmental and regulatory mechanisms in the arginine pathways of *Neurospora crassa* and *Saccharomyces cerevisiae*. *Microbiological Reviews*, **50**, 280-313.

den Hollander, J. A., Ugurbil, K., Brown, T. R. & Shulman, R. G. F. (1981). Phosphorus-31 nuclear magnetic resonance studies on the effect of oxygen upon glycolysis in yeast. *Biochemistry*, **20**, 5871-5880.

Eddy, A. A. (1980). Some aspects of amino acid transport in yeast. In *Microorganisms and Nitrogen Sources*, ed. J. W. Payne, pp. 35-62. John Wiley: Chichester.

Eddy, A. A. (1982). Mechanisms of solute transport in selected eukaryotic microorganisms. *Advances in Microbial Physiology*, **23**, 1-78.

Ewaze, J. O., Moore, D. & Stewart, G. R. (1978). Co-ordinate regulation of enzymes involved in ornithine metabolism and its relation to sporophore morphogenesis in *Coprinus cinereus*. *Journal of General Microbiology*, **107**, 343-357.

Fawole, M. O. & Casselton, P. J. (1972). Observations on the regulation of glutamate dehydrogenase activity in *Coprinus lagopus*. *Journal of Experimental Botany*, **23**, 530-551.

Fincham, J. R. S., Kinnaird, J. H. & Burns, P. A. (1985). The *am* (NADP-specific glutamate dehydrogenase) gene of *Neurospora crassa*. In *Molecular Genetics of Filamentous Fungi*, UCLA Symposia on Molecular and Cellular Biology, New Series volume 34, ed. W. Timberlake, pp. 117-125. Alan R. Liss: New York.

Genetet, I., Martin, F. & Stewart, G. R. (1984). Nitrogen assimilation in mycorrhizas. Ammonium assimilation in the N-starved ectomycorrhizal fungus *Cenococcum graniforme*. *Plant Physiology*, **76**, 395-399.

Harley, J. L. (1978). Ectomycorrhizas as nutrient absorbing organs. *Proceedings of the Royal Society of London, B*, **203**, 1-21.

Henry, S. A., Klig, L. S. & Loewy, B. S. (1984). The genetic regulation and co-ordination of biosynthetic pathways in yeast: amino acid and phospholipid synthesis. *Annual Review of Genetics*, **18**, 207-231.

Hohl, H. R. (1983). Nutrition of *Phytophthora*. In *Phytophthora: its Biology, Taxonomy, Ecology and Pathology*, eds. D. C. Erwin, S. Bartnicki-Garcia & P. H. Tsao, pp. 41-54. American Phytopathological Society: St. Paul, Minnesota, USA.

Hohn, T. M., Lovett, J. S. & Bracker, C. E. (1984). Characterisation of the major proteins, gamma particles, cytoplasmic organelles in *Blastocladiella emersonii* zoospores. *Journal of Bacteriology*, **158**, 253-263.

Holder, A. A., Wootton, J. C., Baron, A. J., Chambers, G. J. K. & Fincham, J. R. S. (1975). The amino acid sequence of *Neurospora* NADP-specific glutamate dehydrogenase. *Biochemical Journal*, **149**, 757-773.

Iaccarino, M., Guardiola, J. & De Felice, M. (1980). Genetics of amino acid transport. In *Microorganisms and Nitrogen Sources*, ed. J. W. Payne, pp. 125-151. John Wiley: Chichester.

Jeanjean, R., Bédu, S., Nieuwenhuis, B. J. W. M. & Hirn, M. (1986). Immunological evidence for the involvement of cell wall proteins in phosphate uptake in the yeast *Saccharomyces cerevisiae*. *Archives of Microbiology*, **144**, 207-212.

Jefferies, T. W., Choi, S. & Kirk, T. K. (1981). Nutritional regulation of lignin degradation by *Phanerochaete chrysosporium*. *Applied & Environmental Microbiology*, **42**, 290-296.

Jennings, D. H. (1964). Changes in the size of orthophosphate pools in mycorrhizal roots of beech with reference to absorption of the ion from the external medium. *New Phytologist*, **63**, 181-193.

Jennings, D. H. (1982). The movement of *Serpula lacrimans* from substrate to substrate over nutritionally inert surfaces. In *Decomposer Basidiomycetes*, British Mycological Society Symposium volume 4, eds. J. C. Frankland, J. H. Hedger & M. J. Swift, pp. 91-108. Cambridge University Press: Cambridge, UK.

Jennings, D. H. (1984a). Polyol metabolism in fungi. *Advances in Microbial Physiology*, **25**, 149-193.

Jennings, D. H. (1984b). Water flow through mycelia. In *The Ecology and Physiology of the Fungal Mycelium*, British Mycological Society Symposium volume 8, eds. D. H. Jennings & A. D. M. Rayner, pp. 143-164. Cambridge University Press: Cambridge, UK.

Jennings, D. H. (1986). Salt relations in cells, tissues and roots. In *Plant Physiology: A Treatise*, vol. IX, *Water and Solutes in Plants*, eds. F. C. Steward, J. F. Sutcliffe & J. E. Dale, pp. 225-379. Academic Press: Orlando, Florida, USA.

Jennings, D. H. (1987a). Presidential Address: The medium is the message. *Transactions of the British Mycological Society*, **89**, 1-11.

Jennings, D. H. (1987b). Translocation of solutes in fungi. *Biological Reviews*, **62**, 215-243.

Jennings, D. H. (1988). The ability of basidiomycete mycelium to move nutrients through the soil ecosystem. In *Field Methods in Terrestrial Ecosystem Nutrient Cycling*, eds. A. F. Harrison, P. Ineson & O. W. Heal, (in press). Elsevier.

Jones, E. W. & Fink, G. R. (1982). Regulation of amino acid and nucleotide biosynthesis in yeast. In *The Molecular Biology of the Yeast Saccharomyces: Metabolism and Gene Expression*, eds. J. N. Strathern, E. W. Jones & J. R. Broach, pp. 181-297. Cold Spring Harbor Laboratory: New York.

Jones, E. W. (1984). The synthesis and function of proteases in *Saccharomyces*: genetic approaches. *Annual Review of Genetics*, **18**, 233-270.

Kalisz, H. M., Moore, D. & Wood, D. A. (1986). Protein utilization by Basidiomycete fungi. *Transactions of the British Mycological Society*, **86**, 519-525.

Kalisz, H. M., Wood, D. A. & Moore, D. (1987). Production, regulation and release of extracellular proteinase activity in Basidiomycete fungi. *Transactions of the British Mycological Society*, **88**, 221-227.

Khan, J. A. & Bhatnagar, R. M. (1977). Studies on solubilization of insoluble phosphates by microorganisms. 1. Solubilization of Indian phosphate rocks by *Aspergillus niger* and *Penicillium* sp. *Fertilizer Technology*, **14**, 329-333.

Kinnaird, J. H. & Fincham, J. R. S. (1983). The complete nucleotide sequence of the *Neurospora crassa am* (NADP-specific glutamate dehydrogenase) gene. *Gene*, **26**, 253-260.

Kinsey, J. A., Fincham, J. R. S., Siddig, M. A. & Keighren, M. (1980). New mutational variants of *Neurospora* NADP-specific glutamate dehydrogenase. *Genetics*, **95**, 305-316.

Knowles, C. J. (1977). Microbial metabolic regulation by adenine nucleotide pools. *Symposia of the Society for General Microbiology*, **27**, 241-284.

Kucey, R. M. N. (1987). Increased phosphorus uptake by wheat and field beans inoculated with a phosphorus-solubilizing *Penicillium bilaji* strain and with vesicular-arbuscular mycorrhizal fungi. *Applied & Environmental Microbiology*, **53**, 2699-2703.

Kulaev, I. S. (1979). *The Biochemistry of Inorganic Polyphosphates*. John Wiley: Chichester.

Kulaev, I. S. & Vagabov, V. M. (1983). Polyphosphate metabolism in micro-organisms. *Advances in Microbial Physiology*, **24**, 83-171.

LéJohn, H. B. (1971). Enzyme regulation, lysine pathways and cell wall structures as indicators of major lines of evolution in fungi. *Nature*, **231**, 164-168.

Lambers, H. (1980) The physiological significance of cyanide-resistant respiration in higher plants. *Plant, Cell & Environment*, **3**, 293-302.

Lea, P. J. & Miflin, B. B. (1974). An alternative route for nitrogen assimilation in higher plants. *Nature*, **251**, 614-616.

Levi, M. P. & Cowling, E. B. (1969). Role of nitrogen in wood deterioration. VII. Physiological adaptation of wood-destroying and other fungi to substrates deficient in nitrogen. *Phytopathology*, **59**, 460-468.

Lewis, D. H. & Harley, J. L. (1965). Carbohydrate physiology of mycorrhizal roots of beech. III. Movement of sugars between host and fungus. *New Phytologist*, **64**, 256-269.

Lichko, L. P., Okorokov, L. A. & Kulaev, S. (1982). Participation of vacuole in regulation of levels of K^+, Mg^{2+} and orthophosphate ions in cytoplasm of the yeast. *Archives of Microbiology*, **132**, 289-293.

Lloyd, D. (1974). *The Mitochondria of Microorganisms*. Academic Press: New York.

Loll, M. J. & Bollag, J.-M. (1983). Protein transformation in soil. *Advances in Agronomy*, **36**, 351-382.

Lopez, S. & Gancedo, J. (1979). Effect of metabolic conditions on protein turnover in yeast. *Biochemical Journal*, **178**, 769-776.

Lowendorf, H. S., Bazinet Jr., G. F. & Slayman, C. W. (1975). Phosphate transport in *Neurospora*. Derepression of a high-affinity transport system during phosphorus starvation. *Biochimica et Biophysica Acta*, **389**, 541-549.

Lowendorf, H. S., Slayman, C. L. & Slayman, C. W. (1974). Phosphate transport in *Neurospora*. Kinetic characterization of a constitutive low-affinity transport system. *Biochimica et Biophysica Acta*, **373**, 369-382.

Lowendorf, H. S. & Slayman, C. W. (1975). Genetic regulation of phosphate transport system II in *Neurospora*. *Biochimica et Biophysica Acta*, **413**, 95-103.

Maclean, D. J. (1982). Axenic culture and metabolism in rust fungi. In The Rust Fungi, eds. K. J. Scott & A. K. Chakravorty, pp. 37-84. Academic Press: London.

Markham, P. & Collinge, A. J. (1987). Woronin bodies of filamentous fungi. *FEMS Microbiology Reviews*, **46**, 1-11.

Martin, F. (1985). ^{15}N-NMR studies of nitrogen assimilation and amino acid biosynthesis in the ectomycorrhizal fungus *Cenococcum graniforme*. *FEBS Letters*, **182**, 350-354.

Martin, F., Stewart, G. R., Genetet, I. & Le Tacon, F. (1986). Assimilation of [15]NH_4^+ by beech (*Fagus sylvatica* L.) ectomycorrhizas. *New Phytologist*, **102**, 85-94.

Marzluf, G. A. (1981). Regulation of nitrogen metabolism and gene expression in fungi. *Microbiological Reviews*, **45**, 437-461.

Maxwell, D. P., Maxwell, M. D., Hanssley, G., Armentrout, V. N., Murray, G. M. & Hoch, H. C. (1975). Microbodies and glyoxylate-cycle enzyme activities in filamentous fungi. *Planta*, **124**, 109-123.

Merrill, W. & Cowling, E. B. (1966). Role of nitrogen in wood deterioration: amounts and distribution of nitrogen in tree stems. *Canadian Journal of Botany*, **44**, 1555-1580.

Messenguy, F. (1987). Multiplicity of regulatory mechanisms controlling amino acid biosynthesis in *Saccharomyces cerevisiae*. *Microbiological Sciences*, **4**, 150-153.

Metzenberg, R. L. (1979). Implication of some genetic control mechanisms in *Neurospora*. *Microbiological Reviews*, **43**, 361-383.

Michalenko, G. O., Hohl, H. R. & Rast, D. (1976). Chemistry and architecture of the mycelial wall of *Agaricus bisporus*. *Journal of General Microbiology*, **92**, 251-263.

Miflin, B. J. & Lea, P. J. (1980). Ammonia assimilation. In *The Biochemistry of Plants: A Comprehensive Treatise*, vol. 5, *Amino Acids and Derivatives*, ed. B. B. Miflin, pp. 169-202. Academic Press: New York.

Millard, P. (1988). The accumulation and storage of nitrogen by herbaceous plants. *Plant, Cell & Environment*, **11**, 1-8.

Moore, D. (1981). Evidence that the NADP-linked glutamate dehydrogenase of *Coprinus cinereus* is regulated by acetyl-CoA and ammonium levels. *Biochimica et Biophysica Acta*, **661**, 247-254.

Moore, D. (1984). Developmental biology of the *Coprinus cinereus* carpophore: metabolic regulation in relation to cap morphogenesis. *Experimental Mycology*, **8**, 283-297.

Moore, D. & Ewaze, J. O. (1976). Activities of some enzymes involved in metabolism of carbohydrate during carpophore development in *Coprinus cinereus*. *Journal of General Microbiology*, **97**, 313-322.

Moore, D., Horner, J. & Liu, M. (1987). Co-ordinate control of ammonium-scavenging enzymes in the fruit-body cap of *Coprinus cinereus* avoids inhibition of sporulation by ammonium. *FEMS Microbiology Letters*, **44**, 239-242.

Nicolay, K., Scheffers, W. A., Briunenberg, P. M. & Kapstein, R. (1983). *In vivo* ^{31}P NMR studies on the role of the vacuole in phosphate metabolism in yeasts. *Archives of Microbiology*, **134**, 270-275.

Nieuwenhuis, B. J. W. M. & Borst-Pauwels, G. W.F. H. (1984). Derepression of the high affinity phosphate uptake system in the yeast *Saccharomyces cerevisiae*. *Biochimica et Biophysica Acta*, **770**, 40-46.

North, M. J. (1982). Comparative biochemistry of the proteinases of eukaryotic microorganisms. *Microbiological Reviews*, **46**, 308-340.

Okorokov, L. A., Lichko, L. P. & Kulaev, I. S. (1980). Vacuoles: main compartments of potassium, magnesium and phosphate ions in *Saccharomyces carlsbergensis* cells. *Journal of Bacteriology*, **144**, 661-665.

Pateman, J. A. & Kinghorn, J. R. (1976). Nitrogen metabolism. In *The Filamentous Fungi*, vol. 2, *Biosynthesis and Metabolism*, eds. J. E. Smith & D. R. Berry, pp. 159-237. Edward Arnold: London.

Paton, F. M. & Jennings, D. H. (1988). Evidence that *Thraustochytrium* is unable to synthesise lysine. *Transactions of the British Mycological Society*, in press.

Paustian, K. (1985). Influence of fungal growth pattern on decomposition and nitrogen mineralisation in a model system. In *Ecological Interactions in Soil*, ed. A. H. Fitter, pp. 159-174. Blackwell Scientific Publications: Oxford.

Paustian, K. & Schnürer, J. (1987a). Fungal growth response to carbon and nitrogen limitation: application of a model to laboratory and field data. *Soil Biology & Biochemistry*, **19**, 621-629.

Paustian, K. & Schnürer, J. (1987b). Fungal growth response to carbon and nitrogen limitation: a theoretical model. *Soil Biology & Biochemistry*, **19**, 613-620.

Rao, P. S. & Niederpruem, D. J. (1969). Carbohydrate metabolism during morphogenesis of *Coprinus lagopus* (*sensu* Buller). *Journal of Bacteriology*, **100**, 1222-1228.

Robertson, B. R. & Button, D. K. (1979). Phosphate-limited continuous culture of *Rhodotorula rubra*: kinetics of transport, leakage and growth. *Journal of Bacteriology*, **138**, 884-895.

Sanders, D. (1986). Generalised kinetic analysis of ion-driven co-transport systems. II. Random ligand binding as a simple explanation of non-Michaelian kinetics. *Journal of Membrane Biology*, **90**, 67-81.

Stewart, G. R. & Moore, D. (1974). The activities of glutamate dehydrogenase during mycelial growth and sporophore development in *Coprinus lagopus* (*sensu* Lewis). *Journal of General Microbiology*, **83**, 73-81.

Straker, C. J. & Mitchell, D. T. (1987). Kinetic characterization of a dual phosphate uptake system in the endomycorrhizal fungus of *Erica hispindula* L. *New Phytologist*, **106**, 129-137.

Tinker, P. B. H. (1975). Effects of vesicular-arbuscular mycorrhizas on higher plants. In *Symbiosis*, Society for Experimental Biology volume 29, eds. D. H. Jennings & D. L. Lee, pp. 325-349. Cambridge University Press: Cambridge, UK.

Veenhuis, M., van Dijken, J. P. & Harder, W. (1983). The significance of peroxisomes in the metabolism of one-carbon compounds in yeasts. *Advances in Microbial Physiology*, **24**, 1-82.

Vogel, H. J. (1964). Distribution of lysine pathway among fungi: evolutionary implications. *The American Naturalist*, **98**, 435-446.

Watkinson, S. C. (1979), Growth of rhizomorphs, mycelial strands, coremia and sclerotia. In *Fungal Walls and Hyphal Growth*, British Mycological Society Symposium volume 2, eds. J. H. Burnett & A. P. J. Trinci, pp. 93-113. Cambridge University Press: Cambridge, UK.

Watkinson, S. C. (1984). Morphogenesis of the *Serpula lacrimans* colony in relation to its function in nature. In *Ecology and Physiology of the Fungal Mycelium*, British Mycological Society Symposium volume 8, eds. D. H. Jennings & A. D. M. Rayner, pp. 165-184. Cambridge University Press: Cambridge, UK.

Wiame, J. M., Grenson, M. & Arst Jr, H. N. (1985). Nitrogen catabolite repression in yeasts and filamentous fungi. *Advances in Microbial Physiology*, **26**, 1-87.

Wolf, D. H. (1980). Control of metabolism in yeast and other lower eukaryotes through action of proteinases. *Advances in Microbial Physiology*, **21**, 267-338.

Wolf, D. H. (1986). Cellular control in the eukaryotic cell through action of proteinases: the yeast *Saccharomyces cerevisiae* as a model organism. *Microbiological Sciences*, **3**, 107-114.

Wolfinbarger Jr., L. (1980a). Transport and utilization of amino acids by fungi. In *Microorganisms and Nitrogen Sources*, ed. J. W. Payne, pp. 63-87. John Wiley: Chichester.

Wolfinbarger Jr., L. (1980b). Transport and utilization of peptides by fungi. In *Microorganisms and Nitrogen Sources*, ed. J. W. Payne, pp. 281-300. John Wiley: Chichester.

Chapter 2

The genetics and biochemistry of nitrate assimilation in ascomycete fungi

A. B. Tomsett

Department of Genetics and Microbiology, The University of Liverpool, P.O. Box 147, Liverpool, L69 3BX, UK

Introduction

All organisms require reduced forms of nitrogen to synthesize proteins and nucleic acids. Inorganic nitrogen is assimilated to the reduced forms by two major pathways, nitrate assimilation and nitrogen fixation. Guerrero, Vega & Losada (1981) have calculated the relative contribution of these two pathways to reduced nitrogen in the biosphere: nitrate assimilation is the major pathway being 100-fold greater than the total contribution of nitrogen fixation. The study of nitrate assimilation therefore, has obvious importance for both ecology and agriculture. Through the combination of genetical, biochemical and more recently, molecular biological techniques, we have developed an excellent understanding of the enzymology and genetic regulation of this pathway.

In this paper, I have been asked to present a general overview of the subject in a limited space. Our knowledge of nitrate assimilation is now sufficiently detailed to fill this entire volume. Even a moderately comprehensive list of the literature would exceed the space available for this article. Inevitably, this review is highly selective in its content: it examines the outstanding contribution made by classical genetic analysis to our understanding of nitrate assimilation, and how biochemistry and molecular biology have built upon its conclusions and hypotheses. Most of this article concerns the study of the two ascomycete fungi *Aspergillus nidulans* and *Neurospora crassa*, although I refer to other organisms where the investigations are more advanced. I have also attempted to

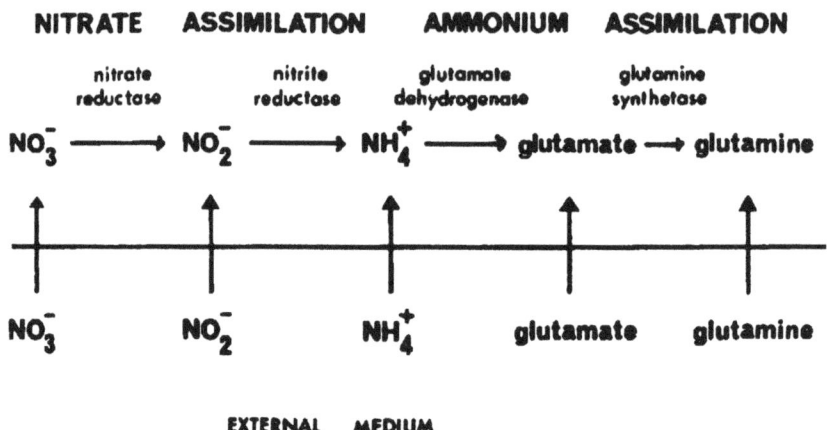

NITRATE   ASSIMILATION   AMMONIUM   ASSIMILATION

Fig. 2.1. The pathways of nitrate and ammonium assimilation.

write this for the non-specialist. For those wishing to pursue this field, there are a number of excellent more extensive reviews (Garrett & Amy, 1978; Cove, 1979; Guerrero *et al.*, 1981; Marzluf, 1981; Dunn-Coleman, Smarrelli & Garrett, 1984; Wiame, Grenson & Arst, 1985).

The pathway of nitrate assimilation

The pathway of nitrate assimilation in the filamentous fungi *A. nidulans* and *N. crassa* involves the complete reduction of nitrate to ammonium, via nitrite (Fig. 2.1). This process occurs in two steps: the reduction of nitrate to nitrite by nitrate reductase (NADPH: nitrate oxidoreductase, E.C. 1.6.6.3.), a reaction involving a two-electron transfer; and the reduction of nitrite to ammonium by nitrite reductase (NAD(P)H: nitrite oxidoreductase, E.C. 1.6.6.4.), a six-electron transfer. The ammonium thus formed is then assimilated into glutamate and glutamine by NADP-linked glutamate dehydrogenase and glutamine synthetase.

Both *A. nidulans* and *N. crassa* can utilise a wide range of nitrogen-containing compounds when supplied as their sole source of nitrogen in the external medium. In broad terms, we can consider three classes of nitrogen-source: inducers, repressors and neutral N-sources. A number of pathways of N-source acquisition in fungi are subject to two genetic control mechanisms, i.e. they are substrate-inducible and product-repressible. For nitrate assimilation, nitrate and/or nitrite will induce enzyme synthesis but only in the absence of the reduced forms of nitrogen, ammonium and glutamine, which act as repressors; compounds such as urea and uric acid are neutral, i.e. they neither induce nor repress nitrate assimilation. For each pathway, substrate-inducibility is pathway-specific but nitrogen metabolite repression controls all the pathways for nitrogen catabolism.

Enzymology

Nitrate reductase is a homodimer of molecular weight 228,000 in *N. crassa* (Garrett & Nason, 1969) and 180,000 in *A. nidulans* (Minagawa & Yoshimoto, 1982). For *N. crassa*, Horner (1983) has reported a subunit size of 145 kD, while the *A. nidulans* subunit is 91 kD (Cooley & Tomsett, 1985). In addition to the two identical polypeptides, nitrate reductase possesses a molybdenum-containing cofactor (see Garrett & Amy, 1978). This cofactor is found in a wide variety of enzymes from diverse phylogenetic sources. In these fungi, the cofactor is also used by xanthine dehydrogenase (E.C.1.2.1.37). Nitrate reductase catalyses an electron-transfer reaction from NADPH to nitrate (Fig. 2.2a). In addition to the molybdenum cofactor, the enzyme contains flavin adenine dinucleotide (FAD) and cytochrome-*b557* (see Dunn-Coleman *et al.*, 1984). Although the physiological activity utilizes the entire electron-transfer sequence, the enzyme has a number of non-physiological partial activities which can be assayed *in vitro*. The initial part of the sequence can be measured by the reduction of certain electron acceptors, e.g. cytochrome-*c* (NADPH-CCR), the latter part by measurement of nitrite production when electrons are passed to the enzyme by reduced methyl viologen (MVH-NaR). These partial activities afford the opportunity of assessing the presence of mutant enzyme molecules.

(a)

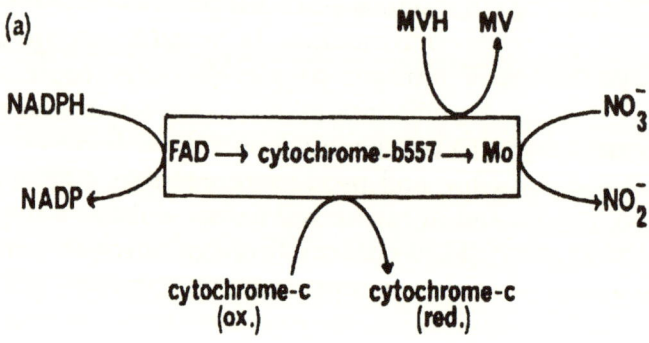

(b)

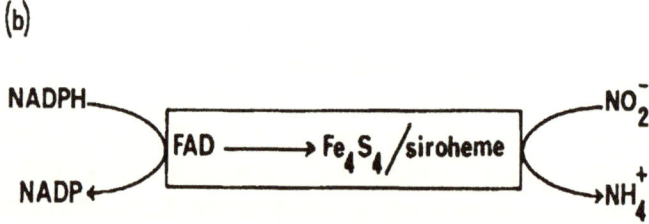

Fig. 2.2. The electron-transfer reactions of (a) nitrate reductase and (b) nitrite reductase.

Nitrite reductase has a molecular weight of 290,000, a homodimer of identical 140,000 molecular weight protomers. The enzyme catalyses a six-electron transfer from NADPH to nitrite, via FAD, an iron-sulphur centre and a siroheme cofactor (Fig. 2.2b; see Dunn-Coleman *et al.*, 1984).

Mutant isolation and characterization

The genetic analysis of nitrate assimilation has been greatly facilitated by having a positive selection method for nitrate-non-utilising mutants, based on the finding that cells lacking nitrate reductase are resistant to the toxic analogue, chlorate (Aberg, 1947). The exact mechanism of chlorate toxicity is unknown. The simplest hypothesis is that nitrate reductase catalyses the conversion of chlorate to chlorite, a highly reactive substance (Aberg, 1947). This mechanism is almost certainly correct, but is insufficient to explain all the data (Cove, 1976a & b, 1979). Nevertheless, resistance to chlorate has been used successfully to select mutants defective in nitrate assimilation in a variety of organisms, including *A. nidulans* and *N. crassa* (Cove, 1976a; Tomsett & Garrett, 1980).

Since the correlation between chlorate resistance and lack of nitrate reductase activity is not absolute, it has been equally important to isolate nitrate non-utilising mutants by other criteria. Several such mutants were obtained in *N. crassa* by a variety of workers (see Garrett & Amy, 1978). In *A. nidulans*, in addition to those isolated without positive selection, the technique of putrescine starvation yields useful numbers of nitrate non-utilising mutants (Cove, 1976a).

A third technique can be used to recognise mutants which have normal or higher levels of nitrate reductase activity. Nitrite in the presence of sulphanilic acid and N-1-naphthylethylenediamine dihydrochloride (NED) produces a pink coloration. When growing on solid media, colonies which can convert nitrate to nitrite will stain pink; colonies which excrete nitrite have a pink halo (see Tomsett & Garrett, 1980). These methods have been used to isolate a wide variety of mutants. Such mutants can be classified on the basis of their ability to utilise certain nitrogen sources, their resistance to chlorate and whether they excrete nitrite (Table 2.1). Using complementation tests, these groups of mutants can be shown to define individual genetic loci as shown in Table 2.1.

Nitrate reductase apoprotein

The nitrate reductase apoprotein is encoded by a single gene, *nia*D in *A. nidulans* and *nit*-3 in *N. crassa* (Cove, 1979; Garrett & Amy, 1978). Mutants in these genes have a unique phenotype, an inability to utilise nitrate whilst exhibiting normal growth on all other nitrogen sources. This phenotype results from a loss of NADPH-nitrate reductase activity which has been demonstrated in a large number of individual mutants (see Cove, 1979; Tomsett & Garrett, 1981). In *A. nidulans*, the genetic evidence that *nia*D encodes the apoprotein comes from temperature-sensitive *nia*D mutants which retain nitrate reductase activity when the organism

Table 2.1. Growth test classification of mutant strains.

Function	A. nidulans locus	N. crassa locus	Growth media							Plate test for nitrite production
			NH$_4^+$	NO$_3^-$	NO$_2^-$	Hypo xanthine	Uric acid	Uric acid +100 mM ClO$_3^-$		
	wild type		+	+	+	+	+	VS		+
NAR structural gene encoding the apoprotein	niaD	nit-3	+	-	+	+	+	most R, few S		-
NIR structural gene encoding the apoprotein	niiA	nit-6	+	-	-	+	+	VS		++
Formation of the molybdenum-containing cofactor required by both NAR and XDH	cnxABC, E, F, G, H	nit-1, 7, 8, 9ABC	+	-	+	-	+	most R, few S		-
Regulator gene necessary for nitrate induction of nitrate assimilation	nirA*	nit-4	+	-	-	+	+	SR		-
Regulator gene responsible for nitrogen metabolite repression of nitrate assimilation and other pathways of nitrogen acquisition	areA*	nit-2	+	-	-	-	-	-		-
Nitrate permease gene	crnA	?	+	+	+	+	+	SR		+

NAR = nitrate reductase; NIR = nitrite reductase; XDH = xanthine dehydrogenase. For nitrogen source utilization in growth media, + indicates growth on that medium, - indicates no growth (nitrogen-starved morphology). For chlorate resistance, R indicates resistance, SR an intermediate resistance characteristic of nirA⁻ and crnA, VS indicates extreme sensitivity, and - that the strain cannot grow. For the nitrite production plate test, + indicates colony stains pink, ++ indicates the colony has a pink halo from excreted nitrite, and - that the colony does not stain. * The nirA and areA genes can mutate to give a variety of mutant phenotypes, these examples are representative of null alleles. Data compiled from Cove (1976a & b), Tomsett & Garrett (1980) and see Wiame et al. (1985).

is grown at 25°C but lack the activity when grown at 37°C. The enzyme from these mutants is more thermolabile than that of the wild-type and hence, has a much shorter half-life *in vitro* (MacDonald & Cove, 1974).

Despite overwhelming genetic evidence that nitrate reductase has only one specific core polypeptide, it is only recently that this has been characterized biochemically. Nitrate reductase has been extensively purified from both *N. crassa* and *A. nidulans*. In *N. crassa*, nitrate reductase was reported to have subunit sizes of 115,000 and 130,000 daltons, but one was presumed to be a degradation product of the other (Pan & Nason, 1978). In *A. nidulans*, the subunit structure was difficult to define. Downey & Focht (1974) isolated a protomer of molecular weight 49,000. Downey & Steiner (1979) later identified proteins which had subunit molecular weights of 49,000, 50,000 and 75,000 daltons. In another study, Steiner & Downey (1982) demonstrated a single polypeptide of molecular weight 54,000, suggesting a homotetrameric enzyme structure for nitrate reductase, whereas Minagawa & Yoshimoto (1982) described the *A. nidulans* nitrate reductase to be 180,000 daltons, consisting of two each of two subunits of 59,000 and 38,000 daltons. These studies of *N. crassa* and *A. nidulans* all involved extensive purification procedures, during which the enzyme was presumably exposed to proteolysis producing the variation in size reported. In the most recent studies, Horner (1983) and Cooley & Tomsett (1985) both employed rapid purification in the presence of higher concentrations of protease inhibitors. For *N. crassa* the subunit size is apparently 145,000 daltons (Horner, 1983), while in *A. nidulans*, a 91,000 dalton peptide is the *nia*D gene product (Cooley & Tomsett, 1985). The fact that proteolysis of nitrate reductase gives the discrete sizes reported above indicates that there may be one or more sites in the core polypeptide which are susceptible to proteolysis. Native enzyme can be purified because the cleaved products remain bound and hence, the enzyme retains full enzyme activity. Upon dissociation by SDS, one or more separate protein fragments are observed. Minagawa & Yoshimoto (1982) report that the 38 kD peptide subunit can be dissociated from the native enzyme by heat treatment, resulting in the loss of overall NADPH-NaR activity and of the NADPH-CCR partial activity but not MVH-NaR. This strongly suggests functional domains for *A.nidulans* nitrate reductase separated by a protease-sensitive site.

Solomonson and his coworkers have examined functional domains in the NADPH-nitrate reductase from *Chlorella vulgaris* (Solomonson *et al.*, 1986). Their model (Fig. 2.3) includes two functional domains connected by a protease-sensitive hinge region. The *Chlorella* enzyme is believed to be a homotetramer, native molecular weight 360 - 380 kD with four 90 - 100 kD subunits.

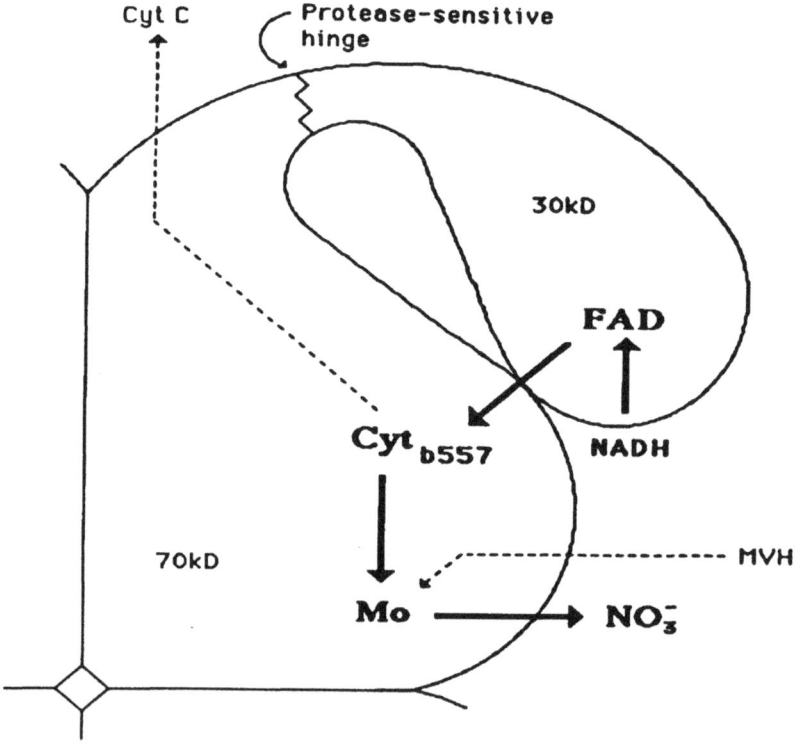

Fig. 2.3. A model for the structure of nitrate reductase from *Chlorella vulgaris* (after Solomonson *et al.*, 1986).

Each 100 kD subunit has a 70 kD domain which is at the core of the molecule and contains the $b5$-type cytochrome and the molybdenum cofactor. The second domain, about 30 kD, is exposed on the surface of the enzyme and contains the FAD/NADH-binding activities. The quaternary structure of the enzyme is maintained by interaction between the 70 kD structures: the 30 kD moieties are linked to the 70 kD molecules by the protease-sensitive hinge and have no role in maintaining the overall structure. Proteolysis releases the 30 kD peptide with consequent loss of the NADH-CCR activity. The 30 kD peptide is thought to be physically closely associated with the 70 kD core.

This model would elegantly explain the data of Minagawa & Yoshimoto (1982): native enzyme cleaved at the hinge *in vivo* or during extensive purification, retains activity by the close physical association between the two domains, but upon heat treatment the 38 kD FAD/NADPH domain is lost leaving the core enzyme (estimated at 126,000 daltons) composed of a dimer of core peptides each 59 kD. Pan & Nason (1978) observed a major 115 kD peptide (and a minor 130 kD peptide) derived from the 145 kD *nit*-3 gene product: this also can be explained by loss of a 30 kD moiety from the native subunit polypeptide. Since the 115 kD and 130 kD polypeptides have the same N-terminal amino acid (Pan & Nason, 1978), it seems likely that the lost 30 kD peptide is from the C-terminal end.

Le & Lederer (1983) have shown, by amino acid sequencing, that the heme-binding domain of *N. crassa* nitrate reductase is a member of the cytochrome-$b5$ superfamily. Analysis of this superfamily reveals 13 conserved amino acid residues. Recently, Caboche and his coworkers (Calza *et al.*, 1987) have isolated cDNA clones complementary to tobacco nitrate reductase mRNA. The amino acid sequence, deduced from the DNA sequence, shows significant homology with different members of the superfamily, including all 13 of the conserved residues. This clearly demonstrates that the cytochrome-$b557$ is encoded within the nitrate reductase structural gene. From the data available, the tobacco and *Neurospora* sequences show only 33 per cent homology in this region.

Calza *et al.* (1987) have also compared their sequence with others in sequence data banks. Downstream (3' on the mRNA) of the cytochrome-*b5* homology, there is a significant homology with human flavoprotein cytochrome-*b5* reductase. This protein catalyses the reduction of cytochrome-*b5*, using NADH as an electron donor and FAD as the redox intermediate; this is analogous to the first partial activity of nitrate reductase. This NADH/FAD domain is separated by 35 amino acids from the cytochrome-*b557* domain and lies toward the C-terminal end of the polypeptide. Unfortunately, the complete sequence for the tobacco nitrate reductase is not yet available. The cDNA contains only about 60% (60 kD polypeptide) of the native polypeptide (100 kD). However, two genomic DNA clones have now been isolated from *Nicotiana tabacum*, each of which can be attributed to one or the other of the ancestors of tobacco, *Nicotiana sylvestris* and *Nicotiana tomentosiformis* (M. Caboche *et al.*, unpublished data). These reveal that each contains two introns which differ in size, but not position, between the two clones. The intron positions map to divide the 1·6 kb cDNA clone into three approximately equal sections. The native DNA sequence appears to start 800 bp (approx.) upstream (5′) of the cDNA clone and extend 1·1 kb (approx.) downstream (3′) of it (M. Caboche *et al.*, unpublished data). However, their data seem to include both the cytochrome-*b557* and the NADH/FAD domains within the same exon, i.e. the third exon downstream. Comparing this with the model of Solomonson *et al.* (1986), and the loss of the C-terminal fragment from *N. crassa* nitrate reductase (Pan & Nason, 1978), we might predict that the molybdenum-cofactor binding site should be located within exon 1 and/or exon 2 (if there is an overall structural analogy in addition to the enzymological (catalytic) similarities). If the model of Solomonson *et al.* (1986) is applicable, one might also expect to find the protease-sensitive hinge region within, or close to, the 35 amino acids separating the cytochrome-*b557* and NADH/FAD domains. This would yield a 39 kD (approx.) C-terminal fragment.

If the protease-sensitive site is a common feature of all NAD(P)H-nitrate reductases, it is tempting to suggest this has an important physiological role. Proteases have been implicated in the turnover of nitrate reductase (see Dunn-Coleman *et al.*, 1984). In *N. crassa*, two nitrate reductase inactivators, both proteolytic, have been described (Sorger, Premakumar & Gooden, 1978;

Walls *et al.*, 1978) and a role for their control of nitrate-reductase has been proposed.

As yet, many of the details of molecular studies of the *nia*D gene of *A. nidulans* and the *nit*-3 gene of *N. crassa* have not been published. The *nit*-3 gene is transcribed to give a 3.4 kb mRNA but as yet the gene has not been sequenced (Fu & Marzluf, unpublished data). The *nia*D gene transcript is about 2.8 - 3 kb; the gene contains 6 introns (J. R. Kinghorn, unpublished data). The *nia*D gene clones have been sequenced but the data are not available at the time of writing this review.

The molybdenum cofactor

Among mutants isolated for their inability to utilise nitrate are a common class which cannot grow on hypoxanthine as sole nitrogen source (see Table 2.1). This phenotype results from a loss of both NADPH-nitrate reductase activity and xanthine dehydrogenase activity (see Cove, 1979; Tomsett & Garrett, 1981). These mutants retain the NADPH-cytochrome-*c* reductase activity of nitrate reductase and lack its MVH-NaR activity which requires the molybdenum-containing cofactor. Pairwise complementation tests between such mutants reveal that they are encoded by several genes. In *A. nidulans*, the initial studies described seven complementation groups (see Cove, 1979): *cnx*E, F, G and H mutants each formed a discrete group and hence these define four separate genes; *cnx*A, B and C mutants show an overlapping complementation pattern, *cnx*B mutants fail to complement mutants in *cnx*A, *cnx*C or of course *cnx*B; *cnx*A and *cnx*C each form a discrete group which complement each other. Since *cnx*A, B and C mutants are all closely linked, it is uncertain whether this locus defines one gene, in which A and C exhibit intragenic complementation, or two genes, *cnx*B being mutant in both (for discussion see Cove, 1979, and Scazzocchio, 1980).

More recently, another *cnx* locus has been described in *A. nidulans* (Arst, Tollervey & Sealy-Lewis, 1982). Unlike the other *cnx* mutants, *cnx*J mutants are unaffected for their utilization of nitrate and hypoxanthine unless tungstate or methylammonium is present. Tungstate is a molybdate analogue, which presumably lowers the concentration of active molybdenum cofactor by competition with molybdate at the level of uptake and/or incorporation into the cofactor (see Arst *et al.*, 1982). Methylammonium, a toxic analogue of ammonium, may also lower cofactor concentrations by triggering nitrogen metabolite repression (see below) of cofactor synthesis (for further discussion, see Arst *et al.*, 1982). Because of this unusual phenotype, Arst *et al.* (1982) have tentatively suggested *cnx*J could regulate cofactor biosynthesis, although no clear evidence for this exists.

The role of the *cnx*ABC, E, F, G and H genes is also unclear. Because mutation at any of these loci results in loss of enzyme activity, all are assumed to have an obligatory function in cofactor synthesis (or its regulation). Mutants in *cnx*E, when grown on high concentrations of molybdate, have some nitrate reductase and xanthine dehydrogenase activity restored (Arst, MacDonald & Cove, 1970). This molybdenum-repairable phenotype led to their suggestion that the *cnx*E gene product is responsible for insertion of molybdenum into the cofactor. The only other evidence for the role of these genes comes from study of temperature-sensitive mutations. MacDonald & Cove (1974) measured the thermostability of nitrate reductase *in vitro*. Temperature-sensitive alleles of *cnx*E and *cnx*F synthesized nitrate reductase with the same half-life as that from wild-type strains. Nitrate reductase from *cnx*H strains was, however, more thermolabile than from wild-type strains. This evidence suggests that the *cnx*H gene product is a polypeptide which is a structural component of the nitrate reductase molecule; whereas, the *cnx*E and *cnx*F gene products are presumably enzymes responsible for the synthesis of cofactor, but are not themselves part of its structure. Arst *et al.* (1982) have also described a cryosensitive *cnx*C allele, which grows normally at 37°C but not at 25°C. This strain has an extremely labile nitrate reductase and hence it is possible that the *cnx*C gene product is also a structural component of the enzyme. A model has been proposed to accommodate the genetic evidence for cofactor biosynthesis and nitrate reductase assembly (Cove, 1979).

In *N. crassa*, four loci have been identified which are responsible for molybdenum cofactor biosynthesis, using the same criteria as for *A. nidulans* (Tomsett & Garrett, 1980). The first gene to be identified, *nit*-1, was isolated as a nitrate/hypoxanthine nonutiliser and has played a major role in the biochemical characterization of the cofactor (see Coddington, 1976, and see below).

Mutation in the three other loci, *nit-7*, *nit-8* and *nit-9ABC*, were selected among chlorate-resistant mutants. Little is known of the role of these *N. crassa* genes. *N. crassa nit-9ABC* shows a similar complementation pattern to *cnx*ABC of *A. nidulans* (Tomsett & Garrett, 1980) but interestingly, unlike *cnx*ABC mutants, alleles of all 3 complementation groups at this locus are molybdenum-re-pairable (Dunn-Coleman, 1984a). Dunn-Coleman (1984b) has isolated a cloned *N. crassa* DNA fragment which, when trans-formed into *Escherichia coli*, can restore *E. coli* nitrate reductase activity to *chl*D mutants. *E. coli chl*D mutants produce a defective molybdenum cofactor. This same DNA fragment also transforms *N. crassa nit-9A* and *nit-9B* to enable growth on nitrate, but not *nit-9C*. As yet, however, no further details of this work have emerged. The discrepancy between the number of loci in *A. nidu-lans* and *N. crassa* almost certainly arises because insufficient ni-trate non-utilizing mutants have been analyzed in *N. crassa*.

Biochemical analysis of the molybdenum cofactor has been pioneered by Rajagopalan and his co-workers. All mo-lybdoenzymes except nitrogenase contain a common cofactor, which is a complex of a novel pterin, called molybdopterin (MPT),

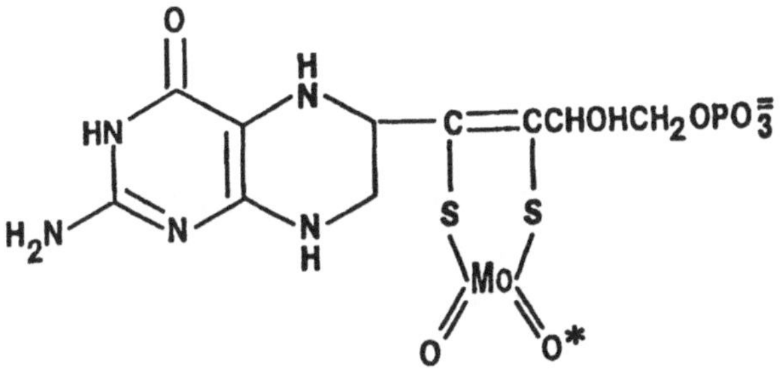

Fig. 2.4. The structure of the molybdenum cofactor (after Johnson & Rajagopalan, 1982). *Sulphur replaces oxygen at this position in xanthine dehydrogenase.

with the molybdenum atom (Johnson, 1980; Johnson *et al.*, 1984). The proposed structure for the molybdopterin with bound molybdenum is shown in Fig. 2.4 (Johnson & Rajagopalan, 1982; Rajagopalan, personal communication).

Until the recent description of the *E. coli chl* M and N loci, the *N. crassa nit*-1 mutant was unique in that it lacks molybdopterin: the other mutants are defective for cofactor biosynthesis but retain some molybdopterin (Kramer, Hageman & Rajagopalan, 1984; Johnson & Rajagopalan, 1987). This has been of great significance in these studies, because molybdopterin from a wide variety of sources can be assayed *in vitro* by its ability to reconstitute nitrate reductase activity in *nit*-1 cell extracts. Thus molybdenum cofactor can be released by denaturation of purified molybdoenzymes, e.g. by guanidine hydrochloride or sodium dodecyl sulphate, and purified by testing fractions with the *nit*-1 reconstitution assay. This has led to the description of the structure shown in Fig. 2.4. Despite the genetic evidence, attempts to identify a polypeptide component of the active cofactor (equivalent to the *cnx*H or *cnx*C gene products of *A. nidulans*) have given negative results. Hageman & Rajagopalan (1985) found no amino acids associated with the cofactor. This evidence is good, but a polypeptide component cannot be totally ruled out since we do not know the nature of the *nit*-1 lesion. It is not inconceivable that *nit*-1 extracts contain such a polypeptide which acts in the reconstitution assay. The purification procedures employed may only measure molybdopterin and hence a polypeptide could be lost at an early stage.

Nitrate and nitrite uptake

The transport of nitrate and nitrite into fungal cells has received relatively little attention in comparison to the enzymology of this pathway. Schloemer & Garrett (1974a) have made a biochemical study of nitrate transport in *N. crassa*. It appears that nitrate uptake is inducible by nitrate or nitrite, even in the presence of ammonium or glutamine; when grown on ammonium as sole nitrogen source, the nitrate uptake system was absent. Curiously, casamino acids repress nitrate uptake, and with the exception of methionine, glutamine and alanine, all individual amino acids prevent nitrate accumulation in the mycelia. These results do not conform to the normal patterns of nitrogen metabolite repression (see below). There is evidence that this is an active transport system which requires transcription and translation. Wild-type and three strains which lack NADPH-nitrate reductase (*nit*-1, *nit*-2, *nit*-3) all displayed similar characteristics for nitrate uptake.

In a parallel study, Schloemer & Garrett (1974b) demonstrated that nitrite uptake in *N. crassa* is induced by nitrate or nitrite, requires transcription and translation, and is an active transport system inhibited by metabolic poisons. They found no evidence, however, for repression or inhibition of this process by ammonium or casamino acids. Coddington (1976) measured nitrite uptake in *nit* mutant strains. He found that *nit*-1 and *nit*-3 strains which can grow on nitrite as sole nitrogen source, did not accumulate nitrite in the mycelia but depleted it from the medium. The *nit*-2 mutant, defective in nitrate and nitrite reductases, totally lacked nitrite uptake, even though it retained nitrate uptake. Mutants of *nit*-4, which lack nitrate and nitrite reductase, accumulate nitrite in the mycelia.

In *A. nidulans*, initial experiments to measure nitrate and nitrite uptake in wild-type, *nia*D, *nii*A and *crn*A mutations used mature mycelia (Tomsett, 1977). These studies demonstrated no difference between wild-type and *crn*A strains but strains lacking nitrate reductase (*nia*D) did not show any nitrate uptake activity, and *nii*A mutants, lacking nitrite reductase, did not show any nitrite uptake activity. This implies that nitrate and nitrite passively diffuse into the mycelium, with diffusion being driven by reduction. Subsequently, Brownlee & Arst (1983) have shown that the *crn*A1 mutation reduces nitrate uptake several-fold in conidia and young mycelia, but it has no effect in older mycelia. They also found that uptake was distinct from, but dependent upon, nitrate reductase activity. Expression of the *crn*A gene is not regulated by *nir*A, which mediates nitrate induction, but is subject to control by the *are*A gene mediating nitrogen metabolite repression.

Although no nitrite uptake gene has formally been described in *A. nidulans*, two mutations have been reported which result in a nitrite hypersensitive phenotype. The *nih*A mutation prevents growth of strains on elevated concentrations of nitrite (Pombeiro, Martinez-Rossi & Rossi, 1983). Since this strain grows normally on normal concentrations of nitrite and can utilize a wide range of nitrogen sources, these authors have suggested that this mutation leads to increased efficiency of nitrite uptake. An alternative explanation, which cannot be eliminated, is that nitrite efflux is reduced or prevented in these strains. The *nih*A locus maps to chromosome I, close to the *pro*A gene. The second nitrite hypersensitive mutation, *nii*C628, also maps to chromosome I, but it is distal to the *bi*A gene (at least 31 centimorgans from *nih*A). The *nii*C628 strains grow with a nitrogen-starved morphology on nitrate or nitrite, but have wild-type levels of both nitrate and nitrite reductases (Cove, 1979). These strains grow normally on all other nitrogen sources but are unable to grown on urea and nitrite. The fact that *nii*C628 has the enzymes to utilise nitrate and nitrite but that nitrite appears to inhibit growth of this mutant again suggests an altered permease function, either increased uptake or decreased efflux.

Nitrite reduction

One of the best biochemical characterizations of nitrite reductase has been undertaken by Garrett and his coworkers in *N. crassa* (Garrett, 1972, 1978; Lafferty & Garrett, 1974). The enzyme has been purified to homogeneity (Greenbaum, Prodouz & Garrett, 1978; Prodouz & Garrett, 1981). It is a homodimer of 140,000 daltons subunits, and contains a siroheme prosthetic group which is thought to serve as the site of binding and reduction of nitrite. Siroheme is also found in assimilatory sulphite reductase. The enzymology of nitrite reductase has been extensively reviewed recently by Dunn-Coleman *et al.*, (1984). Only one structural gene has been identified for nitrite reductase; *nii*A mutants in *A. nidulans* and *nit*-6 mutants of *N. crassa* have identical phenotypes being unable to utilise either nitrate or nitrite as sole nitrogen source (see Cove, 1979; Tomsett & Garrett, 1980). These mutants lack NADPH-nitrite reductase but have normal levels of NADPH-nitrate reductase activity: when nitrate is supplied, they excrete nitrite (see Cove, 1979; Tomsett & Garrett, 1980). There have been no detailed characterizations of mutations affecting siroheme biosynthesis, although a mutant which required both reduced sulphur and reduced nitrogen compounds has been isolated (see Cove, 1979).

The nitrate assimilation gene cluster of *A. nidulans*

The *crn*A, *nii*A and *nia*D genes of *A. nidulans* are tightly linked (Tomsett & Cove, 1979). The genetic organization of this region was determined by deletion mapping. Deletions were selected as spontaneous chlorate-resistant mutants which were unable to utilise either nitrate or nitrite (Cove, 1976a; Tomsett & Cove, 1979). These were easily recognised because *nii*A single mutants are chlorate sensitive and *nir*A⁻ mutants have a distinctive intermediate pattern of resistance. The frequency at which they arose was too high for them to be two independent point mutations. Mutations of *crn*A were initially mapped using heterozygous diploids. However, preliminary studies of bromate toxicity showed that *crn*A mutant strains were more resistant than the wild-type to 1mM bromate when uric acid was present as the nitrogen source. Point mutations involving either the *nii*A or the *nia*D genes do not lead to similar bromate resistance. Analysis of *nii*A *nia*D deletions revealed two groups based upon their bromate resistance. Bromate resistant deletions also lacked part or all of the *crn*A gene, whereas bromate sensitive deletions terminated within the *nii*A gene. The gene order within the cluster was thus defined: *crn*A - *nii*A - *nia*D (Tomsett & Cove, 1979).

Rand & Arst (1977) isolated a mutant, *nis*5, which was tightly linked to the gene cluster but only affected nitrite reductase activity. The *nis*-5 strains have 40% of the wild-type activity when induced, and a low constitutive activity which was insensitive to nitrogen metabolite repression. The mutation results from a translocation which apparently links a new, but inefficient promoter to the *nii*A gene. Deletion mapping showed that *nis*-5 mapped to the central region between the *nii*A and *nia*D genes (Tomsett, 1977) and this was later confirmed as a large portion of the right arm of linkage group II (Arst, Rand & Bailey, 1979). The fact that the *nii*A and *nia*D genes could be separated, but that the *nii*A retained some expression from its normal promoter, is a strong indication that the *nii*A and *nia*D genes are expressed on separate transcripts. Unfortunately, the details of the *nii*A and *nia*D gene sequences were not available to me at the time of writing this review, although the region between *nii*A and *nia*D does contain separ-

ate divergent promoters for these genes (J. R. Kinghorn, personal
communication).

Nitrate induction

The synthesis of the two enzymes of nitrate assimilation is subject
to two control mechanisms: induction by nitrate and/or nitrite and
nitrogen metabolite repression. Both mechanisms are mediated by
the products of positive regulatory genes; that is, proteins bind to
DNA and switch on structural gene expression.

Nitrate/nitrite induction is a pathway-specific regulatory system. In the absence of ammonium or glutamine (the effector of
nitrogen metabolite repression), the pathway of nitrate assimilation is expressed only in the presence of nitrate and/or nitrite. The
$nirA$ gene in *A. nidulans* and the nit-4 gene in *N. crassa* encode
the regulatory protein mediating this process (see Cove, 1979;
Tomsett & Garrett, 1980). The most common class of mutations
in any gene are those which result in the synthesis of an inactive
gene product or no gene product. In $nirA$ and nit-4, most mutations result in an inability to grow on nitrate or nitrite (Cove, 1979;
Tomsett & Garrett, 1980). This can be shown to result from an inability to synthesize normal wild-type levels of both nitrate reductase and nitrite reductase when supplied nitrate or nitrite. Most
retain a low constitutive level of both enzymes on a neutral nitrogen source such as urea whether nitrate/nitrite are present or
not. In the presence of ammonium or glutamine, however, even
this low activity is repressed (Cove, 1979; Tomsett & Garrett,
1981).

In *A. nidulans*, other rare classes of $nirA$ mutants have been isolated (Rand & Arst, 1978; Tollervey & Arst, 1981). The first of
these, $nirA^c1$, was recognised by the plate test for the production
of nitrite, because colonies stain pink much faster than either wild-
type or $niiA$ strains (see Cove, 1979). The reason for this is that
$nirA^c1$ has constitutive high levels of nitrate-reductase and nitrite
reductase when grown on a neutral nitrogen source even when inducer is absent. However, the synthesis of these enzymes in this
strain is still subject to nitrogen metabolite repression. Rand &
Arst (1978) identified a third class of mutation at this locus. The
$nirA^{c/d}101$ mutant has the constitutive synthesis characteristic of
$nirA^c1$ but also leads to strongly derepressed synthesis of nitrate

and nitrite reductase in the presence of ammonium or glutamine. The $nirA^{c/d}101$ strain was derived from $nirA^{c}1$ by mutagenesis and hence carries two mutations within this gene. Tollervey & Arst (1981) isolated the $nirA^{d}106$ mutation which appears to be altered for the derepression site in the $nirA$ gene. Crosses between $nirA^{c}1$ and $nirA^{d}106$ in the appropriate genetic background produce recombinant $nirA^{c}1$ $nirA^{d}106$ progeny, whose phenotype resembles the $nirA^{c/d}101$ mutant. The $nirA$ gene, therefore, contains at least two separable domains: one in which mutations can lead to constitutivity, presumably representing a co-inducer binding domain; another where mutations can result in nitrogen metabolite derepression, which is either a binding site for the $areA$ gene product (see below) or the site which interacts with the initiator/promoter regions adjacent to $areA/nirA$ regulated structural genes.

The $nirA$ gene is not, however, the only gene affecting nitrate induction. Mutations affecting nitrate reductase can also result in altered regulation of nitrate assimilation. Many $niaD$ and cnx mutants in *A. nidulans* have constitutive synthesis for the nitrite reductase and the mutant nitrate reductase (Pateman *et al.*, 1964; see Cove, 1979). This is also true for the equivalent mutants of *N. crassa* (Tomsett & Garrett, 1981). Mutants of $niaD$ (and nit-3) fall basically into two classes: those which are regulated exactly as the wild-type, and those that synthesize the enzymes constitutively (i.e. in the presence and absence of the inducer). This altered regulation is detected by comparison of enzyme levels of both nitrate reductase partial activities and NADPH-nitrite reductase on non-inducing and inducing medium. A few strains remain normally inducible like the wild-type, while others become completely constitutive. Most mutants are found to be at least partly altered with respect to regulation by induction. These observations also apply to the mutants altered for molybdenum cofactor synthesis. Although nitrate/nitrite induction is altered, all these mutants are still subject to nitrogen metabolite repression.

These data demonstrate the dual role played by nitrate reductase in nitrate assimilation. In addition to its catalytic function, the enzyme plays a key role in the regulation of this pathway. Cove (1970) proposed a model to explain this conclusion. The $nirA$ gene product is thought to switch on transcription of the structural genes of both nitrate and nitrite reductase. After translation, the native nitrate reductase is assembled with the molybdenum cofactor into active enzyme. In the absence of nitrate, this nitrate reductase molecule converts the active nirA gene product to an inactive state in which it can no longer stimulate transcription. Once the low level of nitrate reductase had been degraded the cycle would start again. When nitrate was present, nitrate reductase would complex with it and no longer be able to inactivate the *nir*A gene product, leading to further transcription of the structural genes and the induced levels of these enzymes. This model explains all the mutant phenotypes observed; including *nia*D and *cnx* mutants which drastically alter the conformation of native nitrate reductase or prevent its formation, prevent *nir*A gene product inactivation and hence are constitutive, while *nia*D and *cnx* mutants which only affect catalytic activity and not conformation, remain normally regulated. The *nir*A^c mutations prevent nitrate reductase inactivating the *nir*A gene product and hence are constitutive. The *nir*A$^-$ mutants lack the active *nir*A gene product, cannot stimulate high level expression, and hence appear non-inducible. The confirmation of this model awaits complex molecular biology experiments. As the first step towards this, Y. H. Fu & G. A. Marzluf (unpublished data) have examined *nit*-3 mRNA levels; expression requires both nitrogen derepression and nitrate induction. In *nit*-3 mutants, *nit*-3 mRNA is constitutive.

Nitrogen metabolite repression

It is not within the scope of this review to cover all the aspects of nitrogen metabolite repression. These have been eloquently explained in another recent review (Wiame *et al.*, 1985). Instead, I shall give a brief overview of the facts and discuss these in the light of recent data.

The *nit*-2 gene of *N. crassa* and the *are*A gene of *A. nidulans* encode positive-acting regulatory proteins which switch on the expression of a variety of pathways of nitrogen metabolism, including nitrate assimilation. These pathways are only switched on when glutamine is absent; when present, glutamine prevents this switch. Ammonium was formerly thought to be the active effector, but mutations preventing the assimilation of ammonium to glutamine also relieve ammonium repression (Dunn-Coleman, Tomsett & Garrett, 1979; MacDonald, 1982). The most common class of mutations in *nit*-2 and *are*A result in a repressed phenotype: an inability to utilise nitrogen sources other than ammonium or glutamine. A rarer class of mutation, *are*A^d leads to derepressed expression of *are*A regulated pathways: enzymes and permeases in these pathways are synthesized whether glutamine is present or not (but if they require a specific inducer, such as nitrate, this must also be present for synthesis).

The simplest explanation is that the *are*A and *nit*-2 genes encode regulatory proteins which, in the absence of glutamine, bind the promoter/initiator regions of the structural genes which they regulate and stimulate transcription. Glutamine could bind to the regulatory protein preventing the activation of transcription. Grove & Marzluf (1981) isolated non-histone DNA-binding proteins from *N. crassa* nuclei and identified a 22,000 dalton protein which was eluted with glutamine. This protein was present in greatly reduced quantities, or had a different electrophoretic mobility, in nuclei of *nit*-2 mutants. Recently the *nit*-2 gene has been cloned (Stewart & Vollmer, 1986). This gene is transcribed as a 3·5 kb transcript, which is present at considerably higher levels in cells which are limited for nitrogen than those growing under nitrogen-repressed conditions (Fu & Marzluf, 1987). This suggests that *nit*-2 is regulated by other control genes. The *are*A gene has also been cloned and appears to be similarly regulated (Caddick *et al.*, 1986; M. X. Caddick, personal communication).

Conclusions

The genetics and biochemistry of nitrate assimilation in fungi have made outstanding contributions to our knowledge of enzymology, gene expression and gene regulation. By disrupting the normal metabolism of these organisms, detailed information has emerged on the structure of enzymes, their catalytic activity, and the regulation of their synthesis. This has led to models of the interaction of regulatory proteins with DNA and with each other. It is only relatively recently that sophisticated molecular biology has been possible in these fungi. A number of genes have now been cloned and these are currently being sequenced. In the next few years, cloned sequences can be used directly to test the proposals made for the various regulatory interactions. The examination of sequences upstream from structural genes should reveal sequences to which the *areA/nit-2* protein will bind to mediate nitrogen metabolite repression and those to which the *nirA/nit-4* protein binds to mediate nitrate induction. Site-directed mutagenesis should reveal those sites in the *areA/nit-2* protein responsible for binding to DNA and those which bind glutamine. Similar experiments with *nirA/nit-4* could identify whether this protein can bind and be inactivated by nitrate reductase. Certainly in the next few years, we will see these proposals confirmed or refuted by molecular biology.

References

Aberg, B. (1947). On the mechanism of the toxic action of chlorates and some related substances upon young wheat plants. *Kungliga Lantbrukshogskolans Annaler*, **15**, 37-107.

Arst, H. N., MacDonald, D. W. & Cove, D. J. (1970). Molybdate metabolism in *Aspergillus nidulans*. I Mutations affecting nitrate reductase and/or xanthine dehydrogenase. *Molecular and General Genetics*, **108**, 129-145.

Arst, H. N., Rand, K. N. & Bailey, C. R. (1979). Do the tightly linked structural genes for nitrate and nitrite reductases in *Aspergillus nidulans* form an operon? Evidence from an insertional translocation which separates them. *Molecular and General Genetics*, **17**, 89-100.

Arst, H. N., Tollervey, D. W. & Sealy-Lewis, H. M. (1982). A possible regulatory gene for the molybdenum-containing cofactor in *Aspergillus nidulans*. *Journal of General Microbiology*, **128**, 1083-1093.

Brownlee, A. G. & Arst, H. N. (1983). Nitrate uptake in *Aspergillus nidulans* and involvement of the third gene of the nitrate assimilation gene cluster. *Journal of Bacteriology*, **155**, 1138-1146.

Caddick, M. X., Arst, H. N., Taylor, L. H., Johnson, R. I. & Brownlee, A. G. (1986). Cloning of the regulatory gene *areA* mediating nitrogen metabolite repression in *Aspergillus nidulans*. *EMBO Journal*, **5**, 1087-1090.

Calza, R., Huttner, E., Vincentz, M., Rouze, P., Galangau, F., Vaucheret, H., Cherel, I., Meyer, C., Kronenberger, J. & Caboche, M. (1987). Cloning of DNA fragments complementary to tobacco nitrate reductase mRNA and encoding epitopes common to the nitrate reductases from higher plants. *Molecular and General Genetics*, **209**, 552-562.

Coddington, A. (1976). Biochemical studies on the *nit* mutants of *Neurospora crassa*. *Molecular and General Genetics*, **145**, 195-206.

Cooley, R. N. & Tomsett, A. B. (1985). Determination of the subunit size of NADPH-nitrate reductase from *Aspergillus nidulans*. *Biochimica et Biophysica Acta*, **831**, 89-93.

Cove, D. J. (1970). Control of gene action in *Aspergillus nidulans*. *Proceedings of the Royal Society, London, Series B*, **176**, 267-275.

Cove, D. J. (1976a). Chlorate toxicity in *Aspergillus nidulans*: the selection and characterization of chlorate resistant mutants. *Heredity*, **36**, 191-203.

Cove, D. J. (1976b). Chlorate toxicity in *Aspergillus nidulans*: studies of mutants altered in nitrate assimilation. *Molecular and General Genetics*, **146**, 147-159.

Cove, D. J. (1979). Genetic studies of nitrate assimilation in *Aspergillus nidulans*. *Biological Reviews*, **54**, 291-327.

Downey, R. J. & Focht, W. J. (1974). Subunit character of the NADPH-nitrate reductase from *Aspergillus nidulans*. *Microbios*, **11A**, 61-70.

Downey, R. J. & Steiner, F. X. (1979). Further characterization of the reduced NADP: nitrate oxidoreductase in *Aspergillus nidulans*. *Journal of Bacteriology*, **137**, 105-114.

Dunn-Coleman, N. S. (1984a). Biochemical characterization of the molybdenum cofactor mutants of *Neurospora crassa*: *in vivo* and *in vitro* reconstitution of NADPH-nitrate reductase activity. *Current Genetics*, **8**, 581-588.

Dunn-Coleman, N. S. (1984b). Cloning and preliminary characterization of a molybdenum cofactor gene of *Neurospora crassa*. *Current Genetics*, **8**, 589-595.

Dunn-Coleman, N. S., Smarrelli, J. Jr. & Garrett, R. H. (1984). Nitrate assimilation in eukaryotic cells. *International Review of Cytology*, **92**, 1-50.

Dunn-Coleman, N. S., Tomsett, A. B. & Garrett, R. H. (1979). Nitrogen metabolite repression of nitrate reductase in *Neurospora crassa*: effect of the *gln*-1a locus. *Journal of Bacteriology*, **139**, 697-700.

Fu, Y. H. & Marzluf, G. A. (1987). Characterization of *nit*-2, the major nitrogen regulatory gene of *Neurospora crassa*. *Molecular and Cellular Biology*, **7**, 1691-1696.

Garrett, R. H. (1972). The induction of nitrite reductase in *Neurospora crassa*. *Biochimica et Biophysica Acta*, **264**, 481-489.

Garrett, R. H. (1978). Nitrite reduction in fungi. In *Microbiology 1978*, ed. D. Schlessinger, pp. 324-329. American Society of Microbiology Publication, Washington, USA.

Garrett, R. H. & Amy, N. K. (1978). Nitrate assimilation in fungi. *Advances in Microbial Physiology*, **18**, 1-65.

Garrett, R. H. & Nason, A. (1969). Further purification and properties of *Neurospora* nitrate reductase. *Journal of Biological Chemistry*, **244**, 2870-2882.

Greenbaum, P., Prodouz, K. N. & Garrett, R. H. (1978). Preparation and some properties of homogeneous *Neurospora crassa* assimilatory NADPH-nitrite reductase. *Biochimica et Biophysica Acta*, **526**, 52-64.

Grove, G. & Marzluf, G. A. (1981). Identification of the product of the major regulatory gene of the nitrogen control circuit of *Neurospora crassa* as a nuclear DNA binding protein. *Journal of Biological Chemistry*, **256**, 463-470.

Guerrero, M. G., Vega, J. M. & Losada, M. (1981). The assimilatory nitrate-reducing system and its regulation. *Annual Review of Plant Physiology*, **32**, 169-204.

Hageman, R. V. & Rajagopalan, K. V. (1985). Characterization of molybdopterin, the organic portion of the molybdenum cofactor. In *Nitrogen Fixation and CO2 Metabolism*, eds. P. W. Ludden & J. E. Burris, pp. 133-141. Elsevier, New York.

Horner, R. D. (1983). Purification and comparison of *nit*-1 and wild-type NADPH: nitrate reductases of *Neurospora crassa*. *Biochimica et Biophysica Acta*, **744**, 7-15.

Johnson, J. L. (1980). The molybdenum cofactor common to nitrate reductase, xanthine dehydrogenase and sulphite oxidase. In *Molybdenum and Molybdenum-Containing Enzymes*, ed. M. P. Coughlan, pp. 345-383. Pergamon Press, New York.

Johnson, J. L., Hainline, B. E., Rajagopalan, K. V. & Arison, B. H. (1984) The pterin component of the molybdenum cofactor: structural characterization of two fluorescent derivatives. *Journal of Biological Chemistry*, **259**, 5414-5422.

Johnson, J. L. & Rajagopalan, K. V. (1982). Structural and metabolic relationship between the molybdenum cofactor and urothione. *Proceedings of the National Academy of Sciences, U.S.A*, **79**, 6856-6860.

Johnson, M. E. & Rajagopalan, K. V. (1987). Involvement of *chlA, E, M*, and *N* loci in *Escherichia coli* molybdopterin biosynthesis. *Journal of Bacteriology*, **169**, 117-125.

Kramer, S., Hageman, R. V. & Rajagopalan, K. V. (1984). *In vitro* reconstitution of nitrate reductase activity of the *Neurospora crassa* mutant *nit*-1: specific incorporation of molybdopterin. *Archives of Biochemistry and Biophysics*, **233**, 821-829.

Lafferty, M. A. & Garrett, R. H. (1974). Purification and properties of the *Neurospora crassa* assimilatory nitrite reductase. *Journal of Biological Chemistry*, **249**, 7555-7567.

Le, K. H. D. & Lederer, F. (1983). On the presence of a heme-binding domain homologous to cytochrome *b5* in *Neurospora crassa* assimilatory nitrate reductase. *EMBO Journal*, **2**, 1909-1914.

MacDonald, D. W. (1982). A single mutation leads to loss of glutamine synthetase and relief of ammonium repression in *Aspergillus. Current Genetics*, **6**, 203-208.

MacDonald, D. W. & Cove, D. J. (1974). Studies on temperature-sensitive mutants affecting the assimilatory nitrate reductase of *Aspergillus nidulans. European Journal of Biochemistry*, **47**, 107-110.

Marzluf, G. A. (1981). Regulation of nitrogen metabolism and gene expression in fungi. *Microbiological Reviews*, **45**, 437-461.

Minagawa, N. & Yoshimoto, A. (1982). Purification and characterization of the assimilatory NADPH-nitrate reductase of *Aspergillus nidulans. Journal of Biochemistry*, **91**, 761-774.

Pan, S-S. & Nason, A. (1978). Purification and characterization of homogeneous assimilatory NADPH-nitrate reductase from *Neurospora crassa. Biochimica et Biophysica Acta*, **523**, 297-313.

Pateman, J. A., Cove, D. J., Rever, B. M. & Roberts, D. B. (1964). A common cofactor for nitrate reductase and xanthine dehydrogenase which also regulates the synthesis of nitrate reductase. *Nature*, **201**, 58-60.

Pombeiro, S. R. C., Martinez-Rossi, N. M. & Rossi, A. (1983). Nitrate toxicity in *Aspergillus nidulans*: a new locus in a *proAl pabaA6 yA2* strain. *Genetical Research*, **41**, 203-207.

Prodouz, K. N. & Garrett, R. H. (1981). *Neurospora crassa* NAD(P)H-nitrite reductase: studies on its composition and structure. *Journal of Biological Chemistry*, **256**, 9711-9717.

Rand, K. N. & Arst, H. N. (1977). A mutation in *Aspergillus nidulans* which affects the regulation of nitrite reductase and is tightly linked to its structural gene. *Molecular and General Genetics*, **155**, 67-75.

Rand, K. N. & Arst, H. N. (1978). Mutations in *nirA* gene of *Aspergillus nidulans* and nitrogen metabolism. *Nature*, **272**, 732-734.

Scazzocchio, C. (1980). The genetics of the molybdenum-containing enzymes. In *Molybdenum and Molybdenum-Containing Enzymes*, ed. M. P. Coughlan, pp. 487 - 515. Pergamon Press, Oxford.

Schloemer, R. H. & Garrett, R. H. (1974a). Nitrate transport system in *Neurospora crassa*. *Journal of Bacteriology*, **118**, 259-269.

Schloemer, R. H. & Garrett, R. H. (1974b). Uptake of nitrite by *Neurospora crassa*. *Journal of Bacteriology*, **118**, 270-274.

Solomonson, L. P., Barber, M. J., Robbins, A. P. & Oaks, A. (1986). Functional domains of assimilatory NADH: nitrate reductase from *Chlorella*. *Journal of Biological Chemistry*, **261**, 11290-11294.

Sorger, G. J., Premakumar, R. & Gooden, D. (1978). Demonstration *in vitro* of two intracellular inactivators of nitrate reductase from *Neurospora*. *Biochimica et Biophysica Acta*, **706**, 33-47.

Steiner, F. X. & Downey, R. J. (1982). Isoelectric focusing and two dimensional analysis of purified nitrate reductase from *Aspergillus nidulans*. *Biochimica et Biophysica Acta*, **706**, 203-211.

Stewart, V. & Vollmer, S. J. (1986). Molecular cloning of *nit*-2, a regulatory gene required for nitrogen metabolite repression in *Neurospora crassa*. Gene, **46**, 291-295.

Tollervey, D. W. & Arst, H. N. (1981). Mutations to constitutivity and derepression are separate and separable in a regulatory gene of *Aspergillus nidulans*. *Current Genetics*, **4**, 63-68.

Tomsett, A. B. (1977). Structure and function in the *nii*A *nia*D gene region of *Aspergillus nidulans*. Ph.D. Thesis, University of Cambridge, U.K.

Tomsett, A. B. & Cove, D. J. (1979). Deletion mapping of the *nii*A *nia*D gene region of *Aspergillus nidulans*. *Genetical Research*, **34**, 19-32.

Tomsett, A. B. & Garrett, R. H. (1980). The isolation and characterization of mutants defective in nitrate assimilation in *Neurospora crassa*. *Genetics*, **95**, 649-660.

Tomsett, A. B. & Garrett, R. H. (1981). Biochemical analysis of mutants defective in nitrate assimilation in *Neurospora crassa*: evidence for autogenous control by nitrate reductase. *Molecular and General Genetics*, **184**, 183-190.

Walls, S., Sorger, G. J., Gooden, D. & Klein, V. (1978). The regulation of the decay of nitrate reductase: evidence for the existence of at least two mechanisms of decay. *Biochimica et Biophysica Acta*, **540**, 24-32.

Wiame, J. M., Grenson, M. & Arst, H. J. (1985). Nitrogen catabolite repression in yeasts and filamentous fungi. *Advances in Microbial Physiology*, **26**, 1-88.

Chapter 3

Transport and metabolism of sulphur dioxide in yeasts and filamentous fungi

Anthony H. Rose

Zymology Laboratory, School of Biological Sciences, University of Bath, Bath BA2 7AY, Avon, UK

Introduction

From a knowledge of the chemistry of the element sulphur and of its oxidised forms, it can be stated with confidence that yeasts and fungi have been exposed to sulphur dioxide (SO_2) from time immemorial. However, a deeper scientific interest in the reaction of these micro-organisms with SO_2 has recently been evident, largely for two reasons.

First, for centuries SO_2 has been used to preserve beverages and more recently foods. Although it is often stated that the ancient Greeks and Romans disinfected their amphorae with burning sulphur, there is curiously little firm evidence for this practice (Henderson, 1824). We do know, however, that sulphur was included in a mixture of pitch and resin which was used to patch cracks in fermenting jars (Allen, 1961). Sulphur is hardly ever burned nowadays to disinfect, although a visit to a traditional oast house can still provide an experience of this pungent breathtaking vapour (Lloyd-Hind, 1948). The great strides made in inorganic chemistry in the 19th Century led to the availability of a variety of sulphur salts which give rise to SO_2 when dissolved in water. Today, the principal compound used to generate sulphur dioxide and the related anions in the preservation of foods and beverages is sodium metabisulphite ($Na_2S_2O_5$), designated additive E223 in Directives of the European Economic Community (Hanssen & Marsden, 1984). Other compounds frequently employed as sulphiting agents include SO_2, potassium and sodium salts of sulphite and bisulphite and potassium metabisulphite, the common characteristic being their ability to liberate free molecular SO_2 which appears to correlate with their preservative activity.

At the present time, there is, however, deep concern over the use of SO_2 as a beverage and food preservative. This arises in part from the increasing disapproval of adding chemical compounds of any description to beverages and foods. Specific criticism of the use of SO_2 has been prompted largely by the pungent nature of this additive. The merest whiff of gaseous SO_2, when opening a bottle of wine or a packet of food, can bring on an asthmatic attack, which has in one or two instances proved fatal for persons with a history of respiratory problems. The general desire to lower concentrations of SO_2 included in beverages and foods explains in large part the recent upsurge of interest into the physiological basis of the interaction of this food additive with micro-organisms in general. Knowledge of the molecular mechanisms of this interaction, it is argued, could indicate ways in which the lethal effect of SO_2 on microorganisms might be maximised, thereby permitting use of lower concentrations.

The second reason for an intensification of interest in the reaction of yeasts and filamentous fungi with SO_2 stems from the problem of acid rain. Most fossil fuels contain various proportions of sulphur-containing minerals which, when combusted, generate SO_2. When deposited with precipitation, this constitutes acid rain which, especially in parts of Northern Europe and North America, has, partly at least by killing micro-organisms, led to complete or partial sterilization of lakes and rivers.

Effect of pH value

This article is concerned with SO_2 and related ions which accompany this compound in aqueous solution. Throughout, sulphur dioxide is referred to as SO_2, the bisulphite ion as HSO_3^- and the sulphite ion as SO_3^{2-}. It should be stressed, however, that this nomenclature, although one which is very widely used, is at variance with that recommended by I.U.P.A.C. which recommends that the oxoacid $H_2S_2O_5$ be named disulphurous acid, and its salts as disulphites rather than metabisulphites. I.U.P.A.C. also recommend that partially dissociated acids be referred to with the prefix 'hydrogen' replacing 'bi-'. Further information on the I.U.P.A.C. recommendations can be obtained from Fernelius, Leoning &

Adams (1973). An excellent account of the chemistry of the compounds and ions referred to above can be found in the text written by Wedzicha (1984).

While in solution, metabisulphite generates SO_2, HSO_3^- and SO_3^{2-}; the proportion of these species present depends on the pH value of the solution. The equilibria are:

$$SO_2.H_2O \leftrightarrow HSO_3^- \leftrightarrow SO_3^{2-}$$

Although molecular sulphur dioxide will, as already indicated, be referred to in this chapter as SO_2, Raman and infrared spectra of dissolved SO_2 and gaseous or liquid SO_2 have led workers to conclude that dissolved SO_2 exists as $SO_2.H_2O$. Dissociation constants for each of the two equilibria have been determined at low sulphite concentrations of the order of those normally encountered by yeasts and fungi. The reaction leading to ionization of SO_2 has a pK_a value of 1.86 at $25°C$ and zero ionic strength, while the value for the reaction leading to production of the sulphite ion under the same conditions is 7.18 (Wedzicha, 1984). Using these pK_a values, calculations have been made of the proportions of each species present in solution as a function of pH value and tabulated data have been published by King *et al.* (1981. This effect of pH value is vitally important in relation to the mechanism of action of SO_2 with yeasts and filamentous fungi.

Transport of SO2

Passage of sulphite across the microbial plasma membrane appears to be by different mechanisms depending on the species of yeast or mycelial fungus. The currently accepted mechanism for transport of sulphite into *Saccharomyces cerevisiae*, a microbe on which the majority of studies of SO_2 transport have been carried out, is by free diffusion of the molecular form of SO_2 (Stratford & Rose, 1986) which conflicts with the active transport system previously proposed by Macris & Markakis (1974). Stratford & Rose (1986) presented strong evidence in favour of a protein not being involved in SO_2 transport by reporting near vertical Eadie-Woolf plots (Hofstee, 1959) for SO_2 transport at pH 3.0 and 4.0. At these pH values, a high proportion of free SO_2 is present in solution. Values for K_T calculated from kinetic plots of v against v/s (v is the initial velocity of accumulation of sulphite equivalents, and s the extracellular SO_2 concentration) were 3.2 mM and 0.1 mM at pH 3·0 and 4·0 respectively. These K_T values are far in excess of the concentration of SO_2 required to kill *Saccharomyces cerevisiae*. At such concentrations, all active transport systems would break down and yet the initial rate of uptake is theoretically still only half its maximum level, strongly suggesting that a passive transport system predominates under these conditions. This evidence is corroborated by the inability of carbonylcyanide-*m*-chlorophenylhydrazone (CCCP) and dinitrophenol (DNP) (Borst-Pauwels, 1981) to affect initial velocities of SO_2 accumulation. These protonophores are known to dissipate the transmembrane proton gradient and to inhibit protein-mediated transport systems. Further evidence for a lack of active transport of SO_2 came from the finding that exclusion of glucose from reaction mixtures had no effect on initial velocities of accumulation. Similarly, the inability of the glycolytic inhibitor, 2-deoxy-D-glucose, to affect SO_2 accumulation adds fuel to the theory that energy is not required for SO_2 uptake. Yet additional evidence is provided by the absence of an effect of pH value on the process, an effect characteristic of protein-mediated transport. Macris & Markakis (1974) studied the kinetics of radiolabelled SO_2 uptake by *Saccharomyces cerevisiae* var. *ellipsoideus* making some valuable observations on SO_2 toxicity and pH dependence. They found a close correlation between accumulation of radiolabelled sulphite, over the pH range 3·0 - 5·0, and concentration of molecular SO_2 in solution, which was corroborated by Stratford & Rose (1986) and Hinze & Holzer (1985). Evidence strongly suggests that, over this pH range, only the molecular form of SO_2 passes into organisms and, by inference, that *Saccharomyces cerevisiae* does not transport HSO_3^-. In this yeast, the plasma membrane merely acts as a selective barrier to free diffusion of molecular SO_2. For this reason the structure and fluidity of the plasma membrane are likely to influence transport and further investigation in this area is obviously required. A slow transport system for HSO_3^- in *Saccharomyces cerevisiae* has been tentatively suggested which is apparent in the presence of low concentrations of the ion (Stratford & Rose, 1986). As sulphite concentrations are increased, this system rapidly becomes saturated and is masked by diffusion of higher concentrations of molecular SO_2. *Saccharomyces cerevisiae* accumulates SO_2 initially very rapidly, reaching a plateau concentration after about 5 min. Intracellular sulphite concentrations at equilibrium are many times greater than those in the suspension. This can be explained by the dynamic equilibrium between the three forms of SO_2 in solution and the presence of sulphite-binding compounds inside cells (Burroughs & Sparks 1964). A third factor is the difference in pH value between the cell suspension and that inside organisms. Intracellular pH values in *Saccharomyces cerevisiae* are around 6·5 where only 0·0015% of free sulphite exists as molecular SO_2 (King *et al.*, 1981). As the extracellular pH value is below 5·5 in most yeast cultures, molecular SO_2 will diffuse into cells, dissociate to HSO_3^- and H^+, thereby allowing further diffusion into organisms until theoretically the concentration of SO_2 is equal on both sides of the plasma membrane.

Recent reports from Stratford, Morgan & Rose (1987) and Pilkington & Rose (1988) have shown that sulphite transport by *Saccharomycodes ludwigii* and accumulation by *Zygosaccharomyces bailii*, respectively, were very similar to these processes in *Saccharomyces cerevisiae*. Evidence was presented that diffusion alone of molecular SO_2 occurred in both of these yeasts, while both again were shown to be able to accumulate large quantities of sulphite intracellularly.

The few reports that have appeared on sulphite transport in other eukaryotic micro-organisms show that this process differs from that in the yeasts already referred to. Benitez *et al.* (1983) and Garcia *et al.* (1983) directed their studies towards sulphate transport in *Candida utilis* and postulated a common transport system for sulphate, sulphite and thiosulphate. Data presented in these publications yield strong evidence in favour of sulphate transport being an active, energy-dependent process and that sulphate analogues, such as molybdate, selenate and chromate, competitively inhibit sulphate incorporation. In addition, the sulphur oxyanions sulphite, thiosulphate and dithionate cause substantial inhibition. With the exception of dithionate, which acts as a mixed-type inhibitor, all of the other compounds cause competitive inhibition. Uptake of sulphate is also under endogenous control. If sulphite, sulphate or thiosulphate is allowed to accumulate, subsequent uptake of sulphate is specifically inhibited. Thiosulphate and sulphite have also been shown to be extracellular inhibitors of sulphate transport in *Penicillium chrysogenum* (Yamamoto & Segel, 1966). Selenate-resistant mutants of *C. utilis* have been isolated (Garcia *et al.*, 1983). They have a common transport defect showing inability to grow in media with either sulphite, sulphate or thiosulphate as the sole source of sulphur. The wild type grows with any one of these sources. These results imply that sulphite is transported via the common transport system. This contrasts with the situation in *Saccharomyces cerevisiae, Saccharomycodes ludwigii* and *Zygosaccharomyces bailii.* A possible explanation is that transport of the bisulphite ion is biphasic in behaviour and that the common active transport system is the same as that intimated by Stratford & Rose (1986). The experiments of Benitez *et al.* (1983) were carried out at pH 6·1 where molecular SO_2 concentrations are negligible and active transport of bisulphite predominates. This might also explain the absence of sulphite toxicity since only molecular SO_2 is believed to have antimicrobial activity.

Other references to sulphite transport include work on filamentous fungi. Tweedie & Segel (1970) recognized a common active transport system for the ions SO_4^{2-}, $S_2O_3^{2-}$, SeO_4^{2-} and MnO_4^{2-} entering mycelia of *Penicillium* and *Aspergillus* species. In addition, mycelia appear to possess distinct permeases for sulphite or tetrathionate. This is borne out by the ability of sulphate permease-deficient mutants to accumulate either sulphite or tetrathionate. Anion transport in mycelia, as in *C. utilis* (Alonso, Benitez & Diaz, 1984), is regulated by both extracellular and intracellular anion concentrations. Mycelia grown in methionine-rich media accumulate few sulphur anions whereas those grown under sulphur-deficient conditions show a much greater rate of accumulation of substrate.

Evidence for a sulphite-specific permease is still questionable. Uptake of sulphite by sulphate permease-deficient mycelia recorded by Tweedie & Segel (1970) could be equally explained by simple leakage. All transport studies using multi-anionic systems are fraught with problems due to oxidation and cross reaction of anions. Tweedie & Segel (1970) clearly recognized these disadvantages but did not consider the equilibrium position of sulphite. Wherever HSO_3^- exists in solution, some proportion must be present as molecular SO_2 depending upon the pH value. It is possible that molecular SO_2 could diffuse into filamentous fungi as readily as into *Saccharomyces cerevisiae* and other yeasts. Accumulation of sulphite may be misleading in this case and, in fact, merely reflect molecular SO_2 accumulation.

The variety of possible transport mechanisms described does not readily help us to understand the mode of action of sulphite, but only to reveal that its transport is likely to be specific for each organism. Certainly in those yeasts that are a major cause of food spoilage, it seems likely that diffusion of molecular SO_2 is common and is possibly dependent on membrane composition.

Metabolism of SO_2

Sulphite can have a damaging or a beneficial effect upon yeasts and filamentous fungi, depending on the concentration to which the micro-organism is exposed. The remainder of this review deals with the ways in which sulphite, having made contact with a yeast or filamentous fungus, is chemically modified. The first section summarizes the manner in which sulphite can deleteriously interact with biologically important molecules. This is followed by a brief summary of the reactions by which SO_2 acts as a sulphur-containing nutrient for yeasts and filamentous fungi.

Reaction with biologically important molecules

All three molecular species that are found in solutions of sulphite, especially the bisulphite ion, are chemically very reactive. Not surprisingly, a voluminous literature exists on the chemistry of the interaction of these species, most of it dealing with model and low molecular-weight compounds. Molecular mechanisms for the interactions have, in many instances, been the subject of detailed studies, which are comprehensively reviewed in Wedzicha (1984). These studies have revealed that biologically important molecules, both molecularly complex and otherwise, contain chemical groupings that are known to react with sulphite particularly when in the form of the bisulphite ion. What is much more problematical, however, is to infer that, because a biological macromolecule (e.g. a protein or nucleic acid) possesses a grouping that interacts with sulphite, an interaction takes place that leads to inactivation or denaturation of that molecule.

Several enzymes are known to be inhibited by sulphite. Sulphite ions have the ability to react with disulphide bonds in the following reaction:

$$R\text{-}S\text{-}S\text{-}R + SO_3^{2-} \leftrightarrow R\text{-}S\text{-}S\text{-}O_3^- + RS^-$$

The catalytic activity of an enzyme depends on the tertiary conformation of the protein molecule. This conformation is dictated by the primary sequence of amino acid residues in the molecule which in turn governs the manner in which the molecule becomes folded into its tertiary structure. Disulphide bonds forged between juxtaposed cysteine residues help to stabilize the tertiary structure of proteins. A potential point of attack for sulphite on enzyme molecules must be disulphide bridges. However, it cannot automatically be assumed that, because an enzyme molecule contains disulphide bridges, interaction with sulphite takes place leading to inactivation of the enzyme. It is likely, nevertheless, that this reaction is responsible for inactivation of many enzymes.

Sulphite ions also react with a range of coenzymes which are essential for enzyme activity and with prosthetic groups conjugated with biologically active proteins. Inhibition of enzyme activity by sulphite may, therefore, be attributable on occasions to inactivation of a coenzyme or prosthetic group brought about by the oxoanion. Pyridine reacts with sulphite to form an addition compound. An important biological molecule containing a pyridine residue is nicotinamide adenine dinucleotide (NAD^+) and the related $NADP^+$. The reaction with NAD^+ can be represented as follows (Pfleiderer, Sann & Stock, 1960):

$$NAD^+ + SO_3^{2-} \leftrightarrow NADSO_3^-$$

Sulphite can also react with NAD^+ and $NADH^+$ when these coenzymes are bound to dehydrogenase enzymes (Parker, Lodola & Holbrook, 1978).

In addition to NAD^+ and $NADP^+$, a variety of other organic compounds have been shown to be essential for enzyme activity. Unlike NAD^+ and $NADH^+$ those compounds known as prosthetic groups remain attached to the enzyme molecule and are essential for the biological activity of the enzyme. Several of them are inactivated by sulphite. Two prosthetic groups concerned in oxidation-reduction reactions, namely flavins and haems, are both inactivated by sulphite. Other prosthetic groups that are susceptible to sulphite inactivation are the vitamins folic acid, pyridoxal and thiamin. Little is known of the chemical nature of the reaction between folic acid and sulphur oxoanions, although the reaction is thought to involve preferentially the bisulphite ion (Wedzicha, 1984). The reaction between pyridoxal and sulphite almost certainly involves the carbonyl group on the vitamin molecule. A discussion of the reaction between carbonyl compounds and sulphite appears later in this section of the article. Thiamin is cleaved by sulphite to produce pyrimidine sulphonate and a thiazole residue-containing product.

Bisulphite ions can inactivate nucleic acids by undergoing an addition reaction with pyridine residues. The reaction of cytosine and uracil with bisulphite has been closely studied. Residues of the first of these pyrimidines are found in DNA while uracil residues occur in RNA. Reaction of sulphite with DNA presumably terminates polymerization of new DNA on a DNA template, as well as transcription of DNA into RNA. Similarly, interaction of uracil residues in a mRNA molecule would abort the process of translation into polypeptide.

One of the most important and most extensively studied reactions involving sulphite is that with carbonyl groups in organic compounds. The reaction for formation of an adduct with acetaldehyde is:

$$\begin{array}{c} CH_3 \\ \diagdown \\ \diagup \quad C=O \ + \ H^+SO_3^- \ \leftrightarrow \\ H \end{array} \qquad \begin{array}{c} CH_3 \quad OH \\ \diagdown \diagup \\ C \\ \diagup \diagdown \\ H \qquad SO_3^- \end{array}$$

The rate constant for dissociation of the adduct differs for each carbonyl-containing compound, and also with pH value. The data in Table 3.1 illustrate these points, and show the extremely high affinity which acetaldehyde has for sulphite. The table does not include information on the ability of the more common sugars to bind sulphite, although some interesting data in this connection were reported by Ingram & Vas (1950).

Table 3.1. Apparent equilibrium constants of some α-hydroxysulphonates.

Carbonyl compound	Concentration (mM) of		Equilibrium constant at	
	Carbonyl compound	Total SO$_2$	pH 3·0	pH 4·0
acetaldehyde	6·0	4·0	$1·5 \times 10^{-6}$	$1·4 \times 10^{-6}$
2,5-diketogluconic acid	2·0	0·6-7·2	$4·5 \times 10^{-4}$	$4·3 \times 10^{-4}$
galacturonic acid	6·0	4·0	$1·6 \times 10^{-2}$	$2·1 \times 10^{-2}$
2-ketoglutaric acid	2·0	2·0-10·0	$4·9 \times 10^{-4}$	$7·0 \times 10^{-4}$
D-threo-2,5-hexodiulose	2·0	2·0-10·0	$3·4 \times 10^{-4}$	$3·3 \times 10^{-4}$
pyruvic acid	2·0	0·8-5·0	$1·4 \times 10^{-4}$	$2·2 \times 10^{-4}$
L-xylosone	2·0	2·0-10·0	$1·4 \times 10^{-3}$	$1·4 \times 10^{-3}$

Adapted from Burroughs & Sparks (1973).

Finally, there is the interaction of sulphite with molecular oxygen, the stoicheiometric equation for which is:

$$2SO_3^{2-} + O_2 \rightarrow 2SO_4^{2-}$$

This is a much studied reaction, data on which are thoroughly reviewed by Wedzicha (1984). In deducing kinetic data for this equation, the concentration of sulphite is that of SO_3^{2-} since the HSO_3^- ion is much less easily oxidized by oxygen. The reaction can take place spontaneously or through a reaction catalyzed by a transition metal ion. Postgate (1963) cautioned workers who store solutions of sulphite for experimental work against the possibility of oxidation. He recommended storing solutions at -20°C in the presence of 5 mM EDTA to sequester metal ions.

The future

Research into transport and metabolism of SO$_2$ by yeasts and mycelial fungi will continue in the foreseeable future, spurred on for the reasons adduced in the Introduction to this article. Pressure to lower the concentrations of sulphite used to preserve beverages and foods is, if anything, increasing year by year, and brings with it an ever increasing curiosity into the molecular mechanisms of the interaction of yeasts and mycelial fungi to this compound. No less interest is shown in production of acid rain, particularly in countries in the northern hemisphere. When precipitated, vapourised sulphur derivatives are in the form of sulphate. While potentially providing micro-organisms with a source of the element sulphur, the response all too often is to lower the pH value of an environment - lake or soil - to one which will not permit microbial growth. Politically, a solution to this problem has to be found.

References

Allen, H. W. (1961). *A History of Wine*, pp. 17-42. Faber & Faber: London.

Alonso, A., Benitez, J. & Diaz, M. A. (1984). A sulfate, sulfite and thiosulfate incorporating system in *Candida utilis*. *Folia Microbiologica*, **29**, 8-13.

Benitez, J., Alonso, A., Delgado, J. & Kotyk, A. (1983). Sulphate transport in *Candida utilis*. *Folia Microbiologica*, **28**, 6-11.

Borst-Pauwels, G. W. F. H. (1981). Ion transport in yeast. *Biochimica et Biophysica Acta*, **650**, 88-127.

Burroughs, L. F. & Sparks, A. H. (1964). The identification of sulphur dioxide-binding compounds in apple juices and ciders. *Journal of the Science of Food and Agriculture*, **15**, 176-185.

Fernelius, W. C., Leoning, K. & Adams, R. M. (1973). Notes on nomenclature. *Journal of Chemical Education*, **50**, 341-342.

Garcia, A., Benitez, J., Delgado, J. & Kotyk, A. (1983). Isolation of sulphate transport defective mutants of *Candida utilis*; further evidence for common transport system for sulphate, sulphite and thiosulphate. *Folia Microbiologica*, **28**, 1-5.

Hanssen, M. & Marsden, J. (1984). *E for Additives*, pp. 60-65. Thorsons Publishers Ltd.: Wellingborough, Northamptonshire, UK.

Henderson, A. (1824). *The History of Ancient and Modern Wines*, pp. 47-59. Baldwin, Cradock & Joy: London.

Hinze, H. & Holzer, H. (1985). Effect of sulfite or nitrate on the ATP content and the carbohydrate metabolism in yeast. *Zeitschrift für Lebensmitteluntersuchung und -Forschung*, **181**, 87-91.

Hofstee, B. H. J. (1959). Non-inverted versus inverted plots in enzyme kinetics. *Nature*, **184**, 1296-1298.

Ingram, M. & Vas, K. (1950). Combination of sulphur dioxide with concentrated orange juice. I. Equilibrium states. *Journal of the Science of Food and Agriculture*, **1**, 21-27.

King, A. D. Jr., Ponting, J. D., Sanschuck, D. W., Jackson, R. & Mihara, K. (1981). Factors affecting death of yeast by sulphur dioxide. *Journal of Food Protection*, **44**, 92-97.

Lloyd-Hind, H. (1948). *Brewing Science and Practice*, vol. 1, *Brewing Materials*. Chapman & Hall: London.

Macris, B. J. & Markakis, P. (1974). Transport and toxicity of sulfur dioxide in *Saccharomyces cerevisiae* var. *ellipsoideus*. *Journal of the Science of Food and Agriculture*, **25**, 21-29.

Parker, D. M., Lodola, A. & Holbrook, J. J. (1978). Use of the sulphite adduct of nicotinamide adenine dinucleotide to study ionisations and the kinetics of lactate dehydrogenase and malate dehydrogenase. *Biochemical Journal*, **173**, 959-967.

Pfleiderer, G., Sann, E. & Stock, A. (1960). The mechanism of action of dehydrogenases. The reactivity of pyridine nucleotides (PN) and PN-models with sulphite as the nucleophilic reagent. *Berichte der Deutschen Chemischen Gesellschaft*, **93**, 3083-3099.

Pilkington, B. J. & Rose, A. H. (1988). Reactions of *Saccharomyces cerevisiae* and *Zygosaccharomyces bailii* to sulphite. *Journal of General Microbiology*, **134**, 2823-2830.

Postgate, J. R. (1963). The examination of sulphur auxotrophs: a warning. *Journal of General Microbiology*, **30**, 481-484.

Stratford, M. & Rose, A. H. (1986). Transport of sulphur dioxide by *Saccharomyces cerevisiae*. *Journal of General Microbiology*, **132**, 1-6.

Stratford, M., Morgan, P. & Rose, A. H. (1987). Sulphur dioxide resistance in *Saccharomyces cerevisiae* and *Saccharomycodes ludwigii*. *Journal of General Microbiology*, **133**, 2173-2179.

Tweedie, J. W. & Segel, I. W. (1970). Specificity of transport processes for sulfur, selenium and molybdenum anions by filamentous fungi. *Biochimica et Biophysica Acta*, **196**, 95-106.

Wedzicha, B. L. C. (1984). *Chemistry of Sulphur Dioxide in Foods*. Elsevier Applied Science Publishers: London and New York.

Yamamoto, L. A. & Segel, I. W. (1966). The inorganic sulphate transport system of *Penicillium chrysogenum*. *Archives of Biochemistry and Biophysics*, **114**, 523-531.

Chapter 4

Inorganic sulphur oxidation by fungi

M. Wainwright

Department of Microbiology, University of Sheffield, Sheffield S10 2TN, UK

Introduction

The oxidation of reduced sulphur involves three main groups of microorganisms: (a) colourless sulphur bacteria, including members of the genus *Thiobacillus*, (b) photosynthetic sulphur bacteria, and (c) heterotrophs (Wainwright, 1978a). Although the involvement of organisms of groups (a) and (b) in sulphur oxidation in the environment has been generally emphasised, a wide range of heterotrophs, including bacteria and fungi, have also been implicated in the process, particularly in soils (Wainwright, 1984a).

The ability of fungi to oxidize sulphur *in vitro* was recognised during the early part of this century. For example, Waksman (1918), in a review on nutrient cycling by fungi made a passing reference to the ability of *Fusarium* species to oxidize elemental sulphur, while thiosulphate oxidation by fungi was demonstrated by Armstrong (1921). Abbott (1923) also showed that fungi could oxidize sulphur both *in vitro* and when grown in autoclaved soils. Surprisingly however, Starkey (1934) concluded that fungi were incapable of oxidizing thiosulphate. Fungi were also observed contaminating sulphur-enrichment cultures used in the original isolation of *Thiobacillus thiooxidans* (Lipman, Waksman & Joffe, 1922), and Joffe (1922) concluded that fungi, including species of *Aspergillus*, *Fusarium* and *Penicillium* participate in the early stages of sulphur oxidation, prior to the involvement of sulphur bacteria.

Interest in sulphur utilization by heterotrophs, including fungi, eventually declined as the role of the chemolithotrophic sulphur oxidizing bacteria was emphasised (Waksman & Joffe, 1922) since the rates of sulphur oxidation achieved by these bacteria *in vitro* suggested that heterotrophs would play only a minor role in the process in soils. Recent evidence suggests, however, that fungi and heterotrophic bacteria play a more important role in environmental sulphur oxidation than was previously thought, most notably in soils lacking significant populations of chemolithotrophic sulphur oxidizers (Wainwright, 1984a).

Modern studies on fungal sulphur oxidation began with the isolation by Wainwright (1978b & c) of sulphur oxidizing species from vegetation exposed to heavy atmospheric pollution. More recently a number of studies have appeared on the physiology of sulphur oxidation by both yeasts (Kurek, 1979, 1983, 1985) and filamentous fungi (Grayston, Nevell & Wainwright, 1986), and research has also been directed towards determining the role which fungi play in the process in soils (Wainwright, 1984a; Grayston *et al.*, 1986; Lawrence & Germida, 1988a & b).

My aim here is to review the literature on fungal sulphur oxidation, paying particular attention to recent studies which show that fungi may play a more important role in the process in soils than is generally recognized.

Isolation and *in vitro* study of sulphur oxidizing fungi

Fungi capable of oxidizing sulphur to thiosulphate and tetrathionate can be readily isolated from most soils. The isolates can be obtained using conventional media and then screened for their ability to oxidize sulphur using spot-tests for the formation of oxyanions, such as thiosulphate and tetrathionate. Specific media have also been reported for the isolation of sulphur-oxidizing fungi. The first of these media to be described consists of Czapek-Dox medium in which sulphur is precipitated by the reaction between polysulphide and hydrochloric acid (Wainwright, 1978a & b); a modification of a carbon-free medium devised by Wieringa (1966) to isolate chemolithotrophic sulphur-oxidizers. Sulphur oxidizing fungi growing on the surface of this medium oxidize the finely precipitated sulphur and produce a clear halo in the milky-white medium (Fig. 4.1). Colonies producing these haloes are considered presumptive sulphur oxidizers and can be removed and sub-cultured for further study. This medium has been described erroneously as a polysulphide medium and, therefore, has been criticized for containing a sulphur source not commonly added to soils as a fertilizer. This is not the case however, since the reaction between polysulphide and acid forms a fine precipitate of sulphur. Germida (1985) later used a medium for isolating sulphur oxidizers, including fungi, which contains colloidal sulphur as the sulphur source; sulphur oxidation being confirmed by the use of indicators to detect acid production. Lawrence & Germida (1988b) have also recently reported the use of a 'most probable

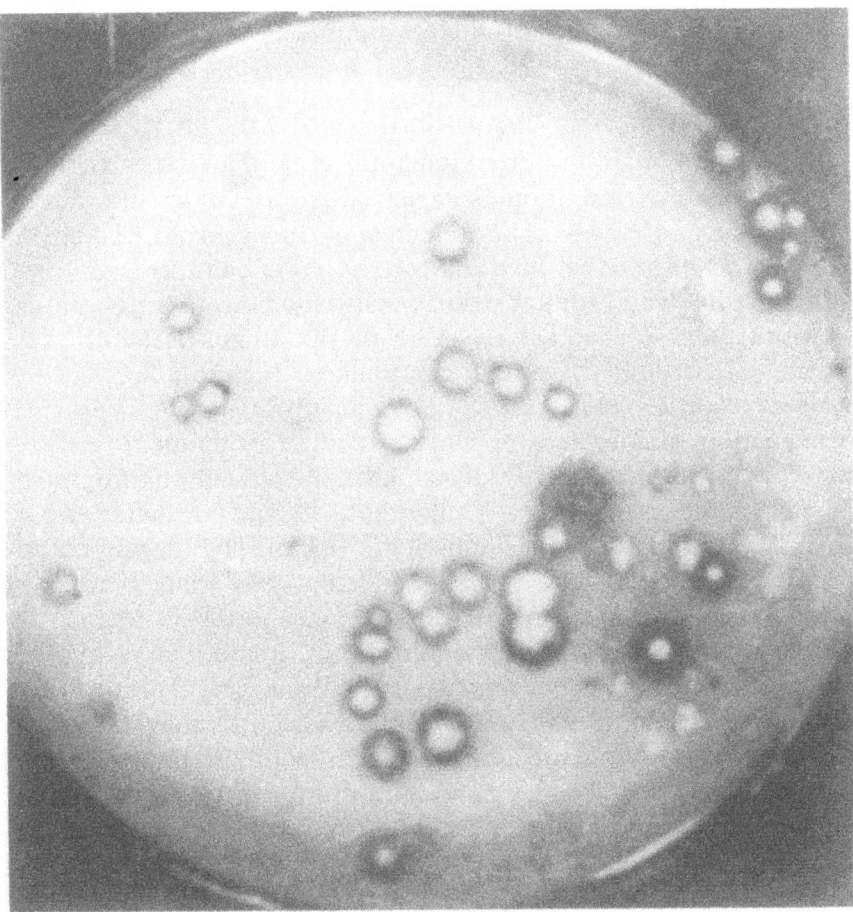

Fig. 4.1. Fungi, isolated from soil, clearing precipitated sulphur when growing on Wainwright's medium (Wainwright, 1978b).

number' method to enumerate heterotrophic thiosulphate sulphur oxidizers, including fungi, in soils.

Once presumptive sulphur oxidizing fungi have been isolated their ability to oxidize elemental and reduced forms of sulphur can be readily demonstrated in shake-cultures, using media such as liquid Czapek-Dox, and the production of thiosulphate and tetrathionate can be measured colorimetrically after cyanolysis. Sulphate is generally measured turbidimetrically. Fungal sulphur oxidation can also be detected by measurements of oxygen uptake; Grayston (1987), for example, showed that oxygen uptake over a 6 hour period was doubled when *Aspergillus niger* oxidized elemental sulphur.

Range of fungi capable of oxidizing sulphur

The ability of fungi to oxidize elemental and reduced forms of sulphur is not restricted to one specialised group, but occurs widely over all classes. However, to date most of the reported studies on the process have been devoted to mesophilic Deuteromycotina, although the list of sulphur oxidizers also includes thermophilous species (Table 4.1), and at least one marine fungus (*Asteriomyces crucicatus*). Free-living soil fungi appear to be particularly active sulphur oxidizers, and some ectotrophic mycorrhizal species and wood rotting basidiomycetes can also oxidize the element (Grayston & Wainwright, 1988). Fungi capable of oxidizing sulphur have been isolated from soils (Wainwright, 1984a; Germida, 1985; Krol, 1983), vegetation (Wainwright, 1978c), H2S-impregnated filter beds (Langenhove, Wuyts & Schamp, 1986) and even from sulphur fungicides (Adams, 1932). The widespread occurrence of the sulphur oxidative ability among fungi isolated from diverse habitats, suggests that these organisms play an important role in sulphur oxidation in nature. Nor does this ability need to be acquired by long term exposure to reduced sulphur, since isolates from culture collections, even without prior exposure to sulphur, can immediately oxidize sulphur when grown *in vitro*. The widespread occurrence among fungi of the ability to oxidize sulphur suggests that it is either a directly useful process, or a fortuitous product of metabolic reactions which are not themselves associated with the ability to oxidize sulphur (see below). Resistance to the fungicidal properties of elemental sulphur is an obvious pre-

Table 4.1. Fungi reported to oxidize sulphur *in vitro*

(a) Deuteromycotina and Zygomycotina

Absidia glauca	*Acremonium*
Alternaria tenuis	*Aspergillus flavus*
Aspergillus fumigatus	*Aspergillus niger*
Aureobasidium pullulans	*Epicoccum nigrum*
Fusarium episphaeria	*Fusarium tricinctum*
Monilia sp.	*Mortierella isabellina*
Mucor flavus	*Penicillium pinetorum*
Trichoderma hamatum	*Trichoderma harzianum*
Trichoderma viride	*Trichoderma sp.*
Zygorhynchus molleri	*Zygorhynchus vuilleminii*

(b) Mycorrhizal fungi

Amanita muscaria	*Hymenoscyphus ericae*
Paxillus involutus	*Pisolithus tinctorius*
Rhizopogon roseolus	*Suillus bovinus*

(c) Cord-forming Basidiomycotina

Hypholoma fasciculare	*Phanerochaete velutina*

(d) Yeasts

Rhodotorula sp.

(e) Thermophilous fungi

Geosmithia argillacea	*Geosmithia emersonii*
Myceliophthora thermophila	

Sources: Germida (1985); Grayston & Wainwright (1988); Wainwright (1984a).

---

requisite if a fungus is to oxidize the element, so that those phytopathogenic fungi which are very sensitive to the fungicidal effect of sulphur cannot, *a priori* participate in the process.

Sulphur oxidation by fungi *in vitro*

The majority of reports on sulphur oxidation by fungi have involved shake culture studies using carbon-rich media e.g. liquid Czapek-Dox. The products of elemental sulphur oxidation by fungi (e.g. *Fusarium solani*) are usually thiosulphate, tetrathionate, and sulphate – the terminal oxidation product (Wainwright & Killham, 1980). It is generally assumed that the following direct polythionate pathway is involved in these transformations, although polythionates may result from side reactions, or even from non-biological oxidation:

$$S^0 \quad \rightarrow \quad S_2O_3^{2-} \quad \rightarrow \quad S_4O_6^{2-} \quad \rightarrow \quad SO_4^{2-}$$

Elemental-S   thiosulphate   tetrathionate   sulphate

Wainwright & Killham (1980) showed that elemental sulphur-oxidation by *F. solani* was optimal at pH 7·0, increased with increasing amounts of added sulphur and doubled as the C:N ratio was increased from 5:1 to 10:1. The process caused a decrease in the pH of the medium (from 6·5 to 2·4 over 10 days) as sulphuric acid was formed. However, most fungi, like *Aspergillus niger*, generally produce only thiosulphate and sulphate when oxidizing elemental sulphur *in vitro* (Fig. 4.2), with oxidation leading to a reduction in the pH of the culture medium to pH 2·0. Thiosulphate oxidation, on the other hand, usually results in a less dramatic effect on medium pH and to the production of tetrathionate as well as sulphate (Fig. 4.3). Fungi achieve differing rates of sulphur oxidation *in vitro*. For example, after 21 days growth with elemental sulphur in Czapek-Dox medium *Penicillium nigricans* produced 430 μg SO_4^{2-} ml^{-1} compared to values for *Aspergillus niger* of 280; *Trichoderma harzianum*, 250; *Mucor flavus*, 28 and *Phanerochaete velutina*, 21 (Grayston, 1987). However, these amounts of sulphate cannot be used directly to list fungi in order of their sulphur oxidative ability, since the values represent net sulphate levels and do not account for differences in the rate of sulphate assimilation between species (Raistrick & Vincent, 1948). It should again be emphasised that these rates of oxidation achieved by fungi are in no way comparable to those seen when thiobacilli oxidize sulphur *in vitro*, although they do compare favourably with rates of oxidation achieved by other heterotrophs (Pepper & Miller, 1978).

When fungi are grown *in vitro* with elemental sulphur they invariably adsorb the sulphur particles onto their mycelial surface, thereby removing them from solution (Fig. 4.4). Particle adsorption by fungi is a widespread phenomenon and since particulate materials like clays are also adsorbed onto hyphae (Wainwright,

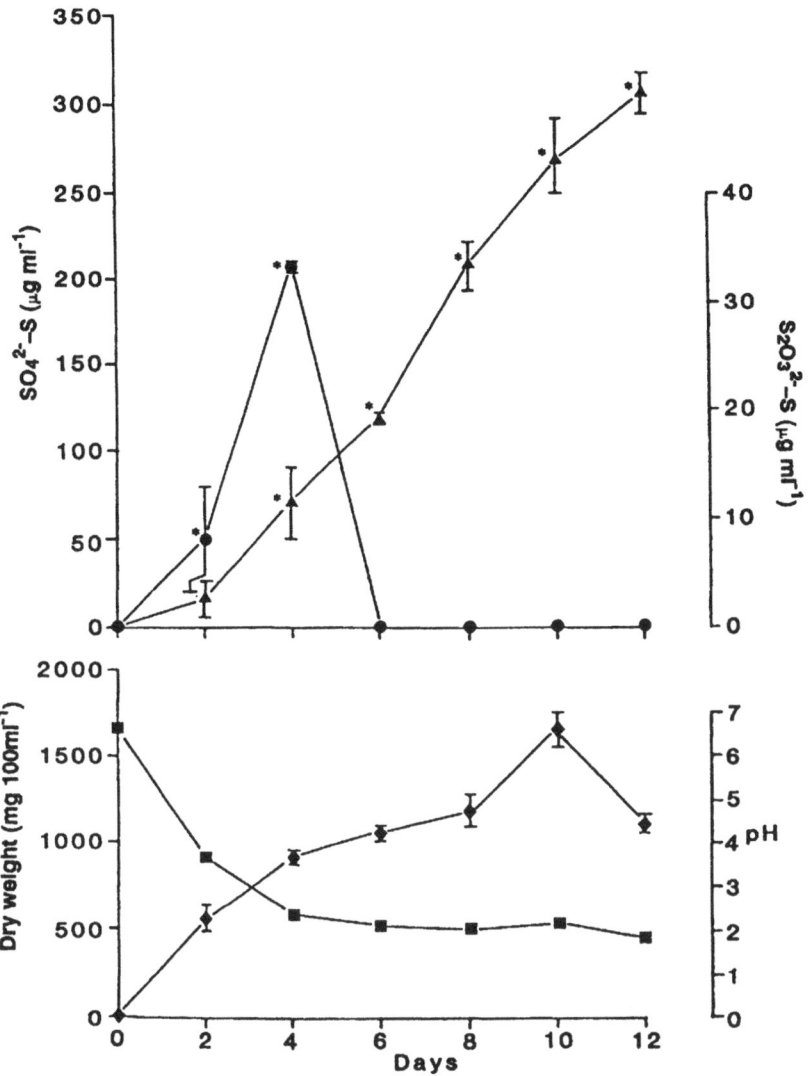

Fig. 4.2. Oxidation of elemental sulphur (1% w/v) to thiosulphate (closed circles) and sulphate (closed triangles) by *Aspergillus niger* *in vitro*; dry weight is shown as diamonds, and medium pH as closed squares. Values are excess over uninoculated controls, and are the means of triplicates. Asterisks indicate values which are significantly increased over the controls at $P = 0.05$. Error bars are ± 1 SD.

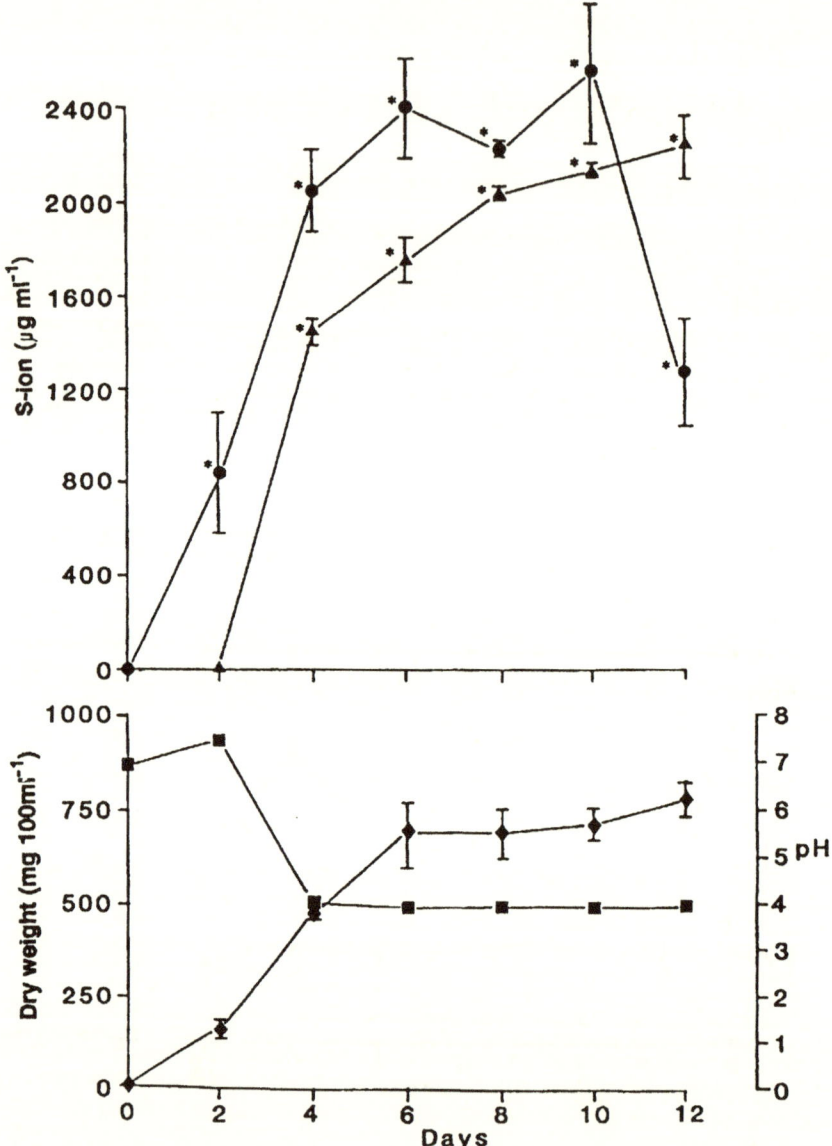

Fig. 4.3. Oxidation of thiosulphate to tetrathionate (closed circles) and sulphate (triangles) by *Aspergillus niger*. Dry weight is shown as diamonds, pH as squares. Asterisks indicate values which are significantly increased over the controls at $P = 0.05$. Error bars are ± 1 SD.

Grayston & DeJong, 1986) the ability of fungi to adsorb sulphur may be unrelated to the fact that they oxidize the element.

Although fungal sulphur oxidation is generally studied using carbon-rich media, low, but measurable, rates of oxidation can occur in media containing relatively small amounts of carbon (10 to 100 μg C ml^{-1}) (Grayston *et al.*, 1986; Wainwright & Grayston, 1988), and some active species such as *A. niger* can clear carbon-free polysulphide medium solidified with colloidal silica. The ability of fungi to oxidize sulphur in media containing low concentrations of carbon is obviously relevant to the question of whether they participate in the process in soils and other environments where supplies of carbon are likely to be limited.

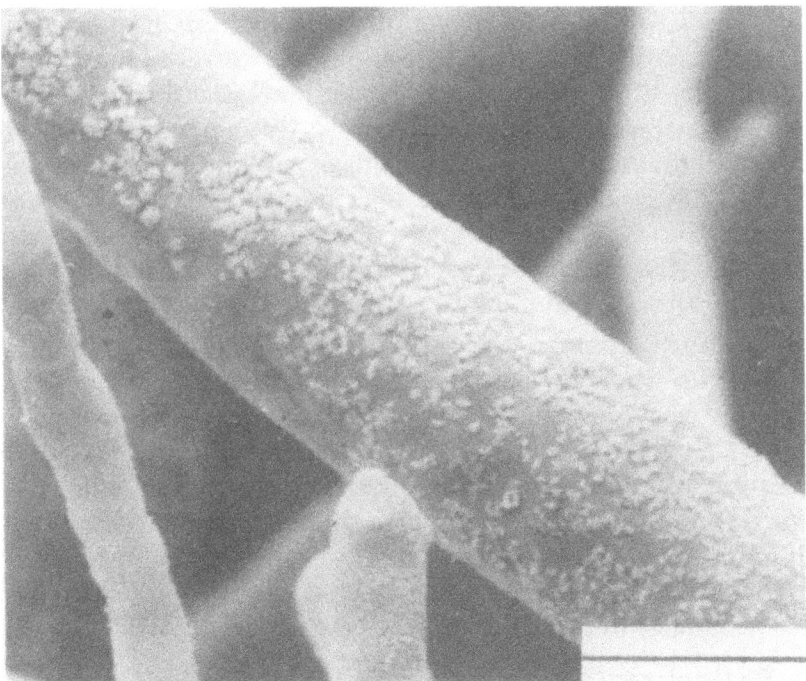

Fig. 4.4. Scanning electron-micrograph (scale bar = 20 μm) of hyphae of *Mucor flavus* adsorbing elemental sulphur (1% w/w) after 7 days growth at 25°C.

A wide range of carbon substrates can support sulphur oxidation by fungi *in vitro* (Grayston, 1987), with rates of oxidation depending upon the carbon source employed. For example, larger amounts of sulphate were produced by both *A. niger* and *T. harzianum* when glucose or sucrose acted as energy source than when amino acids provided the carbon source (Grayston & Wainwright, 1987). In contrast, Krol (1983) reported that the fungi used in her studies achieved higher rates of sulphur oxidation when amino acids, rather than sugars were provided as the carbon source. Cereal straw and acetic acid, the latter being an anaerobic breakdown product of straw, can also support fungal sulphur oxidation, so plant residues should be capable of supplying carbon to support sulphur oxidation by fungi in agricultural soils (Grayston *et al.*, 1986).

Most *in vitro* studies on fungal sulphur oxidation have involved the use of elemental sulphur, thiosulphate or tetrathionate as sulphur source, although Skerman, Dermetjev & Skyring (1957) also showed that some species can oxidize H_2S. Wainwright & Grayston (1986) reported that *A. niger* oxidizes the sulphides of copper, lead and zinc, but not cadmium. Since metal ions were not found free in solution as a result of these oxidation reactions it was concluded that any metal released during sulphide oxidation was rapidly adsorbed by the fungus.

Studies by Germida (1985) suggest that fungi may interact synergistically when oxidizing sulphur on solid media. Similarly, Grayston & Wainwright (1988) showed synergism between *A. niger*, *T. harzianum* and *M. flavus* in liquid culture, with larger quantities of sulphate being initially produced in mixed, compared to single culture. Interactions between sulphur oxidizing fungi and other sulphur oxidizing micro-organisms will probably occur in nature, and maximum rates of oxidation are likely to be achieved by mixed communities of sulphur oxidizers.

Biochemistry of fungal sulphur oxidation

Little is known about the biochemical pathways involved in fungal sulphur oxidation. The process has, however, been shown to be enzymic in *Aureobasidium pullulans* (Killham, Lindley & Wainwright, 1981) and in a species of the yeast *Rhodotorula* (Kurek, 1985). The latter was shown to possess both thiosulphate and sulphite oxidases, with the activity of both enzymes being coupled with ferricyanide and native, but not mammalian, cytochrome-*c*. The thiosulphate enzyme catalyzed the following reaction:

$$2S_2O_3^{2-} + 2Fe(CN)_6^{3-} \rightarrow S_4O_6^{2-} + 2Fe(CN)_6^{4-}$$

While the sulphite oxidase catalyzed the following:

$$2SO_3^{2-} + \text{cytochrome-c } Fe^{3+} + H_2O \rightarrow SO_4^{2-} + $$
$$2 \text{ cytochrome-c } Fe^{2+} + 2H^+$$

Both enzymes were inhibited by thiol-inhibiting, but not chelating agents, and they were both denatured by proteolytic enzymes and sodium deoxycholate. It appears, therefore, that not only proteins, but also phospholipids are important in the activity of this sulphur-oxidizing enzyme complex, and that the sulphite oxidase of *Rhodotorula* is similar to that found in thiobacilli (Kurek, 1985).

The fact that H_2S is often liberated when fungi are grown in culture with elemental sulphur and thiosulphate suggests an alternative means by which sulphate may be produced. Sciarini & Nord (1943) suggested that elemental sulphur serves mainly as a hydrogen acceptor in dehydrogenations and alcoholic fermentations of hexoses and pentoses in *Fusarium lini* thus:

$$S^0 + 2H \rightarrow H_2S$$

$$2H_2S + O_2 \rightarrow 2S + 2H_2O$$

$$2H_2O + H_2S + 4O \rightarrow SO_4^{2-} + 2H_3O^+$$

To date no clear explanation has been put forward to account for why fungi oxidize sulphur. The following possibilities have been suggested, however, and will be considered in turn below:

(1) The process may be fortuitous or incidental to the normal heterotrophic metabolism of the organisms, and may be merely an *in vitro* phenomenon.

(2) Conversely, the process may yield energy, which the organism can presumably use largely chemolithoheterotrophically.

(3) Fungi may oxidize inorganic sulphur to sulphate to meet their assimilatory requirements for sulphur.

(4) Inorganic sulphur oxidation may involve the same metabolic reactions which function in sulphur mineralization.

(5) Sulphur oxidation may provide a means by which the toxic effects of elemental sulphur and heavy metals can be avoided.

(6) Finally, the process may involve free radicals and as a result, may protect fungi from the damaging effects of, for example, superoxide.

The first of these possibilities, namely that the ability of fungi to oxidize sulphur is fortuitous or incidental to the organism's normal metabolism, is a widely held belief and it is generally assumed that no energy is obtained from the process, an assumption which has, however, largely gone unchallenged. Although fungi are generally regarded as being strict heterotrophs there is no *a priori* reason why they should not obtain energy from processes such as sulphur oxidation and nitrification. An inorganic, energy yielding pathway, has been implicated by Aleem, Lees & Lyric (1964) to account for nitrification by fungi, a view which was challenged although not necessarily disproved by Doxtader & Alexander (1966). The usual argument proposed against the view that fungi gain energy from nitrification and sulphur oxidation is that maximum rates of these processes occur towards the end of the fungus growth cycle. However, where this has been observed, carbon rich media were employed, so that the fungus could meet all its energy and carbon requirements by metabolizing sugars without the need to gain energy from inorganic oxidation reactions. The use of such media would tend to mask the relatively small energy gains obtained from the oxidation of reduced nitrogen and sulphur compounds.

The second possibility listed above (i.e. chemolithoheterotrophy) was strengthened when Grayston & Wainwright (1987) and Wainwright & Grayston (1988) showed that thiosulphate stimulated fungal growth at low carbon concentrations (Fig. 4.5), suggesting that they might grow chemolithoheterotrophically gaining energy from thiosulphate oxidation while growing hetero-

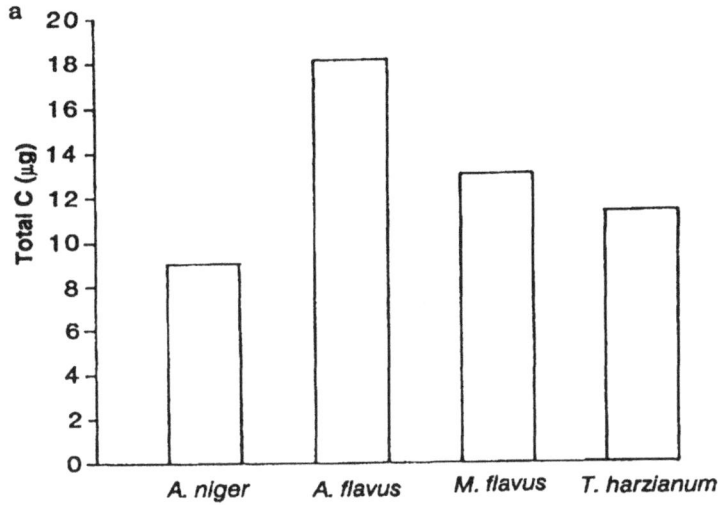

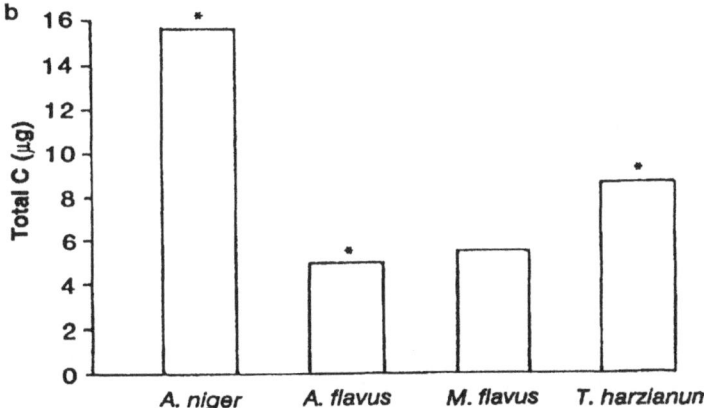

Fig. 4.5. Growth of fungi in media lacking added carbon (a), and (b) increase in fungal growth (7 days at 25°C) in this medium due to thiosulphate (1 mg S ml^{-1}). Asterisks indicate bars which show a significant increase above growth in the absence of thiosulphate at $P = 0.05$. From Wainwright & Grayston, 1988.

trophically on the carbon provided. The result of such a strategy under carbon-limited conditions would be an increase in biomass. Similar thiosulphate induced growth stimulation has also been reported for the mycorrhizal fungus *Glomus caledonium* (Hepper, 1984), and heterotrophic bacteria (Mason & Kelly, 1988). Unfortunately, however, chemolithoheterotrophy is difficult to confirm, particularly by using batch cultures, and proof that fungi are able to grow by oxidizing thiosulphate in this way must await further studies involving the use of chemostats. An aid to such studies would be the recognition that fungi are metabolically more versatile than they are usually assumed to be. For example, they participate in many oxidation-reduction reactions which were formally thought to be mediated solely by bacteria. Evidence has been provided to show that they can fix CO_2 autotrophically and heterotrophically, as well as growing oligotrophically by scavenging traces of both carbon and nitrogen (Wainwright, 1988). It is probable, therefore, that our obsession with growing these organisms on carbon rich media has masked many of the more interesting aspects of their physiology, particularly in relation to transformations of nitrogen and sulphur (Wainwright, 1988).

The fact that fungi can oxidize reduced forms of sulphur in the presence of large amounts of sulphur would appear to dismiss quickly the view, given as point (3) above, that fungi oxidize sulphur solely to meet their assimilatory needs for this element. The possibility (point 4) that sulphur oxidation by fungi involves reactions common to sulphur mineralization processes, while possibly valid, has yet to be tested.

In support of suggestion (5), that the ability of fungi to oxidize sulphur may confer other benefits unrelated to energy gain, Wainwright & Grayston (1983) showed that the thiosulphate and tetrathionate produced during elemental sulphur oxidation can protect fungi from the toxic effects of metals such as mercury. Whether the production of polythionate and thiols, and the associated protective effects of the formation of metal complexes occur in nature is not known. In a similar way, sulphur oxidation may enable fungi to avoid the toxic effect of elemental sulphur and since hydrogen sulphide production appears to play a role here it would clearly be beneficial for a fungus to be able to oxidize both elemental sulphur and H_2S to innocuous sulphate. This would imply that plant-pathogenic fungi which are susceptible to the fungicidal effects of elemental sulphur, are incapable of oxidizing elemental sulphur, a view which is unfortunately difficult to substantiate experimentally.

Finally, Wood (1988) has suggested that nitrification by fungi may involve free radical mechanisms largely because nitrification has been linked with fungal cell lysis. The involvement of such reactions is even more likely to account for fungal sulphur oxidation since the redox potentials for sulphate/reduced forms of sulphur couples are much lower than for nitrate/nitrite or nitrite/ammonium couples (P. M. Wood pers. comm.). An example of the auto-oxidation of sulphite has now been well characterized (McCord & Fridovich, 1969).

The role of fungi in sulphur oxidation in soil

Our interest in the ability of fungi to oxidize sulphur began with a desire to know if these organisms make a major contribution to sulphur oxidation in soils - a process which is often considered to be mediated solely by the thiobacilli. There is, in fact, an increasing awareness of the role which heterotrophs play in sulphur oxidation in soil, particularly where thiobacilli are lacking; and recent evidence suggests that a wide range of fungi and bacteria (Pepper & Miller, 1978), including actinomycetes (Wainwright, Skiba & Betts, 1984), might all play a role in the process.

The general paucity of carbon is usually cited as a major limitation on heterotrophic sulphur oxidation in soils, although it could occur in the rhizosphere or in close proximity to decomposing plant residues where carbon should be more readily available. However, since fungi can grow and oxidize sulphur under oligo-carbotrophic conditions, low levels of carbon in soils may not be such an important limitation as it first appears. Further, actinomycetes can also oxidize sulphur under oligotrophic conditions (Wainwright *et al.*, 1984), so it is clear that carbon deficiency in soils need not necessarily limit the ability of heterotrophs to oxidize sulphur. Also, recent studies by Lawrence & Germida (1988b) have confirmed that heterotrophs play an important role in sulphur oxidation in soil. Using glucose amendment they showed that microbial biomass was produced in direct proportion to an increase in the rate of sulphur oxidation, and with the aid of metabolic inhibitors demonstrated the role of fungi and other heterotrophs in the process.

Although fungi and other heterotrophs readily oxidize sulphur when inoculated into autoclaved soils (Wainwright & Killham, 1980; Grayston et al., 1986; Pepper & Miller, 1978), from which are liberated large amounts of available carbon during sterilization, the process is difficult to demonstrate in non-sterilized soils. Evidence for the ability of fungi to oxidize sulphur while growing in non-autoclaved soils was, however, provided by Wainwright & Killham (1980), who inserted into soil glass capillary pedoscopes containing modified polysulphide medium, a fungal spore germination inhibitor and antibacterial antibiotics. After a period of incubation in the soil, the capillaries were removed and examined for evidence of clearing of the opaque milk-white medium which would indicate sulphur oxidation. As clearing was observed around fungal hyphae, it was concluded that fungi can grow from the soil (the germination inhibitor prevented spores from germinating at the soil-agar interface) and oxidize the sulphur, the inclusion of antibacterial antibiotics avoiding the possibility that sulphur oxidizing bacteria growing on the hyphal surface were responsible for the observed sulphur oxidation. More recent work from my laboratory has shown that ectotrophic mycorrhizal fungi such as *Suillus bovinus* and *Amanita muscaria* are also able to oxidize sulphur *in vitro* (Table 4.1), and limited evidence has also been provided to indicate that these fungi can oxidize the element when growing as mycelial fans in soil (Grayston & Wainwright, 1988). Wood decomposing basidiomycetes e.g. *Hypholoma fasciculare* are also capable of slow rates of oxidation *in vitro*, but no evidence was found by Grayston & Wainwright (1988) to suggest that these fungi actively oxidize the element when growing from inoculated wood blocks into the soil. These fungi receive carbon from a host or food base, but nevertheless they seem to be incapable of oxidizing sulphur at rates achieved by free-living species, at least *in vitro*. This could indicate that the latter group are adapted for growth under the rigours of carbon limitation in soil, and that free-living fungi may have evolved the ability to gain energy by oxidizing sulphur in order to supplement the limited energy supplies normally available to them.

Conclusion

Although the role of the colourless sulphur bacteria in sulphur oxidation has been generally emphasised it is becoming increasingly clear that heterotrophs, including fungi, play a major role in the process. As yet, however, insufficient is known about the biochemical and physiological processes involved in fungal sulphur oxidation to allow us to assess accurately the advantages which these organisms derive from participating in these oxidative reactions. Similarly, while evidence increasingly shows that fungi contribute to the oxidation of sulphur in soils, it is unclear what proportion of the sulphate produced can be attributed to the chemolithotrophs and what proportion derives from sulphur oxidation by fungi and other heterotrophs. Future research will hopefully provide a clearer picture of the part played by fungi in sulphur oxidation and their overall role in the sulphur cycle in nature.

Acknowledgements. The contributions of the following postgraduates to this work are gratefully acknowledged, Susan J. Grayston, Wendy Nevell, K. Killham, Ute Skiba and I. Singleton. Funding from the NERC and SERC is also acknowledged.

References

Adams, J. M. (1932). Bacterial and fungus flora in certain fungicides. *Phytopathology*, **22**, 785-786.

Abbot, E. V. (1923). The occurrence and action of fungi in soils. *Soil Science*, **16**, 207-216.

Aleem, M. I. H., Lees, H. & Lyric, R. (1964). Ammonium oxidation by cell-free extracts of *Aspergillus wentii*. *Canadian Journal of Biochemistry*, **42**, 989-999.

Armstrong, G. M. (1921). Studies in the physiology of the fungi – sulphur nutrition, the use of thiosulphate as influenced by hydrogen ion concentration. *Innals of the Missouri Botanical Garden*, **8**, 237-248.

Doxtader, K. G. & Alexander, M. (1966). Nitrification by growing and replacement cultures of *Aspergillus*. *Canadian Journal of Microbiology*, **12**, 807-815.

Germida, J. J. (1985). Modified sulphur-containing media for studying sulphur-oxidizing micro-organisms. In *Planetary Ecology*, eds. D. E. Caldwell, J. A. Brierley & C. L. Brierley, pp. 333-344. Nostrand Reinhold, New York.

Grayston, S. J. (1987). Sulphur oxidation and nitrification by fungi *in vitro* and in soils. PhD thesis, University of Sheffield, England.

Grayston, S. J., Nevell, W. & Wainwright, M. (1986). Sulphur oxidation by fungi. *Transactions of the British Mycological Society*, **87**, 193-198.

Grayston, S. J. & Wainwright, M. (1987). Fungal sulphur-oxidation: effect of carbon source and growth stimulation by thiosulphate. *Transactions of the British Mycological Society*, **88**, 213-219.

Grayston, S. J. & Wainwright, M. (1988). Sulphur oxidation by soil fungi including some species of mycorrhizae and wood-rotting basidiomycetes. *FEMS Microbiology Ecology*, **53**, 1-8.

Hepper, C. M. (1984). Inorganic sulphur nutrition of the vesicular-arbuscular mycorrhizal fungus *Glomus caledonium*. *Soil Biology and Biochemistry*, **16**, 669-671.

Joffe, J. S. (1922). Preliminary studies on the isolation of sulfur oxidizing bacteria from sulfur floats - soil composts. *Soil Science*, **13**, 161-172.

Killham, K., Lindley, N. D. & Wainwright, M. (1981). Inorganic sulphur oxidation by *Aureobasidium pullulans*. *Applied and Environmental Microbiology*, **42**, 629-631.

Krol, M. (1983). Occurrence in soils and activity of sulphur oxidizing micro-organisms. *Pamietnik Pulawski Prace Iung Zeszyt*, **79**, 45-61.

Kurek, E. (1979). Oxidation of inorganic sulphur compounds by yeasts. *Acta Microbiologica Polonica*, **28**, 169-172.

Kurek, E. (1983). An enzymatic complex active in sulphite and thiosulphate oxidation by *Rhodotorula* sp. *Archives of Microbiology*, **134**, 143-147.

Kurek, E. (1985). Properties of an enzymatic complex active in sulphite and thiosulphate oxidation by *Rhodotorula* sp. *Archives of Microbiology*, **143**, 277-282.

Langenhove, H. van, Wuyts, E. & Schamp, N. (1986). Elimination of hydrogen sulphide from odorous air by a wood bark filter. *Water Research*, **20**, 1471-1476.

Lawrence, J. R. & Germida, J. J. (1988a). A most probable number method (MPN) for the enumeration of heterotrophic thiosulphate producing sulfur oxidizers in soil. *Soil Biology and Biochemistry*, **20**, 577-578.

Lawrence, J. R. & Germida J. J. (1988b). Relationship between microbial biomass and elemental sulfur oxidation in agricultural soils. *Soil Science Society of America Journal*, **52**, 662-667.

Lipman, J. G., Waksman, S. A. & Joffe, J. S. (1921). The oxidation of sulfur by soil micro-organisms. *Soil Science*, **12**, 475-489.

Mason, J. & Kelly, D. P. (1988). Thiosulphate oxidation by obligately heterotrophic bacteria. *Microbial Ecology*, **15**, 123-134.

McCord, J. M. & Fridovich, I. (1969). The utility of superoxide dismutase in studying free radical reactions. *Journal of Biological Chemistry*, **244**, 6056-6063.

Pepper, I. L. & Miller, R. H. (1978). Comparison of the oxidation of thiosulphate and elemental sulphur by two heterotrophic bacteria and *Thiobacillus thiooxidans*. *Soil Science*, **126**, 9-14.

Raistrick, H. & Vincent, J. M. (1948). Studies in the biochemistry of micro-organisms. A survey of fungal metabolism of inorganic sulphates. *Biochemical Journal*, **43**, 90-99.

Sciarini, L. J. & Nord, F. F. (1943). Elementary sulphur as a hydrogen acceptor in dehydrogenations by living Fusaria. *Archives of Biochemisty and Biophysics*, **3**, 261-267.

Skerman, V. B. D., Dermentjev, S. & Skyring, G. W. (1957). Deposition of sulphur from hydrogen sulphide by bacteria and yeast. *Nature*, **179**, 742.

Starkey, R. L. (1934). The production of polythionates from thiosulphate by microorganisms. *Journal of Bacteriology*, **28**, 387-400.

Wainwright, M. (1978a). Microbial sulphur oxidation in soil. *Science Progress, Oxford*, **65**, 459-475.

Wainwright, M. (1978b). A modified sulphur medium for the isolation of sulphur oxidizing fungi. *Plant & Soil*, **49**, 191-193.

Wainwright, M. (1978c). Sulphur oxidizing micro-organisms on vegetation and in soils exposed to atmospheric pollution. *Environmental Pollution*, **17**, 167-174.

Wainwright, M, (1984a). Sulphur oxidation in soils. *Advances in Agronomy*, **37**, 349-396.

Wainwright, M. (1984b). Sulphur oxidation by some thermophilous fungi. *Transactions of the British Mycological Society*, **83**, 721-724.

Wainwright, M. (1988). Metabolic diversity of fungi in relation to growth and mineral cycling in soil. *Transactions of the British Mycological Society*, **90**, 159-170.

Wainwright, M. & Grayston, S. J. (1983). Reduction in heavy metal toxicity towards fungi by addition to media of sodium thiosulphate and sodium tetrathionate. *Transactions of the British Mycological Society*, **81**, 541-546.

Wainwright, M. & Grayston, S. J. (1986). Oxidation of heavy metal sulphides by *Aspergillus niger* and *Trichoderma harzianum*. *Transactions of the British Mycological Society*, **86**, 269-272.

Wainwright, M. & Grayston, S. J. (1988). Fungal growth and stimulation by thiosulphate under oligocarbotrophic conditions. *Transactions of the British Mycological Society*, in press.

Wainwright, M. & Killham, K. (1980). Sulphur oxidation by *Fusarium solani*. *Soil Biology and Biochemistry*, **12**, 555-558.

Wainwright, M., Grayston, S. J. & De Jong, P. (1986). Adsorption of insoluble compounds by mycelium of the fungus *Mucor flavus*. *Enzyme and Microbial Technology*, **8**, 597-600.

Wainwright, M., Skiba, U. & Betts, R. P. (1984). Sulphur oxidation by *Streptomyces* sp. growing in a carbon deficient medium and autoclaved soil. *Archives of Microbiology*, **139**, 272-276.

Waksman, S. A. (1918). The importance of mould action in soil. *Soil Science*, **6**, 137-155.

Waksman, S. A. & Joffe, J. S. (1922). Micro-organisms concerned in the oxidation of sulfur in the soil. II. *Thiobacillus thiooxidans* a new sulfur oxidizing organism isolated from the soil. *Journal of Bacteriology*, **11**, 239-256.

Wieringa, R. T. (1966). Solid media with elemental sulphur for detection of sulphur oxidizing microbes. *Antonie van Leeuwenhoek*, **32**, 183-186.

Wood, P. M. (1988). Monoxygenase and free radical mechanisms for biological ammonia oxidation. In *The Nitrogen and Sulphur Cycles*, Society for General Microbiology Symposium 42, eds. J. A. Cole & S. Ferguson, pp. 65-98. Cambridge University Press, Cambridge, UK.

Chapter 5

Sulphur compounds in fungi

J. C. Slaughter

Department of Brewing and Biological Sciences, Heriot Watt University,
Edinburgh EH1 1HX, UK

Introduction

This chapter is concerned with compounds containing divalent
sulphur at its lowest oxidation state of -2. In its sulphide, thioes-
ter or thioether forms, sulphur plays a significant part in the biochemical function of a wide range of compounds. The high
reactivity of the S-X bond compared to the equivalent O-X bond
is mainly responsible for this and is itself due to the fact that as a
larger atom, sulphur has a lower charge density and so forms
longer, weaker and less polarized bonds than does oxygen. An idea
of the effect of substituting sulphur for oxygen can easily be gained
by comparing the chemical and physical properties of water and
hydrogen sulphide. H_2S has a boiling point of -61°C, is toxic to living organisms at trace levels, reacts readily with metals and is ex-
plosive in air at concentrations between 4 and 46%. A notable
exception to this reactivity rule is that disulphides (-S-S-) are much
more stable than peroxides (-O-O-).

Biosynthesis of sulphide

The fungi can normally utilize inorganic sulphur for all their requirements. The most common source is sulphate which is the
most highly oxidized and stable form of the element. Sulphate is
actively taken up (Bradfield *et al.*, 1970; McCready & Din, 1974)
and reduced to sulphide by a pathway which appears to be quite
well understood and occurs in all organisms investigated which can
utilize sulphate (Fig. 5.1). The reactions absorb a considerable
amount of energy as the free energy difference (ΔG^0) between sulphate and sulphide is about 800 kJ mol^{-1} (Huxtable, 1986). The
first two steps of the pathway can be seen as an 'activation' of sul-

phate. First, ATP sulphurylase (sulphate adenylyl transferase) catalyses reaction between sulphate and ATP to form adenosine-5'-phosphosulphate (APS) and inorganic pyrophosphate. The chemical equilibrium of the reaction is very much in favour of sulphate and ATP. A biosynthetic flux is only possible because pyrophosphate is removed through the action of pyrophosphatase, and the K_m value for APS of the next enzyme in the sequence, APS kinase, is very low. APS kinase catalyses reaction between APS and a further molecule of ATP to form adenosine-3'-phospho-5'-phosphosulphate (PAPS). The characteristics of APS sulphurylase from *Penicillium chrysogenum* have been described (Tweedie & Segal, 1970) but little seems to be known about APS kinase.

The first reduction of the sulphur atom is catalyzed by sulphotransferases using either APS or PAPS as substrate. Variation exists from one organism to another but it appears likely that, in most fungi, PAPS is in the main stream of sulphur reduction. The

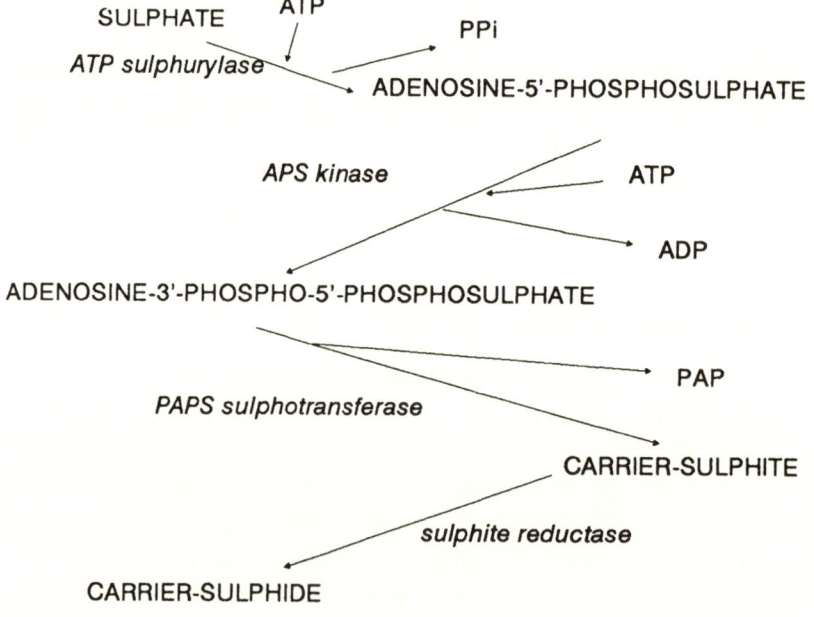

Fig. 5.1. Biosynthesis of sulphide. PAP = adenosine 3', 5'-diphosphate.

sulphate group is transferred to a sulphydryl group on a carrier molecule to form a sulphite ion attached to the carrier through a disulphide bond (a thiosulphonate). The reaction releases either AMP or PAP (adenosine-3'-phospho-5'-phosphate) depending on whether APS or PAPS was the substrate. NADPH is probably required in this step. The carrier is most likely to be a protein but can be a simple thiol and in some organisms glutathione seems to fulfill the function. In the final step the sulphite ion, still bound to the carrier, is reduced to sulphide through the action of sulphite reductase. The source of reducing power is not always clear but in yeast ferredoxin is required in addition to NADPH, and lipoic acid is a cofactor (Cooper, 1983). It is likely that thioredoxin is also involved, as in bacteria and unicellular algae. Furthermore, it may be that in fungi, certainly in *Saccharomyces cerevisiae*, there are two distinct sulphite reductases: one acting on carrier bound sulphite and the other using free sulphite (Umbarger 1978).

Biosynthesis of cysteine and methionine

Reduced sulphur is incorporated into carbon compounds through the reaction of sulphide with various precursors to form the amino acids cysteine and homocysteine (Fig. 5.2). Cysteine is formed by reaction between serine or O-acetylserine and sulphide, in free or bound form, catalyzed by cysteine synthase. O-acetylserine is the preferred substrate (de Robichon-Szulmajster & Surdin-Kerjan, 1971). In an analogous way homocysteine synthase catalyses formation of homocysteine from O-acetylhomoserine and sulphide. Homocysteine can also be formed from cysteine through reaction with O-acetylhomoserine to form cystathionine (cystathionine-γ-synthase) followed by breakdown to yield homocysteine, pyruvate and ammonia (cystathionine-β-lyase). In the reverse reactions from homocysteine to cysteine through cystathionine, homocysteine reacts with serine rather than pyruvate, and α-ketobutyrate is the other product rather than O-acetylhomoserine. The reactions in this direction are catalyzed by two quite separate enzymes, cystathionine-β-synthase and cystathionine-γ-lyase. Methionine is synthesised from homocysteine by methylation of the sulphur atom with N^5-methyltetrahydrofolate as the methyl donor.

Evidence from studies with mutant organisms suggests that though all the pathways mentioned above may be present, they are

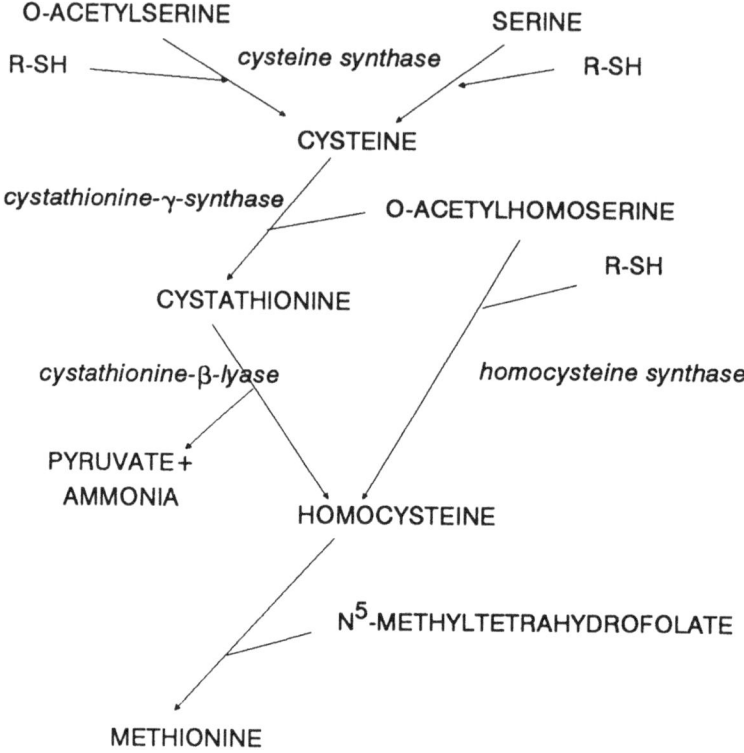

Fig. 5.2. Biosynthesis of cysteine, homocysteine and methionine.

put to different uses in different species. For example, methionine auxotrophs of *Neurospora crassa* can grow when supplied with cystathionine but not when homoserine and sulphide are the substrates, despite the presence of homocysteine synthase (Huxtable, 1986). In *N. crassa* the latter enzyme is possibly involved in sulphur salvage as it functions well with thiomethane (methyl sulphide), a product of cysteine breakdown.

In hemiascomycetous yeasts, it seems likely that the direct sulphydration reaction to homocysteine is the important biosynthetic route (Cherest, Eicher & de Robichon-Szulmajster, 1969). Homocysteine can still be formed from cystathionine but the activity of cystathionine-γ-synthase is very low (Kerr & Flavin, 1970). Savin

& Flavin (1972) have suggested that in yeast the O-acetylhomo-serine sulphydrase (homocysteine synthase) and cystathionine-γ-synthase activities may be properties of a single protein.

Further metabolism of cysteine and methionine

Both cysteine and methionine have several biochemical functions. These are well described in standard textbooks of biochemistry and so only a brief outline will be given here to illustrate the highly significant functions of the reduced divalent sulphur atom.

Proteins

Both cysteine and methionine are incorporated into proteins. Cysteine is particularly important as the sulphydryl group is involved in the chemistry of the catalytic centre of many enzymes and in the binding of metal ions. Two highly developed examples where the function of a protein depends heavily on sulphydryl groups are the thioredoxins, which participate in many redox reactions (Holmgren, 1985) and the metallothioneins which are responsible for binding considerable amounts of metals, such as zinc, within the cell (Hamer, 1986). In addition, formation of the disulphide amino acid, cystine, from two cysteine residues is an important factor in the tertiary structure of proteins. Apart from the peptide bond, the disulphide bond is the only type of covalent linkage between amino acids in proteins.

Peptides

In common with other organisms fungi appear to synthesise the ubiquitous tripeptide, glutathione (γ-glutamylcysteinylglycine). The β-lactam antibiotics, i.e. the penicillins and cephalosporins, are produced by several fungi from the tripeptide δ-(L-aminoadi-pyl)-L-cysteinyl-D-valine. The toxic principles of some fungi are peptides, e.g. *Amanita phalloides* produces phalloidin, a cyclic peptide containing a thioether bond (Wieland & Faulstich, 1978).

S-adenosyl methionine

S-adenosyl methionine is formed from methionine and ATP with release of pyrophosphate and inorganic phosphate. It is an important methylating agent and is also the source of the propylamino groups used in the synthesis of polyamines.

Vitamins and growth factors

Several vitamins and growth factors contain divalent sulphur. The functional group of Coenzyme A is the sulphydryl group of cysteamine, the decarboxylation product of cysteine. The vitamins biotin and thiamine both contain thioether linkages. The exact source of the sulphur is still unclear but methionine appears to be the most likely donor. Lipoic acid contains a disulphide grouping which is reversibly reduced and oxidised in the multienzyme dehydrogenation of pyruvic acid and α-ketoglutaric acid. As mentioned above, lipoic acid is also involved in the biological reduction of sulphur.

Volatile sulphur compounds

In addition to the major sulphur compounds mentioned in the foregoing sections, fungi also produce a limited range of volatile sulphur compounds usually at low concentrations. These appear to represent a more or less accidental loss of sulphur from the biological system and in most cases at least, do not seem to have any function in the intermediary metabolism of the fungus. However, some may have a significance in terms of the life cycle of certain fungi, e.g. they may act as animal attractants for fungi lacking the ability to distribute their own spores (Bellina-Agostinone, D'Antonio & Pacioni, 1987) and they do have significance for man. Sulphur compounds of this type tend to have very low odour thresholds, i.e. they can be detected as smells very easily. For example, in aqueous solution threshold values are 5-10 μg l^{-1} for hydrogen sulphide, 30 μg l^{-1} for dimethyl sulphide and 2 μg l^{-1} for methyl mercaptan. In addition to the low odour thresholds, these sulphur compounds also tend to have very characteristic and easily recognizable smells and tastes and our attitude to a particular fungus may be determined by, amongst other things, its ability to produce volatile sulphur compounds. The topic does not seem to have been investigated systematically but there is considerable information available concerning certain situations where there is an obvious economic significance and these are dealt with in the following paragraphs.

Production by Saccharomyces cerevisiae

The flavour of fermented beverages is a very important characteristic and beer is known to contain a range of sulphur volatiles which can affect the flavour (Garza-Ulloa, 1980). It is not clear whether all these arise from yeast metabolites and some of the mercaptans can certainly be formed chemically during processing, particularly hop boiling. From the point of view of yeast involvement, attention has focussed mainly on hydrogen sulphide and dimethyl sulphide (CH_3-S-CH_3).

Hydrogen sulphide

Most beers contain hydrogen sulphide and, at concentrations just below the flavour threshold, it may be a positive contributor to the 'yeasty' flavour. However, above the flavour threshold, it is always undesirable. In commercial practice brewery fermentations occasionally produce considerable quantities of hydrogen sulphide and Wainwright (1970) demonstrated that the gas was produced abundantly by pantothenate-requiring yeasts in the absence of the vitamin. Pantothenate is an essential part of the Coenzyme A molecule and it was suggested that deficiency of the vitamin in the medium leads to deficiency of Coenzyme A within the cell. As Coenzyme A is the major acylating agent in the cell, one effect of this would be lowered levels of O-acetylhomoserine. This compound is an intermediate in the biosynthetic route to methionine between cysteine and homocysteine (Fig. 5.2). Thus, lack of pantothenate could result in accumulation of cysteine. It was then envisaged that the excess cysteine, a compound which is toxic to cells, would be broken down to pyruvate, ammonia and hydrogen sulphide by the enzyme cysteine desulphydrase (Lawrence & Cole, 1968). In support of these ideas, Tokuyama *et al.* (1973) found that the activity of cysteine desulphydrase was nearly doubled in cells grown in pantothenate-deficient media. They further suggested that some of the external hydrogen sulphide could come directly from the biosynthetic pool of sulphide although they felt this would be a minor source of the compound.

In 1971, de Robichon-Szulmajster & Surdin-Kerjan reported that cysteine synthase used O-acetylserine as substrate in preference to serine. This should mean, using the logic of Wainwright (1970), that pantothenate deficiency would cause a blockage at

Table 5.1. Effect of addition of amino acids to the medium on the production of hydrogen sulphide.

Amino acid	μg Hydrogen sulphide produced	
	1·25 mg l⁻¹ pantothenate	0·125 mg l⁻¹ pantothenate
none	0	6·0
Met	0	0
Cys	17·5	19·0
His	2·4	24
Asp	2·5	23
Glu	11·0	50
Hser	11·0	2·5
Tyr	2·4	50
Ala	2·0	30

All amino acids were present at 5 mM except cysteine where the concentrations were 1 mM in the pantothenate-sufficient medium and 0·5 mM in the pantothenate-deficient medium. Production of hydrogen sulphide from cysteine was strongly concentration dependent and the quoted concentrations are those which led to the highest values. For experimental details see Jordan & Slaughter (1986).

the level of cysteine synthesis as well as at homocysteine synthesis. In that case it is unlikely that the intracellular concentration of cysteine would rise and, rather than hydrogen sulphide being produced by breakdown of cysteine, it seems likely that hydrogen sulphide would arise almost entirely from the biosynthetic pool.

This latter hypothesis has been supported by a number of experiments carried out by Jordan & Slaughter (1986) using a single polyploid strain of *Saccharomyces cerevisiae* (NCYC 1108) and defined glucose/salts media. Under the conditions used, hydrogen sulphide was produced only at pantothenate concentrations less than 0·2 mg l⁻¹ and then in inverse proportion to the concentration of the vitamin. The bulk of the gas was formed towards the end of the growth phase of the cells. Table 5.1 shows the effect of adding single amino acids at 5 mM to media with sufficient pantothenate (1·2 mg l⁻¹) or just limiting pantothenate (0·125 mg l⁻¹). Methionine prevented production of hydrogen sulphide in the pantothenate limiting medium and this is in line with earlier reports that methionine controls sulphate uptake and sulphide pro-

Table 5.2. Effect of amino acids on cysteine desulphydration activity of cell extracts

| Amino acid | Cysteine desulphydration activity (% of the control) | | |
| | Addition to medium | | |
	1·25 mg l⁻¹ pantothenate	0·0125 mg l⁻¹ pantothenate	Addition to assay
Asp	82	133	91
Glu	97	117	91
His	63	118	96
Hser	72	99	85
Tyr	91	85	81
Ala	96	101	90

The amino acids were present at 5 mM in the medium or in the assay as appropriate (Jordan & Slaughter, 1986).

duction (Marzluff, 1972; Breton & Surdin-Kerjan, 1977). In contrast, all the other amino acids stimulated hydrogen sulphide production regardless of whether or not they could act as sources of sulphur themselves. Measurement of the cysteine desulphydrase activity in extracts of yeast cells grown with the addition of non-sulphur amino acids and measurement of the effect of addition of amino acids to the enzyme assay (Table 5.2) failed to show any correlation between the influence of the amino acids on enzyme activity and hydrogen sulphide production. These observations support the idea that hydrogen sulphide results more from metabolic disturbances leading to excess sulphate uptake rather than simple desulphydration of cysteine.

In an extension of this work, the cysteine pool size, hydrogen sulphide production, and cysteine desulphydrase activity were compared in yeast grown in three batches of standard medium varying as follows: 1·25 mg l⁻¹ pantothenate, 0·0125 mg l⁻¹ pantothenate and 0·0125 mg l⁻¹ pantothenate with 5 mM methionine (Table 5.3). It is clear that the cysteine pool, whether measured per mg of yeast or per culture, is inversely proportional to the hydrogen sulphide production. This would be expected if the metabolic blockage came after sulphide formation but before cysteine synthesis. Tokuyama et al. (1973) also noticed that the concentration

J. C. Slaughter

Table 5.3. Influence of medium composition on cysteine pool size, hydrogen sulphide production and cysteine desulphydration activity

Panto-thenate (mg l⁻¹)	Yeast crop (mg dry wt per culture)	Cysteine pool		Hydrogen sulphide production per culture (μmole)	Cysteine desulphydration activity	
		nmole per mg dry yeast	μmole per culture		nmole h⁻¹ mg⁻¹ protein	μmole h⁻¹ per culture
1·25	128	38·0	4·74	0	67	3·43
0·0125	42	8·8	0·37	7·27	140	2·45
0·0125*	47	21·4	1·00	0·30	63	1·18

of cysteine fell under pantothenate deficiency but attributed this to excessive activity of cysteine desulphydrase. However, although the specific activity of cysteine desulphydrase rises about two fold in pantothenate deficiency, the lower amount of growth which occurs under these conditions in comparison to a pantothenate sufficient medium leads to an overall lower catalytic capability in the vitamin deficient culture. The capacity to desulphydrate cysteine is highest in the cultures which do not produce hydrogen sulphide and intermediate in those which produce the most.

This evidence, taken along with the theoretical likelihood of a metabolic blockage in the synthesis of cysteine from O-acetylserine, strongly suggests that biosynthetic sulphide rather than cysteine is the precursor of hydrogen sulphide produced by yeast under these conditions.

Dimethyl sulphide

Dimethyl sulphide is a common flavour active compound in a variety of foodstuffs (Anness & Bamforth, 1982) and the main source is thought to be thermal decomposition of the plant amino acid, S-methyl methionine. Dimethyl sulphide was first noticed in beer by Kepner, Strating & Weurman (1963) but began to excite particular interest when it was found to be associated with lager beer rather than ales (Sinclair, Hall & Thorburn-Burns, 1969). Subsequent work has indicated that decomposition of S-methyl

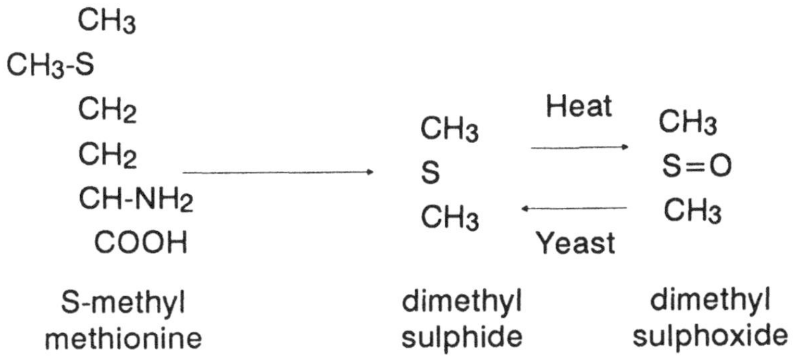

CH3
CH3-S
   CH2
   CH2
   CH-NH2
    COOH

$$CH_3\text{-S-}CH_2\text{-}CH_2\text{-}CH(NH_2)\text{-}COOH \longrightarrow \begin{array}{c} CH_3 \\ S \\ CH_3 \end{array} \xrightarrow[\text{Yeast}]{\text{Heat}} \begin{array}{c} CH_3 \\ S=O \\ CH_3 \end{array}$$

| S-methyl methionine | dimethyl sulphide | dimethyl sulphoxide |

Fig. 5.3. Synthesis of dimethyl sulphide.

methionine during the various heating stages of the brewing process accounts for most of the dimethyl sulphide found in beer (Dickenson, 1983) but in some cases yeast can be responsible for a proportion of the compound (Anness & Bamforth, 1982). Yeast cannot form dimethyl sulphide directly from S-methyl methionine, but Anness & Bamforth (1982) showed that the dimethyl sulphide precursor in fermenting worts is dimethyl sulphoxide which is itself produced from S-methyl methionine during kilning of the malt (Fig. 5.3). Extracts of yeast can reduce dimethyl sulphoxide to dimethyl sulphide using NADPH as a cofactor. It seems unlikely, however, that there is a special dimethyl sulphoxide reductase in yeast and studies have indicated that the reaction is catalyzed by methionine sulphoxide reductase. The affinity of this enzyme is much higher for the methionine derivative than for dimethyl sulphoxide and the catalytic rate is over 600 times faster. The functional enzyme has three components: a methionine sulphoxide reducing protein, thioredoxin and thioredoxin reductase (Bamforth, 1980; Bamforth & Anness, 1981). The normal physiological function of the enzyme is unclear but, presumably, could form part of a sulphur salvage system or possibly be acting to protect the methionine pool against oxidation.

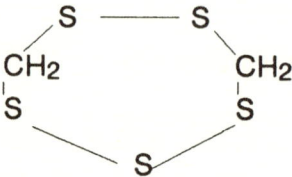

Fig. 5.4. Structure of lenthionine.

Production by other fungi

There do not seem to be many reports on the mechanism of production of volatile sulphur compounds by fungi other than *S. cerevisiae*. However, it is clear that compounds of this type are important in some cases. Over 40 years ago, methyl mercaptan was identified as a product of the wood-rotting fungus *Schizophyllum commune* (Birkinshaw, Findlay & Webb, 1942). This organism can also produce dimethyl sulphide and dimethyl disulphide when grown on a complex medium, and *Scopulariopsis brevicaulis* forms volatile thiols when grown on a bread medium (Challenger & Charlton, 1947). A compound reported as *bis*-methylthiomethane (presumably identical to dimethyl sulphide) was recognised as a product of the white truffle, *Tuber magnatum* (Fiecchi *et al.*, 1967) and thiobismethane (also identical to dimethyl sulphide) has been identified as a component of the odour of the truffle *Tuber aestivum* (Bellina-Agostinone *et al.*, 1987). The unusual compound lenthionine (Fig. 5.4) has been isolated from the edible shiitake mushroom *Lentinula edodes* (Morita & Kobayashi, 1966). The compound appears to be responsible for much of the flavour of the mushroom but is formed during soaking of the dried tissue, which is part of the food preparation process, rather than occurring as a normal metabolite. The natural precursor and the enzyme involved in the process are both unknown but this sort of situation occurs widely in food processing, e.g. the formation of garlic flavour in the crushed bulb due to the formation of allicin from alliin by the action of allinase. No doubt the nose of mycologists has detected volatile sulphur compounds from a wide range of fungi but the individual compounds have not been identified or their function investigated.

Conclusions

There seems to be no feeling that important compounds containing reduced sulphur atoms are waiting to be discovered and there is a clear appreciation of the role and importance of several compounds of this type, e.g. cysteine and methionine, both as intermediary metabolites and combined in proteins. However, the metabolic significance of others, e.g. glutathione, is as uncertain in the fungi as it is in other organisms. Understanding of biosynthesis varies from a good knowledge of the intermediates and enzymes involved, with information on possible control mechanisms, as in sulphide biosynthesis from sulphate, to virtually complete ignorance, as in biotin and thiamine synthesis. Because of the tight, reversible interconversions among the central reduced sulphur compounds the exact metabolic routes followed in real situations are often not clear, even when considerable information is available on the enzymes and intermediates involved. In contrast to the biosynthetic situation, there appears to be a great lack of information on the breakdown of sulphur compounds in fungi, both in terms of general turnover and with regard to the fate of specific compounds, such as methylthioadenosine, formed as products during the normal functioning of metabolism, in this case formation of polyamines from S-adenosyl methionine. The degree to which sulphur is salvaged and the effort necessary to maintain the reduced state in an oxidising world are two other areas where there is little knowledge at present.

References

Anness, B. J. & Bamforth, C. W. (1982). Dimethyl sulphide – a review. *Journal of the Institute of Brewing*, **88**, 244-252.

Bamforth, C. W. (1980). Dimethyl sulphoxide reductase of *Saccharomyces* spp. *FEMS Microbiology Letters*, **7**, 55-59.

Bamforth, C. W. & Anness, B. J. (1981). The role of dimethyl sulphoxide reductase in the formation of dimethyl sulphide during fermentation. *Journal of the Institute of Brewing*, **87**, 30-34.

Bellina-Agostinone, C., D'Antonio, M. & Pacioni, G. (1987). Odour composition of the summer truffle, *Tuber aestivum. Transactions of the British Mycological Society*, **88**, 568-569.

104   J. C. Slaughter

Birkinshaw, J. H., Findlay, W. P. K. & Webb, R. A. (1942). Biochemistry of the wood-rotting fungi. 3. The production of methyl mercaptan by *Schizophyllum commune* Fr. *Biochemical Journal*, **36**, 526-529,

Bradfield, G., Sommerfield, P., Meyen, T., Holly, M., Babcock, D., Bradley, D, Segel, I. H. (1970). Regulation of sulphate transport in filamentous fungi. *Plant Physiology*, **46**, 720-727.

Breton, A. & Surdin-Kerjan, Y. (1977). Sulfate uptake in *Saccharomyces cerevisiae* biochemical and genetic study. *Journal of Bacteriology*, **132**, 224-232.

Challenger, F. & Charlton, P. T. (1947). Studies on biological methylation. Part X. The fission of the mono- and di-sulphide links by moulds. *Journal of the Chemical Society*, 424-429.

Cherest, H., Eicher, F. & de Robichon-Szulmajster, H. (1969). Genetic and regulatory aspects of methionine biosynthesis in *Saccharomyces cerevisiae*. *Journal of Bacteriology*, **97**, 328-336.

Cooper, A. J. L. (1983). Biochemistry of sulphur-containing amino acids. *Annual Review of Biochemistry*, **53**, 187-222.

de Robichon-Szulmajster, H. & Surdin-Kerjan, Y. (1971). Nucleic acid and protein synthesis in yeasts: regulation of synthesis and activity. In *The Yeasts*, eds. A. H. Rose & A. H. Harrison, pp. 335-418. Academic Press: London & New York.

Dickenson, C. J. (1983). Dimethyl sulphide – Its origin and control in brewing. *Journal of the Institute of Brewing*, **89**, 41-46.

Fiecchi, A., Galli-Kienle, M., Scala, A. & Cabella, P. (1967). *Bis*-methylmethane, an odorous substance from white truffle, *Tuber magnatum* Pico. *Tetrahedron Letters*, **18**, 1681-1682.

Garza-Ulloa, H. (1980). Analytical control of sulphur compounds in beer – a review. *Brewers' Digest*, January, 20-25.

Hamer, D. H. (1986). Metallothionein. *Annual Review of Biochemistry*, **55**, 913-951.

Holmgren, A. (1985). Thioredoxin. *Annual Review of Biochemistry*, **54**, 237-271.

Huxtable, R. J. (1986). *Biochemistry of Sulfur*. Plenum Press: New York.

Jordan, B. & Slaughter, J. C. (1986). Sulphate availability and cysteine desulphydration activity as influences on the production of hydrogen sulphide by *Saccharomyces cerevisiae* during growth in a defined glucose-salts medium. *Transactions of the British Mycological Society*, **87**, 525-531.

Kepner, R. E,, Strating, J. & Weurman, C. (1963). Quantitative estimation of esters in beer by gas chromatographic analysis of head space volume. *Journal of the Institute of Brewing*, **69**, 399-405.

Kerr, D. S. & Flavin, M. (1970). The regulation of methionine synthesis and the nature of cystathionine γ-synthase in *Neurospora*. *Journal of Biological Chemistry*, **245**, 1842-1855.

Lawrence, W. C. & Cole, E. R. (1968). Yeast sulphur metabolism and the formation of hydrogen sulphide in brewery fermentations. *Wallerstein Laboratory Communications*, **31**, 95-115.

Marzluff, G. A. (1972). Control of synthesis, activity and turnover of enzymes of sulfur metabolism in *Neurospora crassa*. *Archives of Biochemistry and Biophysics*, **150**, 714-724.

McCready, R. G. L. & Din, G. A. (1974). Active sulphate transport in *Saccharomyces cerevisiae*. *FEBS Letters*, **38**, 361-363.

Morita, K. & Kobyashi, S. (1966). Isolation and synthesis of lenthionine, an odorous substance of Shiitake, an edible mushroom. *Tetrahedron Letters*, **6**, 573-577.

Savin, M. A. & Flavin, M. (1972). Cystathionine synthesis in yeast: an alternative pathway for homocysteine synthesis. *Journal of Bacteriology*, **122**, 299-303.

Sinclair, R., Hall, R. D. & Thorburn-Burns, D. (1969). The analysis of volatile sulphur compounds in beer. *Proceedings of the 12th European Brewery Convention*, 427-444.

Tokuyama, T., Kuraishi, H., Aida, K. & Vemura, T. (1973). Hydrogen sulphide evolution due to pantothenic acid deficiency in the yeast requiring this vitamin, with special reference to the effect of adenosine triphosphate on yeast cysteine desulphydrase. *Journal of General and Applied Microbiology*, **19**, 439-466.

Tweedie, J. W. & Segal, I. H. (1970). Specificity for transport processes for sulfur, selenium and molybdenum anions by filamentous fungi. *Biochimica et Biophysica Acta*, **196**, 95-106.

Umbarger, H. E. (1978). Amino acid synthesis and its regulation. *Annual Reviews of Biochemistry*, **47**, 533-606.

Wieland, T. & Faulstich, H. (1978). Amatoxins, phallotoxins, phallolysin and antamamide - biologically active components of poisonous *Amanita* mushrooms. *Critical Reviews in Biochemistry*, **5**, 185-260.

Wainwright, T. (1970). Hydrogen sulphide production by yeast under conditions of methionine, pantothenate or vitamin B_6 deficiency. *Journal of General Microbiology*, **61**, 107-119.

Chapter 6

Phosphonates: antifungal compounds against Oomycetes

M. D. Coffey & D. G. Ouimette

Department of Plant Pathology, University of California at Riverside, CA 92521-0122, USA

Introduction

While phosphorus is normally considered as an essential element necessary for the growth of microorganisms and microbial communities in nature (see Jennings, Chapter 1), recent studies have demonstrated that some simple phosphorus compounds have powerful and selective antifungal properties. Historically, one of the more frustrating elements in the long search for anti-fungal compounds has been the lack of discovery of molecules with good selective activity against Oomycete plant pathogens in higher plants. Necrotrophic pathogens in this class include many *Pythium* and *Phytophthora* species which cause damping off, root rot, crown and foot rot diseases as well as postharvest rots of various fruits and vegetables. In addition, the downy mildew fungi represent an important group of biotrophic Oomycete parasites causing serious diseases of many valuable food plants. The discovery, in the 1970s, of the phenylamide fungicides, and particularly metalaxyl, marked an important milestone in the improved chemical control of many of these diseases. At about the same time, another class of anti-Oomycete compounds was also discovered in France by researchers at Rhône-Poulenc Agrochimie. These compounds are phosphonates and they include salts of inorganic phosphonic (phosphorous) acid and of short chain organic derivatives, notably aluminium tris-*O*-ethyl phosphonate (fosetyl aluminium) (Fig. 6.1). This latter compound possessed some unique properties apparently unlike those of other known antifungal agents. The systemic activity of fosetyl aluminium proved to be remarkable since foliar applications frequently led to control of soil-borne diseases caused by Oomycetes. Further, fosetyl aluminium displayed very low activity *in vitro* against the majority of culturable Oomycete fungi, including different *Pythium* and *Phytophthora* species. Finally, the compound, despite its very simple structure, displayed antifungal properties which were unusually selective, possibly suggestive of a site-specific mode of action. For instance, whilst it provides effective disease control against *Plasmopora viticola* on grapevines and *Pseudoperonospora humuli* on hops, activity against *Peronospora hyoscyami* on tobacco is poor. Perhaps even more remarkable is the behaviour of this compound against diseases caused by some *Phytophthora* species. Control of many diseases caused by *Phytophthora citricola*, *P. cinnamomi*, *P. citrophthora*, *P. fragariae*, *P. cactorum*, and *P. palmivora* is generally excellent but, in contrast, little or no control of diseases caused by *P. megasperma* f.sp. *glycinea* or *P. infestans* has been possible.

The low activity of fosetyl aluminium *in vitro* against culturable Oomycetes has led some scientists to the viewpoint that this compound triggers a resistance reaction in the host plant (Bompeix, Fettouche & Saindrenan, 1981; Guest, 1984a). However, a careful review of some of the more pertinent patent literature led us to an alternative and still controversial conclusion, namely that phosphonate fungicides have important direct antifungal activity against their target pathogens. Crucial to this opinion was our realization that the first phosphonate fungicides with anti-Oomycete activity to be discovered by the Rhône-Poulenc scientists, were not organic derivatives such as fosetyl aluminium, but in fact simple salts of phosphonic (phosphorous) acid (Thizy *et al.* 1978). Subsequently, our own studies with the potassium salt of phosphonic acid led to the conclusion that this compound has significant antifungal activity against many *Phytophthora* species, but has little or no activity against the majority of other fungi tested (Coffey & Bower, 1984; Fenn & Coffey, 1984).

The phosphonate story, which is slowly emerging, is novel, unexpected and challenges currently held concepts. The purpose of this review is to present some of the new information now available on phosphonate chemistry and biology as it relates to specific target effects on the growth and development of *Phytophthora* species, its behaviour in higher plants and probable fate in the soil environment.

Phosphonates

The simplest phosphonate compounds are phosphorous acid [P(OH)$_3$] and phosphonic acid [H-PO(OH)$_2$]. These two inorganic oxy acids are in tautomeric equilibrium (Guthrie, 1979), but due to the strength of the phosphoryl group (P=O) this equilibrium overwhelmingly favours phosphonic acid (Fig. 6.1). Previous literature has tended to use the term phosphorous acid

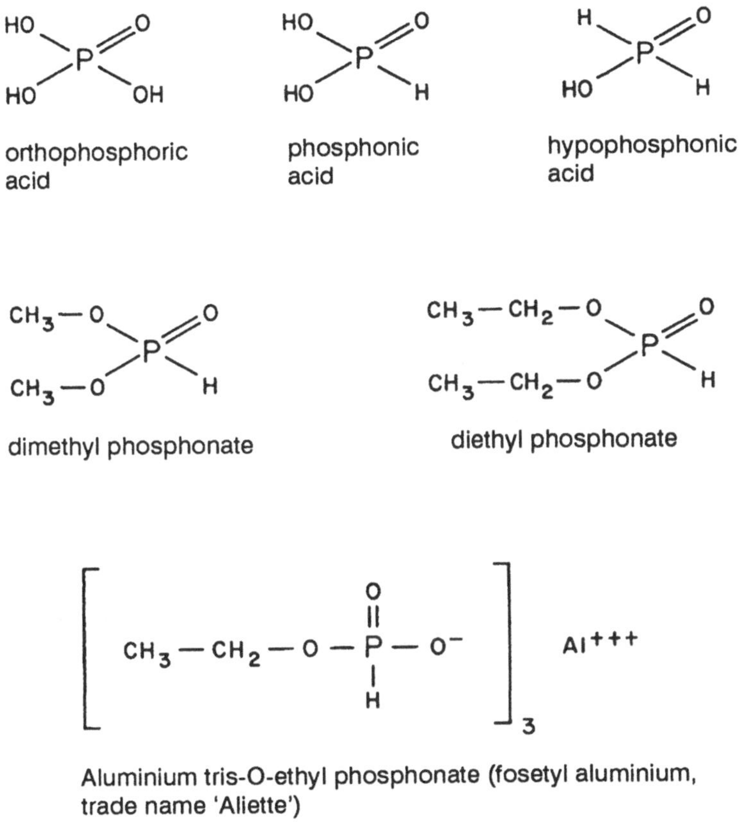

orthophosphoric acid

phosphonic acid

hypophosphonic acid

dimethyl phosphonate

diethyl phosphonate

Aluminium tris-O-ethyl phosphonate (fosetyl aluminium, trade name 'Aliette')

Fig. 6.1. Structures of phosphorus oxy acids, including phosphonic acid, and alkyl-substituted phosphonates.

indiscriminately when referring to a tautomeric mixture consisting predominantly of phosphonic acid. Phosphonic acid is dibasic ($pK_1 = 1.3$, $pK_2 = 6.7$) and has one hydrogen bound to phosphorus rather than oxygen. This hydrogen exchanges rapidly with the deuterium of heavy water and this rate can be increased further by the addition of HCl (Martin, 1959). In marked contrast, the sodium and potassium salts of phosphonic acid exhibit negligible or very slow deuterium exchange, respectively. This property has been used to advantage in preparing relatively stable isotopes of phosphonate salts where tritium is bound exclusively to the phosphorus atom (Fenn & Coffey, 1985, 1988). Phosphonate salts such as sodium or potassium phosphonate have strong buffering action in the pH 5·5 to 7·5 range, and in solution are stable to oxidation over a pH range of 1·5 to 7·6 for many months (Robertson & Boyer, 1956a).

Alkyl esters of phosphorous and phosphonic acid are termed phosphites and phosphonates, respectively (Cohen & Coffey, 1986). Again, the tautomeric equilibrium favours the phosphonate esters. A striking feature of phosphonic acid, inorganic phosphonate salts and simple alkyl esters such as dimethyl phosphonate (Fig. 6.1) is their high solubility in water. Aluminium tris-O-ethyl phosphonate (fosetyl aluminium) is currently the best known example of an alkyl phosphonate which is a systemic fungicide with good activity against some serious diseases caused by Oomycetes (Bruin & Edgington, 1983; Cohen & Coffey, 1986). It has a solubility in water of 12% and is routinely used as a 10% formulation for injection into tree trunks for disease control (Darvas, Toerien & Milne, 1984).

Most early studies on phosphonate activity against pathogenic fungi concentrated on fosetyl aluminium since this was the commercial product. Little or no direct antifungal activity was detected until concentrations exceeded 200 $\mu g \ ml^{-1}$ even with Oomycete species known to be affected *in planta* (Bompeix *et al.*, 1981; Farih, Tsao & Menge, 1981; Guest, 1984b). Consequently, research had tended to focus on the possibility of indirect effects of the compound, perhaps through stimulation of the host defence system (Bompeix *et al.*, 1980; Guest, 1984a). However, two important facts were largely overlooked. First, Rhône-Poulenc Agrochimie had already provided technical data with the release of their experimental compound, indicating that fosetyl aluminium was degraded to phosphonic acid. Second, the patent literature indicated that phosphonic acid had strong antifungal properties specifically against Oomycetes (Thizy *et al.*, 1978). Interestingly, Bompeix *et al.* (1980) did suggest that phosphonic acid was the active product of degradation of fosetyl aluminium in plant tissues. However, the possibility that phosphonic acid was directly antifungal was not investigated until much later (Bompeix & Saindrenan, 1984; Fenn & Coffey, 1984).

Direct antifungal activity
Mycelial growth

Potassium phosphonate was investigated for its antifungal activities against a range of fungi grown on liquid and solid media (Coffey & Bower, 1984; Fenn & Coffey, 1984; Dolan & Coffey, 1988). The compound was active against a number of *Phytophthora* species and isolates in the range of 2 to 90 μg ml^{-1} based on the concentration needed to cause 50% inhibition (EC$_{50}$) of mycelial growth (Table 6.1). The most sensitive species were *Phytophthora cinnamomi*, *P. citricola*, *P. citrophthora*, and *P. palmivora* with EC$_{50}$ values of 5 to 10 μg ml^{-1}. At the other extreme were several with EC$_{50}$ values of about 90 μg ml^{-1}, especially isolates of *P. megasperma* from lucerne (Table 6.1). However, by far the most insensitive species was *P. infestans*, where inhibition of mycelial growth at 200 μg ml^{-1} ranged from 71·2 to only 30·4% for various isolates (Coffey & Bower, 1984). Similar observations with different *Phytophthora* species were also made by Bompeix & Saindrenan (1984).

The direct activity of phosphonate salts against different *Pythium* species has not been investigated in the same detail as *Phytophthora*. Of four species examined, an isolate of *Pythium aphanidermatum* was the most sensitive to potassium phosphonate, with an EC$_{50}$ value of about 140 μg ml^{-1} on 0·5% cornmeal agar (Fenn & Coffey, 1984). Other fungi, including Deuteromycotina, Ascomycotina, Zygomycotina and Basidiomycotina, were not affected by potassium phosphonate.

Thus phosphonate demonstrates high selectivity against some culturable Oomycete pathogens, notably certain *Phytophthora*

Table 6.1. EC$_{50}$ values for inhibition of mycelial growth of representative *Phytophthora* species and isolates.

Species	Isolate	Host	EC$_{50}$[a] (μg ml^{-1})
P. citrophthora	M143	citrus	5·2
P. palmivora	P376	cocoa	6·6[b]
P. citricola	P1287	avocado	6·8
P. citricola	P1273	avocado	7·0
P. cinnamomi (A1)	Pc97	camellia	9·0
P. cinnamomi (A2)	Pc356	avocado	9·9
P. palmivora	P736	cocoa	$\approx$10·0[b]
P. citrophthora	P1163	citrus	10·4
P. cinnamomi (A2)	Pc402	avocado	11·9
P. capsici	P1091	capsicum	18·5
P. megasperma f.sp. *glycinea*	P405	soybean	22·3
P. capsici	P1314	capsicum	30·6
P. parasitica	M134	citrus	30·9
P. capsici	P1319	capsicum	34·7
P. megasperma	P1253	chickpea	91·2
P. megasperma	P1316	lucerne	88·9
P. infestans	P1300	potato	224·4[c]

Data from Coffey & Bower, 1984; Dolan & Coffey, 1988. [a]The majority of isolates were grown on a defined medium containing 0·84 mM potassium phosphate; [b]*P. palmivora* was grown on 0·5% cornmeal agar and [c]*P. infestans* was grown on rye-seed medium.

---

species within the Peronosporales, while no such activity can be detected among a wide range of non-Oomycete fungi.

Sporulation

The life cycle of a pathogenic Oomycete fungus such as *Phytophthora* involves not only mycelial growth in its host, but also the formation of various spore types. Sporulation is a vital component of such life cycles and disease progress depends critically on the quantity and quality of such processes. Interference with the life cycle, by inhibition of germination and/or sporulation processes provides a window of opportunity for an antifungal compound. Production of zoosporangia is especially sensitive to potassium phosphonate. The EC$_{50}$ values for inhibition of sporangial development range from 0·1 μg ml^{-1} with an isolate of *Phytophthora palmivora* (Dolan & Coffey, 1988) to 1·4 and 1·8 μg ml^{-1}, respectively, for *P. citricola* and *P. cinnamomi* (Coffey & Joseph, 1985). EC_{90} values for the same isolates of *P. citricola* and *P. cinnamomi* were 5.5 and 3.8 μg ml^{-1} respectively. Oospore production by the same isolate of *P. citricola* was also sensitive, being inhibited by as little as 1 μg ml^{-1} of phosphonate. Oospore and especially chlamydospore formation in *P. cinnamomi* were not as sensitive to phosphonate; around 50 μg ml^{-1} caused significant inhibition of their production (Coffey & Joseph, 1985).

In contrast to spore production, germination of fungal propagules was much less affected. The one exception was indirect germination of zoosporangia by cytoplasmic cleavage to form motile zoospores. Such zoospore release with *Phytophthora cinnamomi* and *P. citricola* was inhibited by 50% in the presence of 6 μg ml^{-1} phosphonate (Coffey & Joseph, 1985).

Production of resistant strains

The development of laboratory strains of *Phytophthora* species with high resistance to phosphonate provides important evidence supporting the hypothesis of direct and selective inhibition of these fungi by phosphonate. By use of the chemical mutagen *N*-methyl-*N'*-nitro-*N*-nitrosoguanidine (MNNG), Bower & Coffey (1985) succeeded in producing strains of *Phytophthora capsici* with high resistance to phosphonate. One strain of *P. capsici* (P1361), derived from a single-zoospore wild type isolate (P1319), possessed an EC_{50} on 0.5% cornmeal agar of 415 μg ml^{-1} phosphonate compared to a value of 37 μg ml^{-1} for P1319 (Table 6.2). Studies by Fenn & Coffey (1988) led to the production of further mutant strains of both *P. capsici* and *P. parasitica* var. *nicotianae* with resistance to phosphonate (Table 6.2). The choice of *P. capsici* and *P. parasitica* var. *nicotianae* was based on the fact that these species had been used by Bompeix *et al.* (1981) and Guest (1984a), in their studies supporting an indirect role for phosphonate and/or fosetyl aluminium via stimulation of host defence mechanisms. In tomato leaflets, the phosphonate-resistant mutant of *P. capsici* (P1361) demonstrated an EC_{50} of >348 μg ml^{-1} for lesion development (Fenn & Coffey, 1985). Likewise with tobacco seedlings, a mutant of *P. parasitica* var. *nicotianae*, with an EC_{50} *in vitro* on 0.5% cornmeal agar of 665 μg ml^{-1} phosphonate, was only partially controlled when the phosphonate level in the plant

Table 6.2. EC_{50} values (μg ml^{-1}) for resistance to phosphonate on 0·5% cornmeal agar of mutant strains of *Phytophthora palmivora*, *P. capsici* and *P. parasitica* var. *nicotianae*.

Phytophthora species and isolate	Wild type	Mutant strains
P. palmivora (P376)	7	118
		130
P. capsici (P1319)	37	415
P. capsici (P1503)	57	311
		462
		815
P. parasitica (P1495)	<50	270
		301
P. parasitica (P1352)	47	514
		535
		650
		665

Data taken from Fenn & Coffey (1988).

---

tissue was 753 μg g^{-1} fr. wt. In contrast, the wild type isolate was almost completely controlled by a phosphonate concentration of 279 μg g^{-1} fr. wt. (Fenn & Coffey, 1988).

One possible criticism of much of the work on the mode of action of the phosphonate fungicides, potassium phosphonate and fosetyl aluminium, is that neither *P. capsici* nor *P. parasitica* var. *nicotianae* are very sensitive to these compounds (Table 6.1). To overcome this criticism we chose to obtain mutants of *P. palmivora* (P376), an isolate selected because of its high sensitivity to phosphonate *in vitro* (Table 6.2, Fig. 6.2). A range of mutants was obtained *in vitro* using MNNG with EC_{50} values varying from 37 to 130 μg ml^{-1} phosphonate on 0·5% cornmeal agar (Dolan & Coffey, 1988). Using tomato as a host plant, it was possible to demonstrate that the range of responses obtained *in vivo* paralleled those found *in vitro* (Fig. 6.2).

Finally, there has been a single report of resistance to fosetyl aluminium occurring in the field with *P. cinnamomi* (Vegh *et al.*, 1985). This finding has yet to be corroborated by other workers,

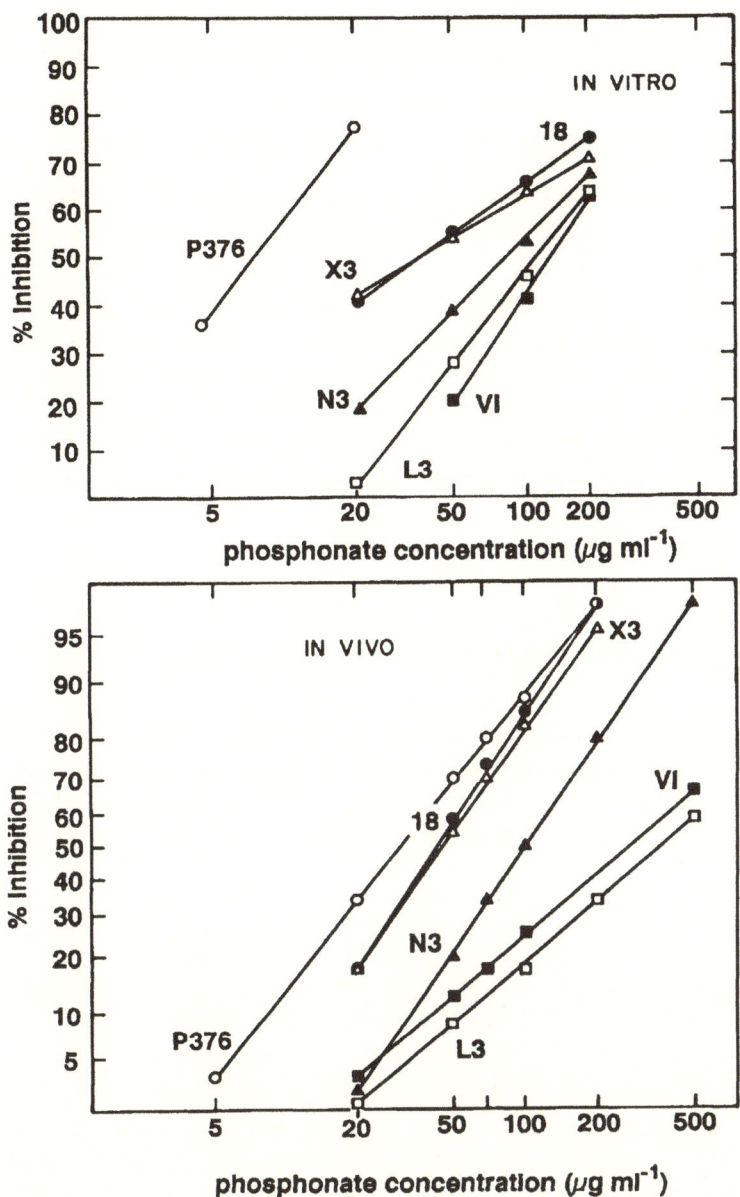

Fig. 6.2. Comparison of (a) *in vitro* and (b) *in vivo* responses to potassium phosphonate of a wild type (P376) and five resistant mutants of *Phytophthora palmivora* (from Dolan & Coffey, 1988).

but in view of the relative ease with which resistant strains of some *Phytophthora* species can be produced in the laboratory, it is not entirely unexpected.

Phosphonate behaviour in higher plants

Metabolism and uptake of organic phosphonates

Although it is known that fosetyl aluminium is degraded to phosphonate in plant tissues (Piedallu & Jamet, 1985; Saindrenan, Darakis & Bompeix, 1985), good quantitative data on levels and persistence in plant tissues has not been available. Recently, Fenn & Coffey (1988) found that tomato leaflets treated with 400 μg ml^{-1} fosetyl aluminium for 48 h contained only 14 μg g^{-1} fr. wt. ethyl phosphonate and 358 μg g^{-1} fr. wt. phosphonate (Table 6.3). Results with whole tomato plants provide a similar data base (Luttringer & de Cormis, 1985). Ethyl phosphonate was only detectable for a few hours following a soil application of 8 mg ml^{-1} and subsequent uptake by the plant roots. Phosphonate accumulated over 7 days to a final concentration of about 150 μg g^{-1} fr. wt. in the leaves, but only about 50 μg g^{-1} fr. wt. in stems.

The levels of ethyl phosphonate and phosphonate were determined in avocado seedlings following either soil or foliar applications of fosetyl aluminium (Ouimette & Coffey, 1989). With soil treatment, ethyl phosphonate was detectable at low levels (2-3 μg g^{-1} fr. wt.) in roots and stems after 1 week. The phosphonate le-

Table 6.3. Phosphonate and ethyl phosphonate concentrations in detached tomato leaflets floating on solutions of fosetyl aluminium for 48 hours

Fosetyl aluminium in floating solution (μg ml^{-1})	Phosphonate concentration μg g^{-1} fr. wt.	Ethyl phosphonate concentration μg g^{-1} fr. wt.
0	0	0
80	49	0·5
160	88	3·0
240	181	4·0
320	222	10·0
400	358	14·0

Data taken from Fenn & Coffey (1988).

Table 6.4. Concentration of phosphonate (HPO_3^{2-}) in tissues of avocado seedlings up to 8 weeks after soil or foliar application with either fosetyl aluminium or potassium phosphonate at equivalent HPO_3^{2-} concentrations

| | | μg HPO_3^{2-} g^{-1} fr. wt. tissue | | | | | |
| | | Soil application[a] | | | Foliar application[b] | | |
Fungicide	Week	Roots	Stems	Leaves	Roots	Stem	Leaves
Fosetyl	1	208	191	57	8	43	140
aluminium	6	706	1084	86	15	121	73
	8	488	543	105	12	95	49
Potassium	1	356	221	221	15	21	42
phosphonate	6	512	784	72	11	130	74
	8	213	382	47	14	75	19

[a]Potassium phosphonate and fosetyl aluminium were applied at a volume of 500 ml per seedling at 2·1 and 3·0 mg ml^{-1} respectively.
[b]Potassium phosphonate and fosetyl aluminium were applied twice, with a 24 hour interval between applications, at 2·1 and 3·0 mg ml^{-1}, respectively. Data from Ouimette & Coffey, 1989.

---

vels in roots and stems were 208 and 191 μg g^{-1} fr. wt., respectively. After 8 weeks the level in roots was 488 μg g^{-1} fr. wt. phosphonate (Table 6.4). With foliar application, no ethyl phosphonate residues were detected in avocado tissues even at 1 week. Concentrations of phosphonate in foliage were 140 μg and 49 μg g^{-1} fr. wt. at 1 and 8 weeks, respectively (Table 6.4). Levels in root tissues remained fairly constant at 8 to 15 μg g^{-1} fr. wt. over the experimental period. Interestingly, much more phosphonate was present in root tissues following soil, as compared to foliar, application of fosetyl aluminium (Table 6.4). The levels of phosphonate in avocado tissues was found to be quite similar when potassium phosphonate was used at equivalent rates to fosetyl aluminium (Table 6.4; Ouimette & Coffey, 1989). The data are consistent with the hypothesis that fosetyl aluminium is rapidly degraded in higher plant tissues to phosphonate and that it persists in this form for at least several months.

Other short chain alkyl-substituted phosphonates such as dimethyl and diethyl phosphonate (Fig. 6.1) have been shown to provide disease control in plants against various Oomycetes (Ab-blard, Gaulliard & Lacroix, 1983; Ouimette & Coffey, 1985). Pepper plants were treated with 5 mM solutions of either potassium phosphonate, monoethyl, diethyl or dimethyl phosphonate as root drenches for 24 h and then analyzed for their phosphonate content (Ouimette & Coffey, 1988). In addition to residues of the respective parent compound, the organic phosphonate treatments contained high levels of inorganic phosphonate, indicating that partial metabolism had occurred within 24 h. Plants treated with dimethyl phosphonate contained the highest levels of phosphonate, 397 μg g^{-1} fr. wt. in roots, which was slightly higher than the amounts found in roots of plants treated with potassium phosphonate.

Factors affecting the degradation of different organic phosphonates to inorganic phosphonate are unknown. In addition, there are no data available on the influence of different alkyl substitutions on their uptake and subsequent distribution in the plant. Most likely they enter the plant by diffusion, since it is difficult to envisage selective uptake processes being involved. Once in the plant tissue, rapid degradation to phosphonate near the site of entry could generate a chemical gradient, perhaps facilitating loading into the symplast and eventually transport in the phloem. The possibility also exists that phosphonate might be taken up actively into plant tissues as is orthophosphate, since the two molecules possess a very similar spatial arrangement of atoms (Fig. 6.1). In mammalian and yeast systems, however, phosphonate has been shown to exert little effect on phosphate-mediated metabolism such as cell respiration (Robertson & Boyer, 1956b). It was concluded that phosphonate and phosphonate esters were essentially inert, even at relatively high ionic strengths (0.067 M–0.2 M). This fundamental biological inertness relative to phosphate might be explained by differences in electrostatic charge distribution (Robertson & Boyer, 1956b). The persistence of phosphonate in higher plants is also consistent with this hypothesis. The ability of phosphonate to mimic phosphate in certain plant metabolic processes, particularly active uptake by cells, needs to be investigated.

Influence of phosphate on phosphonate activity

Bompeix *et al.* (1980) were the first to study the influence of phosphate concentrations on the efficacy of fosetyl aluminium using tomato leaflets and *Phytophthora capsici* as a pathogen. They observed that 250 μg ml^{-1} fosetyl aluminium no longer controlled *P. capsici* on detached leaflets floated on a solution in which the sodium phosphate level had been increased from 10 to 67 mM. However, neither pre- nor post-application of the fungicide-treated leaflets with 67 mM phosphate influenced fungicide efficacy. In contrast, increasing the potassium phosphate concen-

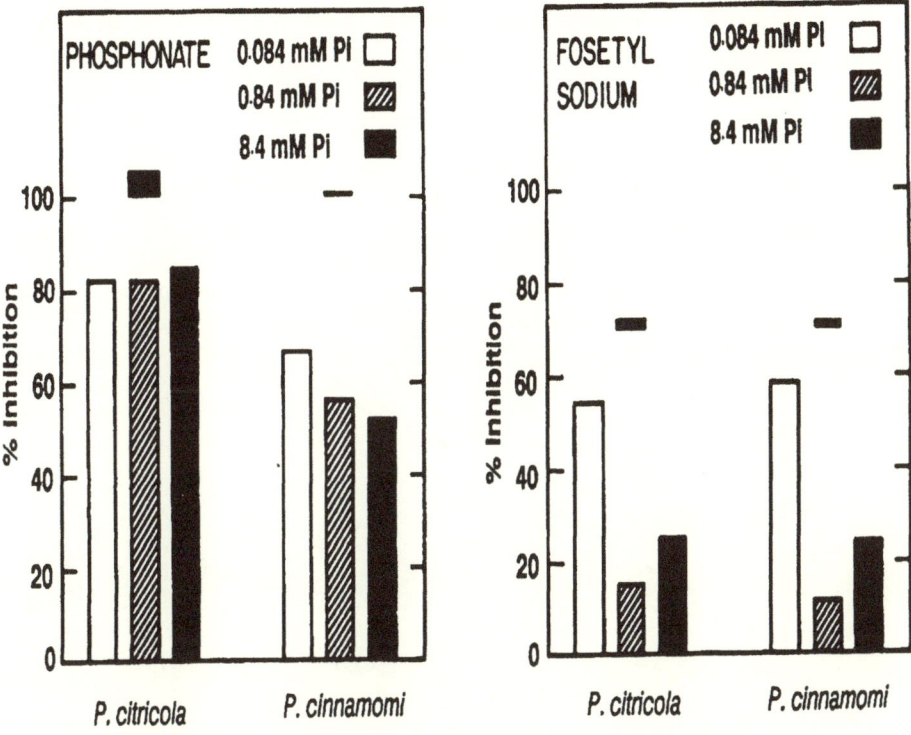

Fig. 6.3. Influence of different potassium phosphate levels (Pi) on the inhibition in defined liquid medium of *Phytophthora citricola* (P1273) and *P. cinnamomi* (Pc356) by (a) 47 μg ml^{-1} potassium phosphonate and (b) 132 μg ml^{-1} fosetyl sodium (from Fenn & Coffey, 1984).

tration of the external medium from 0 to 10 mM with *P. palmivora* on tomato seedlings actually enhanced the antifungal effects of both fosetyl sodium and potassium phosphonate (Dolan & Coffey, 1988). Fenn & Coffey (1984) examined the effect of potassium phosphate levels on the efficacy of both fosetyl sodium and potassium phosphonate directly against *P. cinnamomi* and *P. citricola* in liquid culture. Fosetyl sodium was chosen in preference to fosetyl aluminium since the latter compound formed a precipitate, probably of aluminium hydroxide, when the pH of the solution was adjusted to a more suitable physiological level, typically from 3.1 to 6.2-6.5. It was found that an increase in phosphate from 0.084 to 0.84 mM reduced the efficacy of fosetyl sodium (132 μg ml^{-1}) towards both pathogens in a dramatic fashion (Fig. 6.3). In con-

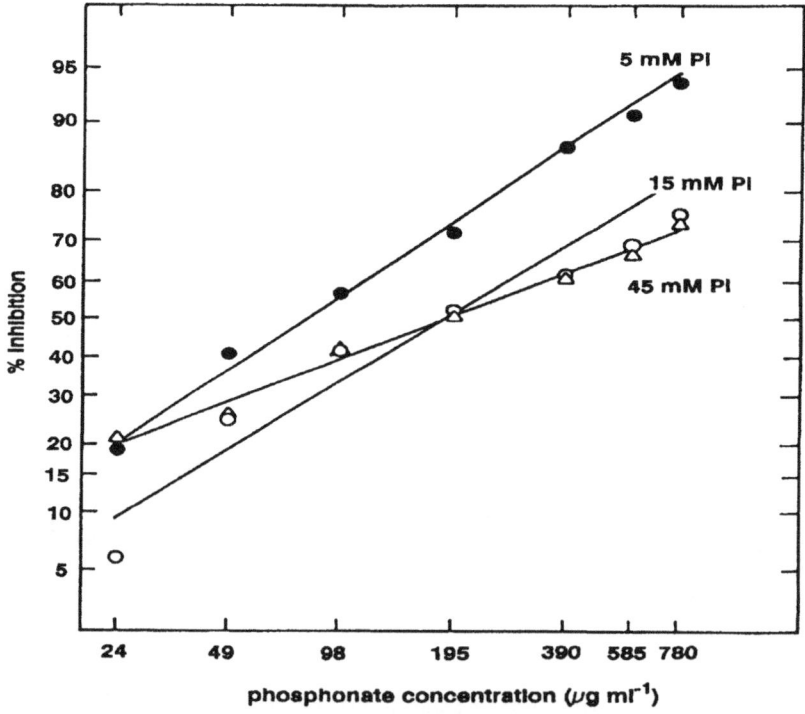

Fig. 6.4. Influence of 5, 15 and 45 mM potassium phosphate (Pi) on the *in vitro* response to potassium phosphonate of *Phytophthora capsici*, isolate P1319 (from Fenn & Coffey, 1988).

trast, there was absolutely no response to an increase in phosphate level to 8·4 mM, with potassium phosphonate (47 μg ml^{-1}) against *P. citricola*. The effect on efficacy of phosphonate against *P. cinnamomi* was slight, reducing inhibition of fungal growth from 67% at 0·084 mM to 52% at 8·4 mM phosphate concentration (Fig. 6.3). Recently, the effect of phosphate concentration on *P. capsici* (P1319)(Fig. 6.4) and *P. parasitica* var. *nicotianae* (P1352) was examined with respect to its influence on phosphonate efficacy (Fenn & Coffey, 1988). The EC$_{50}$ values for inhibition of mycelial growth of *P. capsici* with 5, 15 and 45 mM potassium phosphate concentrations were 77, 184 and 186 μg ml^{-1} phosphonate (Fig. 6.4). Corresponding EC$_{50}$ values for *P. parasitica* var. *nicotianae* at 5, 15 and 45 mM phosphate levels were 55, 123 and 78 μg ml^{-1}. Summarizing these effects, it can be seen that increasing the level of phosphate can have an effect on the fungicidal efficacy of fosetyl aluminium and fosetyl sodium (Bompeix *et al.*, 1980; Fenn & Coffey, 1984), though it may change efficacy in either a positive or negative direction, or alternatively its influence may be negligible.

A key issue in all these considerations of the phosphate:phosphonate interaction is whether phosphate levels in plant tissues are sufficient to reduce the direct antifungal activity of phosphonate against the target pathogen. Proponents of an indirect mode of action for both fosetyl aluminium and phosphonate argue that phosphate levels in plant tissues are too high to account for a direct mode of action of these compounds against the pathogen (Bompeix *et al.*, 1980; Saindrenan *et al.*, 1988). Fenn & Coffey (1988) have produced data indicating that the level of inhibition by phosphonate of two *Phytophthora* species, *P. capsici* and *P. parasitica* var. *nicotianae*, is similar both *in vitro* and *in vivo* at comparable phosphate concentrations (Table 6.5). It was also shown that the responses of resistant strains of these same fungi to phosphonate was quite similar in culture and in plant tissues. In addition, values for phosphonate levels in plant tissues may not accurately reflect the concentrations to be found in fungal tissues. Barchietto, Saindrenan & Bompeix (1988) demonstrated that *P. citrophthora* grown in a synthetic liquid medium could take up phosphonate against a concentration gradient and that this process was inhibited by 10 or 25 mM sodium azide. Consequently,

Table 6.5. Comparison of *in vivo* and *in vitro* antifungal activity of potassium phosphonate (HPO_3^{2-}) towards growth of HPO_3^{2-}-sensitive and resistant isolates of *Phytophthora capsici* and *P. parasitica* var. *nicotianae*.

Isolate[a]	Growth medium[b]	HPO_3^{2-} concentration[c]	HPO_4^2 concentration (mM)	% inhibition
P. capsici	Tomato leaflet	78-110	8·9	61
(P1319)	*In vitro* on SM	77	5·0	50
P. capsici	Tomato leaflet	484-642	8·9	37
(P1361)	*In vitro* on SM	344	5·0	50
P. parasitica	Tobacco stem	279	3·2	96
(P1352)	*In vitro* on CMA	279	< 1·0	86
P. parasitica	Tobacco stem	753	3·2	62
(P1755)	*In vitro* on CMA	665	< 1·0	50

[a]P1319 and P1352 are HPO_3^{2-}-sensitive parental isolates. P1361 and P1755 are HPO_3^{2-}-resistant mutants obtained by treating zoospores of the parental isolates with the chemical mutagen MNNG. [b]The *in vitro* antifungal activity of HPO_3^{2-} was determined by measuring radial growth on either a synthetic medium (SM) or on 0·5% cornmeal agar (CMA) at various concentrations of HPO_3^{2-}. [c]Concentrations of HPO_3^{2-} are expressed as $\mu g\ g^{-1}$ fr. wt. of plant tissue or $\mu g\ ml^{-1}$ of culture medium. Data taken from Fenn & Coffey (1988).

phosphonate levels determined in plant tissues may not accurately reflect those present in fungal tissues.

Influence of amino-oxy acetic acid on phosphonate efficacy

Bompeix *et al.* (1981) showed that 300 μM 2-amino-oxy acetic acid (AOA) could negate the effect of 250 $\mu g\ m^{-1}$ fosetyl aluminium in inhibiting lesion development of *P. capsici* on detached tomato leaflets. Since AOA is known to be an inhibitor of phenylalanine ammonia-lyase (PAL), it was hypothesized that the mode of action of fosetyl aluminium might involve a stimulation of host defence mechanisms. This was based on the assumption that AOA, by interference with PAL activity, blocked phenylpropanoid biosynthesis and that the latter was essential for host defence. Contradictory evidence was obtained in this study, however, since L-2-amino-oxy-3-phenylpropionic acid (AOPP), did not interfere with the antifungal activity of fosetyl aluminium (Bompeix *et al.*, 1981). AOPP is known to be a potent inhibitor of PAL activity *in vitro*, much more so than AOA. One additional factor that was overlooked by Bompeix *et al.* (1981) was the need for a control consisting only of AOA. Fenn & Coffey (1985) attempted to repeat these results concentrating on the effect of AOA on fosetyl aluminium and potassium phosphonate activity against *P. capsici*. They found that 80 μM AOA alone caused a 28% increase in lesion length of *P. capsici* on tomato leaflets. At higher concentrations of fosetyl aluminium (260 μg ml^{-1}) or potassium phosphonate (180 μg ml^{-1}), AOA had no influence on lesion development. At lower concentrations of these two fungicides (130 μg and 90 μg ml^{-1} respectively), 80 μM AOA did increase lesion size by 62% and 48%. Since leaflets treated with AOA alone were

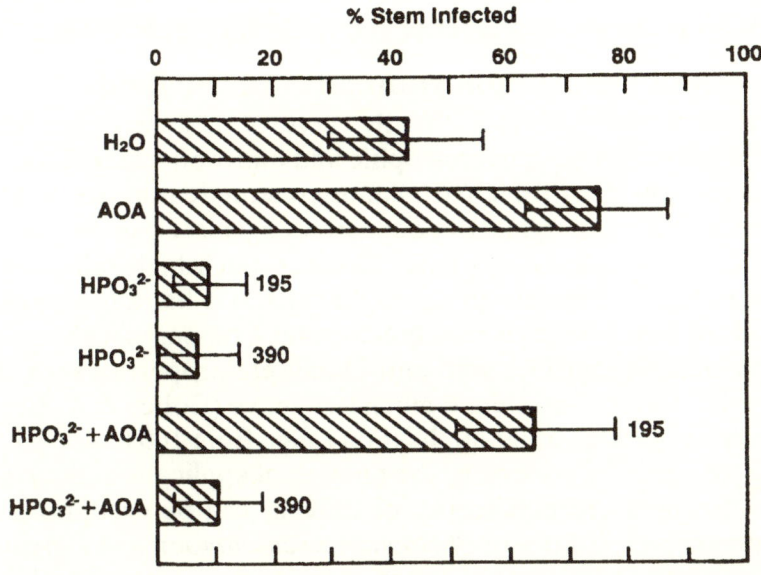

Fig. 6.5. Effect of 0·08 mM 2-aminooxy acetic acid (AOA) and/or potassium phosphonate (HPO$_3^{2-}$) with respect to control of *Phytophthora parasitica* var. *nicotianae* (P1352) on tobacco seedlings. Concentrations of HPO$_3^{2-}$ used were 195 and 390 μg ml^{-1} (From Fenn & Coffey, 1988).

made more susceptible to *P. capsici*, this result might be explained by assuming that more fungicide was required to achieve a similar level of control. In addition, 80 μM AOA was shown to reduce the *in vitro* uptake of phosphonate into *P. capsici* by 62% after 3 h. Consequently, the primary effect of AOA against phosphonate is more likely to be through its effect on uptake by the fungus (Fenn & Coffey, 1985).

These results have been repeated with *Phytophthora parasitica* var. *nicotianae* on tobacco seedlings (Fenn & Coffey, 1988). AOA used alone markedly increased the susceptibility of tobacco to *P. parasitica* var. *nicotianae* (Fig. 6.5). Higher levels of potassium phosphonate (390 μg ml^{-1}) controlled *P. parasitica* var. *nicotianae* and were not affected by the presence of 80 μM AOA. Lower levels of the fungicide (195 μg ml^{-1}) were affected by AOA, however (Fig. 6.5). AOA inhibited *in vitro* uptake of phosphonate by *P. parasitica* var. *nicotianae* by 80% over 4 h. The combination of enhanced host susceptibility and reduced uptake of phosphonate by the fungus offer a feasible explanation of the results obtained with AOA in these two studies (Fenn & Coffey, 1985; 1988).

Behaviour of phosphonates in the field

Fosetyl aluminium provides effective control of some important diseases caused by both necrotrophic and biotrophic members of the Peronosporales (Table 6.6). An interesting feature of its behaviour is its more marginal or poor activity against closely-related Oomycete species causing similar diseases. Such high selectivity within a single taxonomic group might suggest it has a biochemical mode of action which is very site-specific. Compared with some other selective fungicides with anti-Oomycete properties such as metalaxyl, fosetyl aluminium is much less active (Cohen & Coffey, 1986); rates used for foliar application range from 100 to 400 g hl^{-1} and 2-10 kg ha^{-1}. Conversely, the number of applications necessary are low and the persistence of efficacy can be exceptional. With pineapples, control of *Phytophthora cinnamomi* and *P. parasitica* was achieved with a single pre-planting dip which was effective for 18 months (Rohrbach & Schenck, 1985). In the case of avocado root rot caused by *P. cinnamomi*, as little as two applications, as a trunk injection, resulted in some circumstances in a reversal of the symptoms of tree decline and restoration of good

Table 6.6. Examples of the relative efficacy of fosetyl aluminium against selected fungal pathogens within the Peronosporales (Oomycetes).

Pathogens	Efficacy rating		
	High	Moderate	Low
Phytophthora spp.	P. cactorum	P. capsici	P. infestans
	P. cinnamomi	P. palmivora	P. megasperma f.sp. glycinea
	P. citricola	P. parasitica	P. megasperma f.sp. medicaginis
	P. citrophthora		
	P. fragariae		
Pythium spp.	Py. splendens	Py. aphanidermatum	Many Pythium spp.
Downy mildews and Albugo spp.	Peronospora parasitica	Bremia lactucae	Peronospora hyoscyami
	Pseudoperonospora humuli		
	Albugo candida		

root and shoot growth in the following season (Darvas *et al.*, 1984; Pegg *et al.*, 1985). Practical exploitation of the unique phloem-mobile properties of phosphonate fungicides has been both rapid and innovative. A good example is the use of foliar sprays or trunk injections to control root rots caused by various *Phytophthora* species. Currently, however, basic information on the mechanisms which regulate phloem transport of phosphonates is lacking.

Application of fosetyl aluminium as a drench to soils for control of *Pythium* or *Phytophthora* diseases has also been utilized, sometimes with good results. Fosetyl aluminium is rapidly degraded to phosphonate in soils. In a study with three different soil types the half-lives of the fungicide were 0.32, 0.73 and 0.77 days at 20°C (Piedallu & Jamet, 1985). Ouimette & Coffey (1989) found no trace of fosetyl aluminium present in soil 2 weeks after application of 3 g l^{-1}. Phosphonate was detectable in this same soil for 4-6 weeks after application of fosetyl aluminium, though it was not detectable after 8 weeks. Adams & Conrad (1953) demonstrated that phosphonate was oxidised by soil microorganisms to phosphate. Many different soil microorganisms including bacteria, fungi and actinomycetes can apparently utilize phosphonate as a phosphorus source (Casida, 1960; Malacinski & Konetzka, 1966). However, the rate of conversion of phosphonate to phosphate in soils would appear to be quite slow (MacIntire *et al.*, 1950), with a half-life of about 16 weeks recorded for one clay loam soil (Adams & Conrad, 1953). Since phosphate reduction can occur naturally under anaerobic conditions (Tsubota, 1959), it has been suggested that phosphonate is part of a natural phosphorus cycle, with general features similar to that of the nitrogen and sulphur cycles (Alexander, 1977). Since many bacteria have retained the genetic ability to oxidise phosphonate, even though they preferentially utilize phosphate, it has been suggested that they must be frequently exposed to it in nature (Malacinski & Konetzka, 1967).

Conclusions

The active ingredient of fosetyl aluminium, a unique phloem-mobile xenobiotic possessing remarkable antifungal activity against some important Oomycete plant pathogens, is phosphonate. This simple anion is believed to be an important component of the phosphorus cycle in soils (Alexander, 1977). Many microorganisms have retained the enzyme systems necessary to oxidise it to phosphate. What is different about the biology or metabolism of some Oomycetes that causes phosphonate to be inhibitory to their growth? A simple inability by Oomycetes to utilize phosphonate as a phosphorus source (Fothergill & Child, 1964) would appear unlikely to provide an explanation of this selective inhibition. Neither higher plants, nor mammals (Robertson & Boyer, 1956b), can utilize phosphonate, and yet it exhibits no toxic effects. We believe that the answer to the ability of phosphonate to inhibit selectively certain Oomycetes, may lie in unique features of their phosphorus metabolism. Wang & Bartnicki-Garcia (1973) have demonstrated that novel phosphoglucans are abundant in the storage vacuoles of some *Phytophthora* species at certain stages in their life cycle, notably during sporangium development. Polyphosphates are common in filamentous fungi and yeasts, but have not been detected in Oomycetes (Dietrich, 1976). Could it be that Oomycetes store their phosphate in a unique fashion, quite unlike that of other fungi, or higher organisms? Are specific metabolic steps in this storage process highly sensitive to phosphonate? Certainly, the ability of extremely low concentrations of phosphonate (1 to 2 μg ml^{-1}) to inhibit zoosporangium production by *Phytophthora* (Coffey & Joseph, 1985; Dolan & Coffey, 1988) may indicate selective interference with key biosynthetic events. It is our belief that phosphonate will prove not only to be a valuable tool in the control of important plant diseases, but may also be used experimentally to unmask new and important properties concerning both phloem translocation in plants and unique features of phosphorus metabolism in Oomycetes.

Acknowledgements. This research was supported in part by a grant from the California Avocado Commission. The authors are very grateful to BettyAnn Merrill who typed the text and to Linda Bobbitt and John Kitasako for preparing the figures.

References

Alexander, M. (1977). Microbial transformations of phosphorus. In *Introduction to Soil Microbiology*, pp. 347-348. John Wiley & Sons, New York.

Abblard, J., Gaulliard, J.-M. & Lacroix, G. (1983). Fungicidal compositions. *United States Patent Number 4,382,928.*

Adams, F. & Conrad, J. P. (1953). Transition of phosphite to phosphate in soils. *Soil Science*, **75**, 361-371.

Barchietto, T., Saindrenan, P. & Bompeix, G. (1988). Uptake and utilisation of phosphonate ions by *Phytophthora citrophthora* and *Nectria haematococca* in relation to their selective toxicity. *Pesticide Science*, **22**, 159-167.

Bompeix, G., Fettouche, F. & Saindrenan, P. (1981). Mode d'action du phoséthyl Al. *Phytiatrie-Phytopharmacie*, **30**, 257-272.

Bompeix, G., Ravisé, H., Raynal, G., Fettouche, F. & Durand, M. C. (1980). Modalités de l'obtention des nécroses bloquantes sur fueilles détachées de tomate par l'action du tris-*O*-éthyl phosphonate d'aluminium (phoséthyl d'aluminium), hypothèses sur son mode d'action *in vivo*. *Annales de Phytopathologie*, **12**, 337-351.

Bompeix, G. & Saindrenan, P. (1984). *In vitro* antifungal activity of fosetyl Al and phosphorous acid on *Phytophthora* species. *Fruits*, **39**, 777-786.

Bower, L. A. & Coffey, M. D. (1985). Development of laboratory tolerance to phosphorous acid, fosetyl-Al, and metalaxyl in *Phytophthora capsici*. *Canadian Journal of Plant Pathology*, **7**, 1-6.

Bruin, G. C. A. & Edgington, L. V. (1983). The chemical control of diseases caused by zoosporic fungi. In *Zoosporic Plant Pathogens*, ed. S. T. Buczacki, pp. 193-232. Academic Press, London & New York.

Casida, L. E. (1960). Microbial oxidation and utilization of orthophosphite during growth. *Journal of Bacteriology*, **80**, 237-241.

Coffey, M. D. & Bower, L. A. (1984). *In vitro* variability among isolates of eight *Phytophthora* species in response to phosphorous acid. *Phytopathology*, 74, 738-742.

Coffey, M. D. & Joseph, M. C. (1985). Effects of phosphorous acid and fosetyl-Al on the life cycle of *Phytophthora cinnamomi* and *P. citricola*. *Phytopathology*, 75, 1042-1046.

Cohen, Y. & Coffey, M. D. (1986). Systemic fungicides and the control of oomycetes. *Annual Review of Phytopathology*, **24**, 311-338.

Darvas, J. M., Toerien, J. C. & Milne, D. L. (1984). Control of avocado root rot by trunk injection with phosethyl-Al. *Plant Disease*, **68**, 691-693.

Dietrich, S. M. C. (1976). Presence of polyphosphate of low molecular weight in Zygomycetes. *Journal of Bacteriology*, **127**, 1408-1413.

Dolan, T. E. & Coffey, M. D. (1988). Correlative *in vitro* and *in vivo* behavior of mutant strains of *Phytophthora palmivora* expressing different resistances to phosphorous acid and fosetyl-Na. *Phytopathology*, **78**, 974-978.

Farih, A., Tsao, P. H. & Menge, J. A. (1981). Fungitoxic activity of efosite aluminum on growth, sporulation, and germination of *Phytophthora parasitica* and *P. citrophthora*. *Phytopathology*, **71**, 934-936.

Fenn, M. E. & Coffey, M. D. (1984). Studies on the *in vitro* and *in vivo* antifungal activity of fosetyl-Al and phosphorous acid. *Phytopathology*, **74**, 606-611.

Fenn, M. E. & Coffey, M. D. (1985). Further evidence for the direct mode of action of fosetyl-Al and phosphorous acid. *Phytopathology*, **75**, 1064-1068.

Fenn, M. E. & Coffey, M. D. (1988). Quantification of phosphonate and ethyl phosphonate in tobacco and tomato tissues and significance for the mode of action of two phosphonate fungicides. *Phytopathology*, **78**, in press.

Fothergill, P. G. & Child, J. H. (1964). Comparative studies of the mineral nutrition of three species of *Phytophthora*. *Journal of General Microbiology*, **36**, 67-78.

Guest, D. I. (1984a). Modification of defence responses in tobacco and capsicum following treatment with fosetyl-Al [aluminium tris (*O*-ethyl phosphonate)]. *Physiological Plant Pathology*, **25**, 125-134.

Guest, D. I. (1984b). The influence of cultural factors on the direct antifungal activities of fosetyl-Al, propamocarb, metalaxyl, SN 75196 and Dowco 444. *Phytopathologische Zeitschrift*, **111**, 155-164.

Guthrie, J. P. (1979). Tautomerization equilibria for phosphorous acid and its ethyl esters, free energies of formation of phosphorous and phosphonic acid and their ethyl esters, and pKa values for ionization of the P-H bond in phosphonic acid and phosphonic esters. *Canadian Journal of Chemistry*, **57**, 236-239.

Luttringer, M. & de Cormis, L. (1985). Absorption, dégradation et transport du phoséthyl-Al et de son métabolite chez la tomate (*Lycopersicon esculentum* Mill.). *Agronomie*, **5**, 423-430.

MacIntire, W. H., Winterberg, S. H, Hardin, L. J., Sterges, A. J. & Clements, L. B. (1950). Fertilizer evaluation of certain phosphorus, phosphorous, and phosphoric materials by means of pot cultures. *Agronomy Journal*, **42**, 543-549.

Malacinski, G. & Konetzka, W. A. (1966). Bacterial oxidation of orthophosphite. *Journal of Bacteriology*, **91**, 578-582.

Malacinski, G. M. & Konetzka, W. A. (1967). Orthophosphite-nicotinamide adenine dinucleotide oxidoreductase from *Pseudomonas fluorescens*. *Journal of Bacteriology*, **93**, 1906-1910.

Martin, R. B. (1959). The rate of exchange of the phosphorus bonded hydrogen in phosphorous acid. *Journal of the American Chemistry Society*, **81**, 1574-1576.

Ouimette, D. G. & Coffey, M. D. (1985). *In vivo* efficacy of five phosphite compounds against *Phytophthora capsici* on pepper plants. *Phytopathology*, **75**, 1330 (abstract).

Ouimette, D. G. & Coffey, M. D. (1988). Quantitative analysis of organic phosphonates, phosphonate, and other inorganic anions in plants and soil using high performance ion chromatography. *Phytopathology*, **78**, 1150-1155.

Ouimette, D. G. & Coffey, M. D. (1989). Phosphonate levels in avocado (*Persea americana* Mill.) seedlings and soil following treatment with fosetyl-Al or potassium phosphonate. *Plant Disease*, **72**, in press.

Pegg, K. G., Whiley, A. W., Saranah, J. B. & Glass, R. J. (1985). Control of *Phytophthora* root rot of avocado with phosphorus acid. *Australian Journal of Plant Pathology*, **14**, 25-29.

Piedallu, M.-A. & Jamet, P. (1985). Cinétique de disparition et dégradation du phoséthyl-Al dans le sol en conditions contrôlées. In *Fungicides for Crop Protection: 100 Years of Progress*, ed. I. M. Smith, pp. 297-300. British Crop Protection Council, Croydon, UK.

Robertson, H. E. & Boyer, P. D. (1956a). Orthophosphite as a buffer for biological studies. *Archives of Biochemistry and Biophysics*, **62**, 396-401.

Robertson, H. E. & Boyer, P. D. (1956b). The biological inactivity of glucose 6-phosphite, inorganic phosphites and other phosphites. *Archives of Biochemistry and Biophysics*, **62**, 380-395.

Rohrbach, K. G. & Schenck, S. (1985). Control of pineapple heart rot, caused by *Phytophthora parasitica* and *P. cinnamomi*, with metalaxyl, fosetyl Al, and phosphorous acid. *Plant Disease*, **69**, 320-323.

Saindrenan, P., Barchietto, T., Avelino, J. & Bompeix, G. (1988). Effects of phosphite on phytoalexin accumulation in leaves of cowpea infected with *Phytophthora cryptogea*. *Physiological and Molecular Plant Pathology*, **32**, 425-435.

Saindrenan, P., Darakis, G. & Bompeix, G. (1985). Determination of ethyl phosphite, phosphite and phosphate in plant tissues by anion-exchange high-performance liquid chromatography and gas chromatography. *Journal of Chromatography*, **347**, 267-273.

Thizy, A., Pillon, D., Debourge, J.-C. & Lacroix, G. (1978). Fungicidal compositions containing phosphorous acid and derivatives thereof. *United States Patent Number 4,075,324*.

Tsubota, G. (1959). Phosphite reduction in the paddy field. *Soil Plant Food*, **5**, 10-15.

Vegh, I., Leroux, P., LeBerre, A. & Lanen, C. (1985). Détection sur *Chamaecyparis lawsoniana* 'Ellwoodii' d'une souche de *Phytophthora cinnamomi* Rands résistante au phoséthyl-Al. *Pépiniériste, Horticulteurs, Maraïchers. Revue Horticol*, **262**, 19-21.

Wang, M. C. & Bartnicki-Garcia, S. (1973). Novel phosphoglucans from the cytoplasm of *Phytophthora palmivora* and their selective occurrence in certain life cycle stages. *Journal of Biological Chemistry*, **248**, 4112-4118.

Chapter 7

Phosphorus and nitrogen fluxes between plant and fungus in parasitic associations

D. R. Walters

*Department of Plant Sciences, West of Scotland College, Auchincruive,
Nr Ayr KA6 5HW, UK.*

Introduction

Infection with pathogenic fungi results in considerable alterations in host plant physiology and metabolism, including profound effects on the uptake and translocation of nutrients (see Paul, Chapter 8). In certain host/pathogen interactions these effects are accompanied by changes in the utilization and metabolism of some nutrients (eg. nitrogen-containing compounds and phosphorus), which may in turn affect their uptake from the rooting medium and subsequent transport throughout the plant (see Walters, 1985). Although in many plant/parasite systems, altered nutrient uptake and translocation may be the indirect result of other changes in host physiology (eg. transpiration), such changes, if they result in transport of nutrients to the fungal domain, will prove of benefit to the pathogen. Thus, altered rates of xylem and phloem transport within the diseased plant may lead to nutrient and metabolite accumulation at sites of fungal infection (Scott, 1972) and the fungus may be able to draw upon these compounds to support its growth and sporulation. In this chapter the phenomenon of nutrient accumulation at infection sites is first examined. The movement of nutrients, mainly nitrogen and phosphorus, from the host to the fungus is then considered. Because most of the available information relates to biotrophic fungal pathogens, major emphasis will be placed on plants infected with rusts or powdery mildews.

Phosphorus and nitrogen in infected leaves
Changes in whole leaf concentrations

Leaves infected with rust or powdery mildew fungi show considerable changes in nutrient concentrations (Paul, Chapter 8). Elevated phosphate concentrations have often been reported. For
example, rust-infected leeks showed increases of 21% (Roberts &
Walters, 1988a), barley infected with brown rust showed increases
of 20% (Ahmad *et al.*, 1982) and in various powdery mildew infections increased accumulations of 20-25% have been reported
(Comhaire, 1963; Fric, 1978; Walters & Ayres, 1981). However,
changes in nitrogen concentrations in infected tissues do not always take the form of an increase. Raised total nitrogen concentrations have been reported in rusted groundsel (Paul & Ayres,
1988) and rusted leek (Roberts & Walters, 1988a), whereas
Ahmad *et al.* (1982) found a reduced total nitrogen concentration
in rusted barley. Study of alterations in total nitrogen will, however, mask the changes occurring in concentrations of individual
nitrogenous compounds. Information on individual nitrogenous
compounds is important since changes will reflect responses of the
host to infection and may in turn provide information concerning
the nutrient requirements of the pathogen.

Recent work on rust-infected leeks has indicated that most of
the increase in total nitrogen is probably due to increases in amino
acid and nitrate concentrations, rather than ammonium, especially
since ammonium concentration was reduced in infected leaves
(Roberts & Walters, 1988b). Piening (1972) also found an increased nitrate concentration in rust-infected rye and although no
change was observed in nitrate in powdery mildew infected barley
leaves, amino acid and amide concentrations increased (Sadler &
Scott, 1974).

Changes in discrete regions of the infected leaf: nutrient accumulation at infection sites

A common feature of infection with biotrophic fungi is the accumulation of metabolites at infection sites. Gottlieb & Garner
(1946), who were the first clearly to demonstrate this phenomenon, showed that radioactive phosphorus accumulated in wheat
leaves infected with the stem rust fungus (*Puccinia graminis* f.sp.

Table 7.1. Concentrations of free amino acids (μmoles g^{-1} f.wt.) in pustules of *Uromyces phaseoli* var *typica* on Pinto bean leaves, in regions surrounding pustules (3 and 5 mm)* at primary sporulation, and in healthy control Pinto bean leaves.

Amino acids	Pustule	First ring	Second ring	Control
Cysteic acid	0·3	0·2	0·1	0·1
Aspartic acid	4·1	2·3	1·1	1·0
Threonine	1·1	0·3	0·2	0·3
Asparagine	4·4	10·9	8·7	1·6
Glutamic acid	5·2	2·4	1·7	1·8
Glutamine	11·0	3·1	1·8	0·5
Cystine	2·3	2·1	1·1	0·5
Leucine	0·4	0·1	0·1	0·2
Tyrosine	0·6	0·3	0·2	0·1
Lysine	1·2	0·9	0·3	0·4
Arginine	2·0	2·6	0·8	0·3

* Discrete regions were sampled as follows: pustules — discs of 1 mm diameter containing mainly stroma and host cells invaded by parasite; first ring — discs of 3 mm diameter containing the 'green-island' zone; second ring — discs of 5 mm diameter, from beyond the 'green island' zone. (Adapted from Raggi, 1974).

tritici), the label being concentrated in pustule regions. Since that time several workers have demonstrated nutrient accumulation at infection sites. Thus, Shaw & Samborski (1956) found substantial accumulations of labelled phosphorus in pustule regions of stem rust on wheat and of powdery mildew on barley. Phosphorus accumulation was also observed at *Uromyces* infection sites on bean leaves (Yarwood & Jacobsen, 1955). Later work by Raggi (1974, 1976) showed that amino acids and amides also accumulated in pustule areas on bean leaves infected with *Uromyces phaseoli* (Table 7.1) and on peach leaves infected with the powdery mildew *Sphaerotheca pannosa persicae*.

Two factors appear to be important in the accumulation of nutrients at infection sites: (1) an increased rate of nutrient transport towards the pustule area, and (2) a reduction in nutrient movement away from these regions. Thus, rusted primary leaves of bean were shown to import as much as 40 times more assimilate than controls (Livne & Daly, 1966), while substantial reductions in assimilate movement out of infected leaves has been demonstrated in wheat infected with yellow rust (Doodson, Manners & Myers, 1965) and in rusted bean (Zaki & Durbin, 1965; Livne & Daly, 1966). It should be noted however, that evidence for increased phloem import into infected monocot leaves is lacking. Thus, although nutrient accumulation at infection sites on monocot leaves might reflect a redistribution of material within the infected leaf or import of nutrients in the xylem, no data exists to support the view that phloem import into infected leaves of say, grasses or cereals, takes place. Work on brown rust-infected barley has shown that when $^{14}CO_2$ is fed to a non-infected leaf, very little label appears in infected leaves (Farrar, 1984). Similarly in leeks infected with rust, the construction of a carbon budget for individual leaves has shown that there is no assimilate movement into infected leaves (Roberts, 1987).

The mechanisms underlying the movement of nutrients to infection sites have been the subject of much research. Some workers have suggested that the vigorously growing fungal mycelium represents a powerful sink for the diversion of nutrients from other parts of the plant (Thrower, 1965). Indeed, partly because of their ability to 'create' sink effects, plant growth substances like auxins and cytokinins have been implicated in the mobilization of nutrients towards infection sites (Shaw & Hawkins, 1958; Pozsar & Kiraly, 1966; Brian, 1967). Furthermore, nutrient accumulation at infection sites of biotrophic pathogens, together with retention of chlorophyll and of the capacity for protein synthesis, are probably indicative of the maintenance of a 'juvenile' condition at and around infection sites. Evidence of such juvenility takes the form of the so-called 'green-islands', which are typical of rust and powdery mildew infections of many hosts. Interest in the possible involvement of cytokinins in green-island formation on infected leaves has been partly due to their ability to induce 'green-island' formation on detached leaves. The increased cytokinin activity, at least in rusted bean, was suggested to be of host origin (Dekhuijzen & Staples, 1968), but rust and powdery mildew spores are known to contain high levels of cytokinins (Kiraly, El-Hammady

& Pozsar, 1967). Some authors have suggested that localized delay of senescence and other phenomena associated with green-island formation on infected leaves may be the result of release of cyto-kinins by the parasite (Kiraly, Pozsar & El-Hammady, 1966). If this is the case, then it would represent the flux of an important nitrogen-containing compound from the fungus to the host. Al-though all of the research on cytokinins in fungally-infected plants has involved the use of bioassays, and as such is open to some criti-cism, it is interesting to note that using monoclonal antibodies raised against zeatin riboside and dehydrozeatin riboside, in-creases in cytokinins have been detected in rusted leek leaves, the largest increases occurring in and around pustules (Roberts, 1987).

Polyamine accumulation at infection sites

Cytokinins are not the only substances capable of delaying se-nescence and causing green-island formation. Among the substan-ces that possess this property, is a group of simple aliphatic amines called polyamines. The most common of these compounds are pu-trescine (a diamine), spermidine (a triamine) and spermine (a tetra-amine). These compounds are known to be essential for growth in most organisms and among their many proposed func-tions are included the *in vivo* regulation of senescence (Galston, 1983; Slocum, Kaur-Sawhney & Galston, 1984; Walters, 1987). As nitrogenous compounds, the polyamines represent major pools of tissue nitrogen, and their formation and possible mobilization have large potential effects on fluxes of the element in host-bio-troph relations. Greenland & Lewis (1984) were the first to examine the possibility that polyamines may be involved in the delay in senescence and green-island formation in a plant infected with a biotrophic fungus. They showed that barley leaves infected with brown rust accumulated spermidine and, to a lesser extent, spermine, with peak accumulation coinciding with fungal sporula-tion (Fig. 7.1). Soon thereafter, putrescine, spermidine and sper-mine were found to accumulate in mildewed barley leaves (Walters, Wilson & Shuttleton, 1985), the largest increases occur-ring in and around fungal pustules (Walters & Wylie, 1986) (Fig. 7.2).

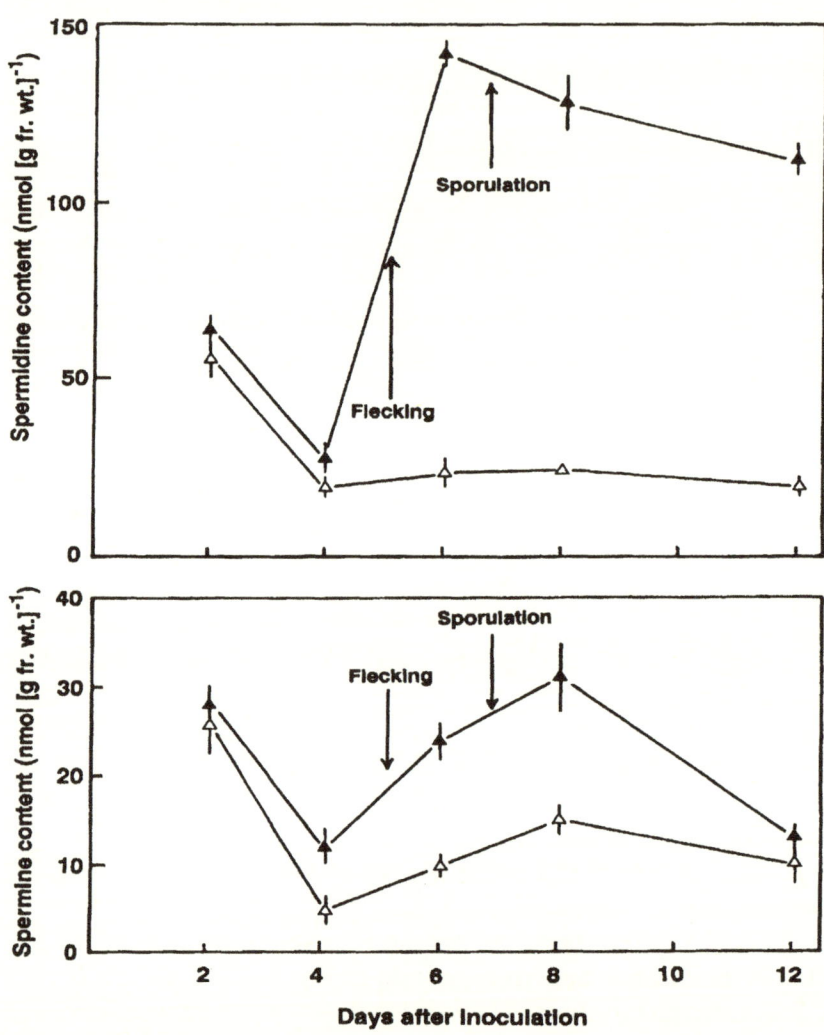

Fig. 7.1. Changes in the concentrations of (a) spermidine and (b) spermine plus an unidentified amine, in healthy (open triangles) primary leaves of barley and comparable leaves infected with *Puccinia hordei* (closed triangles). Vertical bars represent twice the standard errors of the means; where none are shown the limits are within the size of the symbols. (Adapted from Greenland & Lewis, 1984).

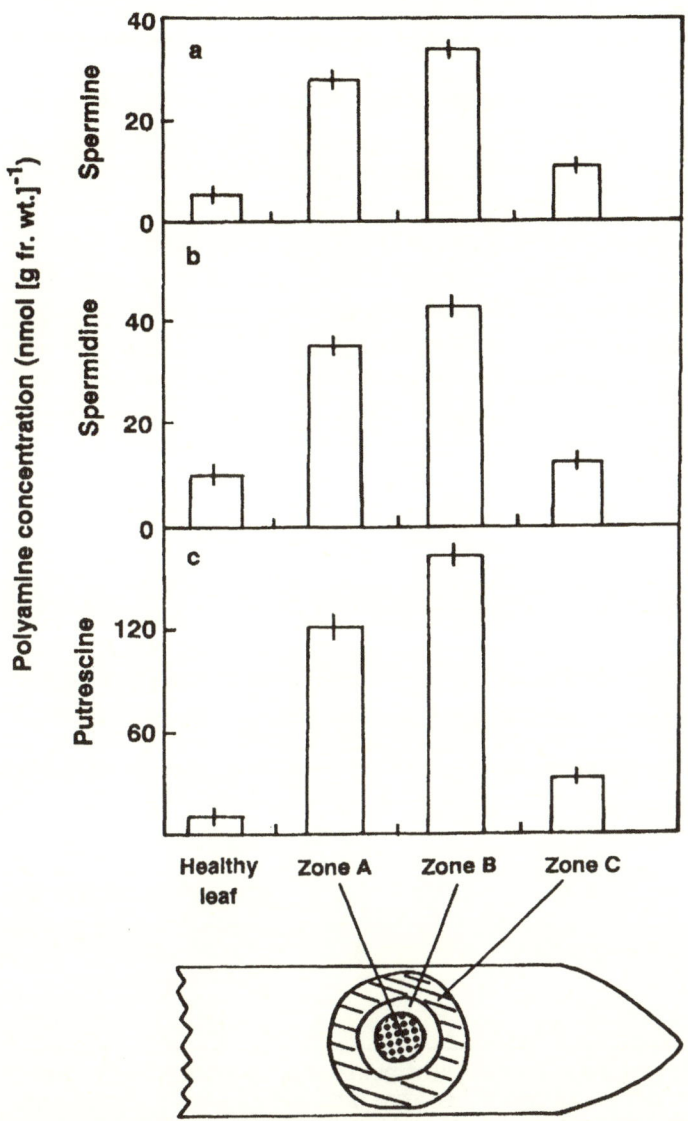

Fig. 7.2. The concentrations of spermine (a), spermidine (b) and putrescine (c) in healthy leaves and in discrete regions of mildewed barley leaves. Zone A, pustule; Zone B, region up to 5 mm from pustule boundary; Zone C, region up to 5 mm from boundary to Zone B. Vertical bars represent SEM of four replicates. (From Walters & Wylie, 1986).

However, on mildewed barley, green-islands are not visible until late on in infection at which stage pustules and surrounding regions remain green while the remainder of the leaf senesces. Green-islands can be induced to form more rapidly if senescence is hastened by detaching the leaf. In this way, recent work has detected substantial increases in the concentrations of the three polyamines in green-island tissues on mildewed barley leaves (S. E. Coghlan & D. R. Walters, unpublished). Because the peak in polyamine accumulation occurs at about the time of fungal sporulation, it is possible that much of the increase in amines could be located in the fungus. Indeed, recent work using autoradiographic localization of ornithine decarboxylase (ODC), the rate limiting

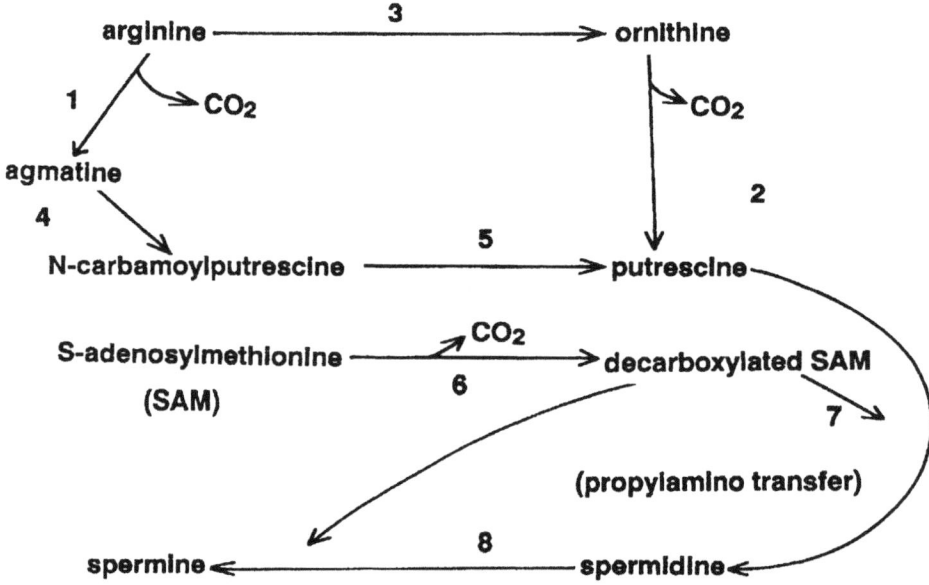

Fig. 7.3. Biosynthetic pathways of putrescine, spermidine and spermine. Putrescine can be formed either from ornithine in a reaction catalyzed by ornithine decarboxylase or from arginine in a reaction catalyzed initially by arginine decarboxylase. Spermidine is formed by the addition of a propylamino group to putrescine; the addition of another propylamino moiety to spermidine leads to spermine. 1 = arginine decarboxylase; 2 = ornithine decarboxylase; 3 = arginase; 4 = agmatine iminohydrolase; 5 = N-carbamoylputrescine aminohydrolase; 6 = S-adenosylmethionine decarboxylase; 7 = spermidine synthase; 8 = spermine synthase.

enzyme in fungal polyamine biosynthesis and one of two enzymes (ODC and arginine decarboxylase, ADC) involved in plant amine biosynthesis (Fig. 7.3), has shown that most of the increased ODC in brown rust infected barley appears to be located in the fungal material of the pustules, much less being found elsewhere (Bailey, Bower & Lewis, 1987). The situation in mildewed barley is rather different. Here, increases in both ODC and ADC were found in mildewed leaves, especially in green-islands (S. E. Coghlan & D. R. Walters, unpublished). Moreover, removal of surface fungal growth, leaving haustoria within epidermal cells, reduced, but did not eliminate, the increased polyamine concentrations and enzyme activities; it would thus appear that in mildewed barley, increased polyamine accumulation is of both host and fungal origin.

In the course of the above studies of green-islands in mildewed barley, it became apparent that substantial changes were occurring in the concentration of another polyamine, cadaverine. Cadaverine, like putrescine, is a simple diamine and can be synthesized in plants from lysine in a reaction catalyzed by lysine decarboxylase. Cadaverine concentration was found to decrease in mildewed leaves, including green-islands, but it virtually disappeared from senescing areas well away from pustules (S. E. Coghlan & D. R. Walters, unpublished). A reduction in cadaverine biosynthesis or an increased cadaverine breakdown should result in a build-up of its precursor lysine, which could then be used for other purposes. It is known that some rusts take up lysine from their hosts (see Mendgen, 1981; p. 147) and it is possible that any lysine accumulating as a result of decreased cadaverine synthesis or enhanced breakdown, could be transported toward mildew pustules and be taken up by the fungus.

Inhibition of fungal polyamine biosynthesis

Since fluxes of nitrogen, in the form of polyamines, between host and fungus, may be key processes in the nutrient relations of the association, much interest surrounds the possibility that disease control could be achieved by inhibition of their synthesis. Most fungi appear to be dependent upon ODC for polyamine biosynthesis, thus, providing the fungus cannot obtain all of its polyamine requirement from its host, specific inhibition of fungal

D. R. Walters

Table 7.2. Mean number of lesions induced by uredospores of *Uromyces phaseoli* on unifoliate leaves of bean plants exposed to α-difluoromethylornithine (DFMO) before or after inoculation.

DFMO concentration (mM)	Lesions cm^{-2}	
	DFMO treatment Preinoculation	DFMO treatment Postinoculation
None	59 ± 4	61 ± 2
0.025	34 ± 1**	28 ± 2**
0.050	29 ± 4**	17 ± 2**
0.10	14 ± 2**	5 ± 1**
0.25	2 ± 1**	2 ± 1**
0.50	0**	0**

Each value is the mean $\pm$ SEM, based on six replicates (one leaf per plant); ** denotes significant differences from controls at the 1% level. (Adapted from Rajam, Weinstein & Galston, 1985).

ODC might appear to offer a unique means of inhibiting fungal growth. Although we have no information concerning either the ability of biotrophic fungi to take up polyamines from plant cells or of the fungal requirements for host polyamines, it would seem likely that necrotrophic fungi might take up polyamines from the contents of host cells killed in advance of colonization. Disease control via inhibition of polyamine biosynthesis is possible because of the existence of several ODC specific inhibitors; e.g. α-difluoromethylornithine (DFMO) — a specific, irreversible inhibitor of ODC. It has been shown that excellent control of rust infection of *Phaseolus vulgaris* (Rajam, Weinstein & Galston, 1985) and *Vicia faba* (Walters, 1986) can be achieved using pre-inoculatory or post-inoculatory sprays of DFMO (Table 7.2; Fig. 7.4). Weinstein *et al.* (1987) confirmed this effect of DFMO on several rust diseases of wheat, while work at Auchincruive (West & Walters, 1988) has shown that powdery mildew infection of barley can be controlled using several different inhibitors of ODC or other inhibitors of polyamine biosynthesis.

It is interesting to note that work so far on hemibiotrophs and necrotrophic fungi has highlighted differences in the responses to

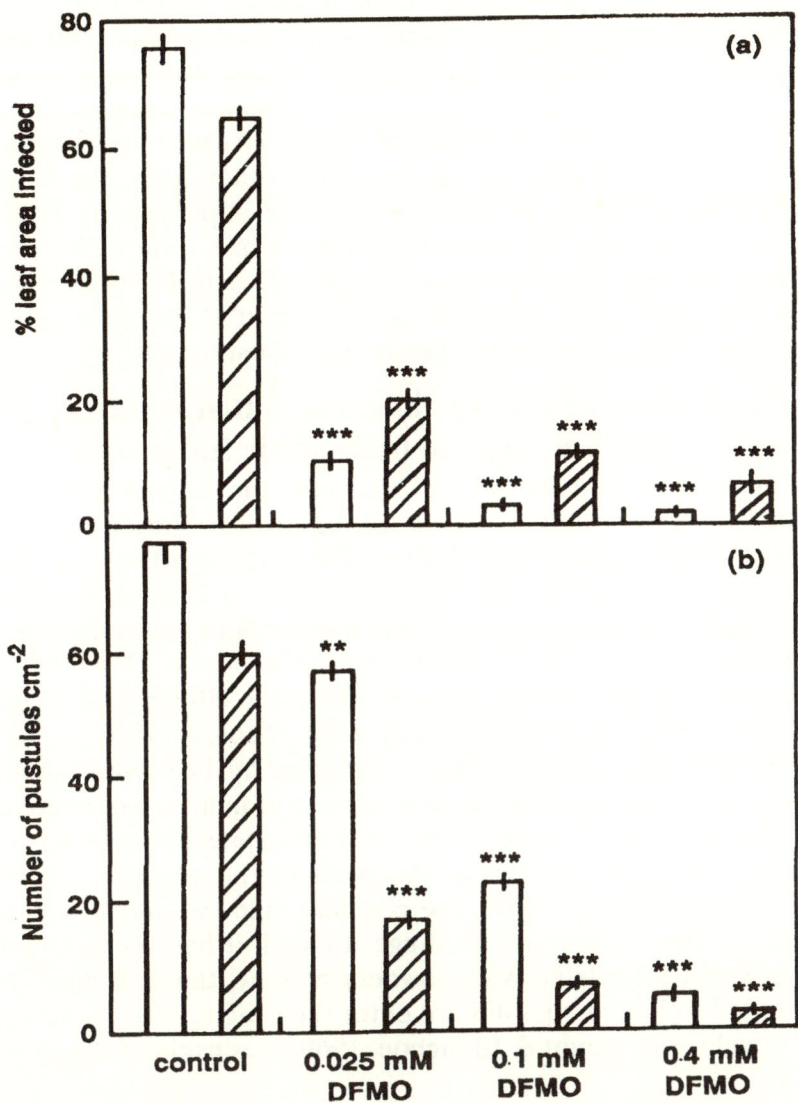

Fig. 7.4. Effect of post-inoculatory (open bars) or pre-inoculatory (hatched bars) treatment with α-difluoromethylornithine (DFMO) on (a) the percentage leaf area infected and (b) pustules per cm^2 on leaves of broad bean inoculated with rust, *Uromyces viciae-fabae*. Vertical bars represent the standard error of the means and significant differences are shown at $P \leq 0.05$ *; $P \leq 0.01$ **; $P \leq 0.001$ ***. (From Walters, 1986).

inhibitors of polyamine biosynthesis in different types of fungi. Thus, although very good control of rust and powdery mildew fungi can be achieved with this approach, some fungi, e.g. *Pyrenophora teres*, require much higher inhibitor concentrations for effective control of growth, while polyamine biosynthesis inhibitors appear to have no effect on the growth of other fungi, e.g. *Fusarium culmorum* (H. M. West & D. R. Walters, unpublished). These differences in fungal response to inhibitors may reflect differences in sensitivity of, for example, fungal ODCs to DFMO and/or they may reflect differences in the abilities of different fungi to take up DFMO (or other inhibitors) from host tissues.

Role of intercellular hyphae and haustoria in N, P and S uptake in biotrophic fungi

For many powdery mildew fungi, the only contact with host cells and therefore the only structure through which nutrient uptake can occur is the haustorium. Indeed, Gay and his colleagues (see Manners & Gay, 1983; Gay, 1984) have convincingly shown that haustoria are the main means by which assimilates are transported to powdery mildew fungi. According to Manners & Gay (1983), haustoria offer several advantages over intercellular hyphae for nutrient uptake. In particular, stimulation of solute efflux over an area surrounding the haustorium is provided by the establishment of an isolated apoplast and by modification of membrane function at localized areas of the host cell. It would appear that by virtue of structural and functional modification of the host plasmalemma in the haustorial region, the powdery mildew controls efflux from the host to its own advantage. Indeed, it has been shown, using wheat inoculated with powdery mildew, that ^{32}P and ^{35}S presented to host tissues did not enter the fungus until haustoria had developed (Mount & Ellingboe, 1969; Slesinski & Ellingboe, 1971).

In plants infected with rusts or Oomycetes little information is available concerning the relative contributions to uptake of haustoria and intercellular hyphae. It has been shown that certain solutes supplied to host tissues are incorporated into fungal material before haustoria are formed whereas other compounds are not incorporated until haustorial establishment has occurred (Onoe, Tani & Haito, 1973; Andrews, 1975). In a microautoradiographic study of uptake of tritiated lysine during the early stages of rust infection on bean, Mendgen (1979) showed that maximum uptake occurred via mature haustoria and very little occurred via intercellular hyphae. Interestingly, only the plasmalemma of pea powdery mildew haustoria possess ATPase activity. The plasma membranes of intercellular hyphae of bean rust possess no detectable ATPase activity (Spencer-Phillips & Gay, 1981), and very little activity is found in intercellular hyphae of *Albugo candida* (Woods & Gay, 1983). Gay (1984) has suggested that intercellular hyphae of the bean rust may thus be unable to take up nutrients at concentrations lower than those obtained internally. However, according to Farrar (1984), there will be movement of nutrients to fungal surfaces in the apoplast, in which case it seems important to know how much fungal growth can be supported in this manner. Hancock & Huisman (1981) have calculated that the concentration of nutrients in intercellular spaces is sufficient to sustain fungal growth. More recently, by calculating the surface areas of haustoria and intercellular hyphae, Kneale & Farrar (1985) were able to show that all of the hexose needed to support growth of brown rust on barley could be taken up through either fungal structure. However, this does not rule out the possibility that haustoria may be necessary for the uptake of specific substances e.g. amino acids like lysine.

Uptake of phosphorus by biotrophic fungi

While there are several reports of phosphorus accumulation at infection sites of biotrophically infected leaves (see above), it is important to know whether the fungus takes up this accumulated phosphorus. Martin & Ellingboe (1978) showed that *Erysiphe graminis tritici* took up labelled phosphorus from wheat leaves as early as 10 h after inoculation, and this continued as long as fungal growth occurred. Martin (1974, cited in Ellingboe, 1978) found that transfer of radioactive phosphorus to the fungus was ten times greater in an incompatible host/parasite interaction than in a compatible interaction. Ellingboe (1978) has suggested that because an incompatible interaction between the plant and the fungus places great stress on the host, it might cause a substantial movement of phosphorus to sites of interaction, especially since phosphorus is important in cell energy metabolism. Earlier work by Hsu

& Ellingboe (1972) had shown that uptake of ^{35}S by barley mildew depended upon successful primary infections and genotypes with high infection efficiency (i.e. those which produced more haustoria and elongating secondary hyphae) had high uptake rates. Conversely, mildew genotypes with lower infection efficiencies had lower uptake rates.

Similar experiments with rust infections do not appear to have been carried out, probably because of the difficulty in separating fungus from host. In fact, few data exist on the transfer of phosphorus from host to parasite. Ahmad, Farrar & Whitbread (1984) made a detailed study of phosphorus fluxes from barley to the brown rust fungus on the day of sporulation. They found that phosphorus retranslocation from first leaves was greatly reduced by infection and as a result, phosphorus concentration in infected leaves increased substantially. It was considered that accumulation was largely in the fungus. By calculating a phosphorus budget for healthy and infected leaves, these authors were able to show that phosphorus fluxes from the host to the rust could be sustained largely because the leaf was supplied by way of the xylem. Interestingly, very little change was observed in the concentrations of phosphorus in the xylem sap from healthy and rusted barley plants. Because the flux of phosphorus from the host symplast to the apoplast was greater than phosphorus transfer to the fungus, there appeared to be no need to invoke pathogen-induced membrane leakiness to provide intercellular rust hyphae with phosphorus (Ahmad et al., 1984). In fact, later work showed that brown rust infection of first leaves of barley did not result in gross impairment of host membrane integrity (Ahmad, Farrar & Whitbread, 1985).

Ahmad et al. (1984) found that brown rust stored most of its phosphorus as polyphosphate. Earlier work by Bennett & Scott (1971) had shown that stem rust infected primary leaves of wheat accumulated eight times more polyphosphate than controls and most of the polyphosphate appeared to be located in the stem rust uredospores (Table 7.3). Apparently, production of brown rust spores on barley leaves used phosphorus of xylem origin and the fungus did not mobilize its own polyphosphate store (Ahmad et al., 1984). These authors have speculated that polyphosphate may represent a store of phosphorus which the fungus can 'leak' to the host once it begins to senesce. In view of the importance of inor-

Table 7.3. Amounts of orthophosphate and 'soluble' polyphosphate in non-infected and rusted wheat leaves and in uredospores and saprophytic mycelium of *Puccinia graminis tritici*.

Source of material	Orthophosphate	'Soluble' polyphosphate
	\multicolumn{2}{c}{(mg P g^{-1} f.wt.)}	
Non-infected wheat leaves[b]	0·08	0·03
Rusted wheat leaves[b,c]	0·19	0·29
Uredospores	0·22	2·44
Infected leaves after removal of uredospores from sori	0·17	0·11
Saprotrophic mycelium	0·14	0·50

[a]Cold acid-soluble, not adsorbed to charcoal, acid labile, shown by paper chromatography to be almost entirely polyphosphate with a trace of pyrophosphate.
[b]Primary leaves of wheat 17 days from sowing.
[c]Ten days after inoculation, uredospores constituted 7·5% of the fresh weight of the infected leaf. (From Bennett & Scott, 1971).

ganic phosphate in controlling the rate of photosynthesis and perhaps also in controlling sink activity, the fungal polyphosphate store may represent a means of maintaining host function in a disease lesion while the rest of the leaf senesces (Ahmad *et al.*, 1984).

Uptake of nitrogen by the fungus

As previously described, infection can cause large changes in nitrogen in host tissues. Thus, twelve days after inoculation of barley with mildew, total nitrogen in mildewed areas of the leaf increased considerably (Bushnell & Gay, 1978) and most of this increase appeared to be in the mildew fungus (Fig. 7.5). Various workers have shown that soluble nitrogen compounds e.g. aspartic acid, glutamic acid, and their corresponding amides, asparagine and glutamine, can accumulate in powdery mildew infected tissues (Sadler & Scott, 1974). Bushnell & Gay (1978) have postulated that powdery mildews take up and utilize these host amino acids and amides. Recent work by Murray & Ayres (1986) supports this premise. They found that the mildew accumulated organic nitrogen at a time when there was no nitrate reduction in infected shoots. Prior to sporulation, mildew infection resulted in only a minor alteration in the rate of nitrogen recirculation to the roots and in the flux of nitrogen to the shoots via the xylem. They thus suggested that early in infection the mildew draws on normal supplies of nitrogen available in the shoot. However, the recycling of nitrogen to the roots stops after fungal sporulation, and nitrogen which is normally recycled becomes available for the mildew. It

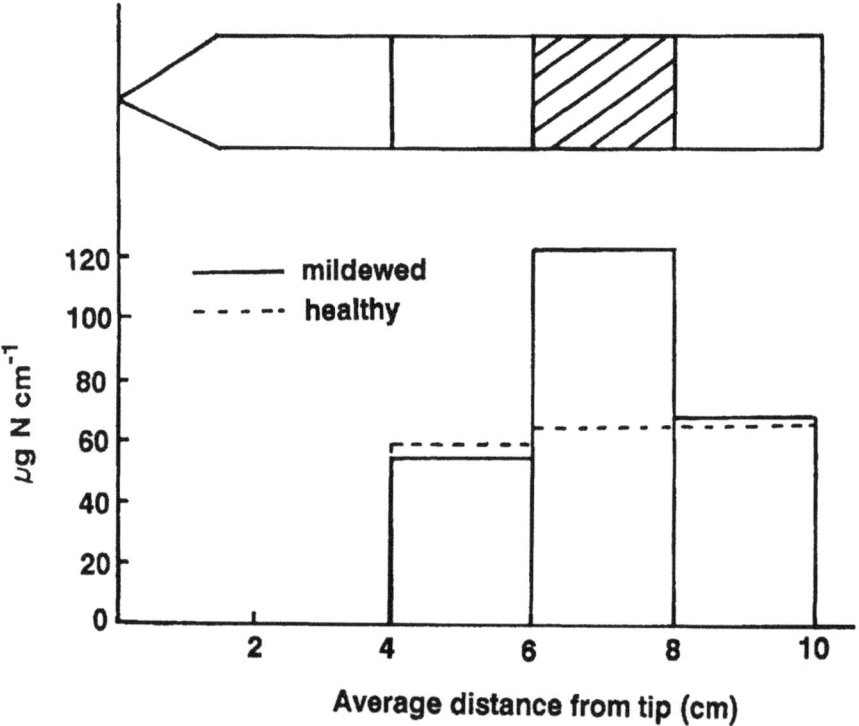

Fig. 7.5. Accumulation of nitrogen in mildewed tissues as shown by the total nitrogen content of mildewed and adjacent non-mildewed segments of attached barley leaves twelve days after inoculation with *Erysiphe graminis* f.sp. *hordei*. All spores were included in the mildewed sample. Nitrogen content of the host tissues in mildewed segments is relatively unchanged. Hatched area represents powdery mildew infection. (From Bushnell & Gay, 1978).

would appear that the fungus cannot obtain all of its nitrogen in this way and it must draw upon a supply of reduced nitrogen either from the xylem (as suggested for phosphorus uptake by brown rust, see p. 144) or by redirecting the flow of nitrogen within the shoot. Unfortunately, we have very little data on transfer of nitrogen-containing compounds from hosts to mildew and, because mildew fungi cannot be grown axenically, we know little of their nutritional requirements.

Interestingly, it has been suggested that alterations in purine metabolism are important in facilitating the successful establishment of powdery mildew infection on barley (Butters, Burrell & Hollomon, 1985). The powdery mildew fungus appears unable to synthesize purines *de novo* (Hollomon, 1979) and so it must obtain these compounds from the host plant. Butters *et al.* (1985) found that although purines were taken up from the leaf surface and metabolized by the powdery mildew fungus, it was difficult to determine what purines were transported to the mildew mycelium through the haustoria. They were able to show that $[^3H]$-adenosine supplied to the host, was taken up by the fungus and extensively deaminated to inosine. Apparently, the substantial activity of the enzyme adenosine deaminase, present in the fungus and in infected leaves but not in healthy leaves and which is responsible for the conversion of adenosine to inosine, would help to maintain an adequate supply of purines for fungal growth and development (Butters *et al.*, 1985).

Some data are available on the uptake of nitrogenous compounds, mainly of amino acids and amides, by rust fungi from their hosts. Jager & Reisener (1969) found that lysine and arginine were probably transported intact into the stem rust fungus from wheat. Amino acids such as alanine and glycine appeared to be metabolized prior to uptake and incorporation into spores. Autoradiographic analysis of $[^3H]$-lysine uptake by resistant and susceptible bean/rust interactions showed that although lysine uptake by rust haustoria occurred in both types of interaction, only in the compatible combination did uptake occur in the very young developing haustorium (Mendgen, 1979). In an incompatible interaction between the cultivar Golden Gate Wax and rust, labelled lysine uptake was much less than that in the compatible interaction, although haustoria were still labelled when cell contents of both partners had become disorganised and broken down. Mendgen (1979) considered that the label had simply diffused from host cell to haustorium via damaged membranes. However, not all incompatible interactions are the same. In another incompatible interaction, lysine uptake was constant and similar to that observed in the compatible combination, suggesting that here, nutrient uptake by the haustorium remains normal until other metabolic processes result in death of the fungus and the host cell (Mendgen, 1979). Experiments by Burrell & Lewis (1977) on infection of coltsfoot (*Tussilago farfara* L.) with the rust *Puccinia poarum*, indicated that serine and alanine were probably taken up directly by the rust, whereas aspartic acid, glutamic acid and glutamine were not readily taken up by the parasite. Indeed, these authors considered serine and alanine to be important in the successful establishment and growth of the rust on its host. Bushnell (1984), on the other hand, considered glutamine to be the most important of the amino acids or amides to be utilized by rusts. Thus, it is known to accumulate in rust infections, is readily translocated in plants and together with ammonia and asparagine, can be an important product of proteolysis (Lea & Miflin, 1980). Glutamine is also a favoured source of nitrogen for axenically grown rusts and it is a precursor in fungal chitin synthesis (e.g. Raggi, 1974).

Conclusions

In their 1946 paper describing the effects of rust infection on phosphorus distribution in wheat leaves, Gottlieb & Garner stated that '... Despite the many years which have elapsed since the early work of Marshall Ward (1901; 1902) on the physiology of obligate parasitism, this complex is still in many respects an enigma'. Today, some 42 years later, this statement is still true. In spite of much excellent work in certain areas, our knowledge of nutrient movement from host to parasite is still fragmentary and many questions remain unanswered. Moreover, very little detailed information exists on the movement of solutes into hemibiotrophic and necrotrophic fungi. This is surprising in view of the obvious importance of nutrient fluxes in infected tissues (and indeed, in whole infected plants) in controlling the growth potential of pathogenic fungi. However, because of the difficulty in separating and isolating host and fungus in infected tissues, studying the movement of nutrients and metabolites from host to pathogen is not an easy task. Nevertheless, it is clear that more detailed experiments on nutrient fluxes in infected plants are required (viz the work of Ahmad *et al.*, 1984 and Murray & Ayres, 1986) and especially on fluxes between host and parasite. Given the importance of polyamines in the growth and development of various organisms, including fungi, and the potential for controlling fungal infections by inhibiting polyamine biosynthesis, it would be useful to know more about fungal polyamine requirements in host tissue. Further, it is desirable to know whether fungal pathogens are able to take up polyamines from their hosts or whether precursors are absorbed and used in polyamine biosynthesis in fungal tissue. Thus, a study of solute fluxes between host and fungus in parasitic associations is not solely of academic interest, since a better understanding of such phenomena may be useful in the search for much needed novel crop protection agents.

References

Ahmad, I., Owera, S. A. P., Farrar, J. F. & Whitbread, R. (1982). The distribution of five major nutrients in barley plants infected with brown rust. *Physiological Plant Pathology*, **21**, 335-346.

Ahmad, I., Farrar, J. F. & Whitbread, R. (1984). Fluxes of phosphorus in leaves of barley infected with brown rust. *New Phytologist*, **98**, 361-375.

Ahmad, I., Farrar, J. F. & Whitbread, R. (1985). Membrane integrity in leaves of barley infected by brown rust: an examination using tracer efflux and *in vivo* chlorophyll fluorescence. *New Phytologist*, **99**, 107-115.

Andrews, J. H. (1975). A freeze-etch study of membranes of *Plasmodiophora*-infected and non-infected cabbage root hairs. *Canadian Journal of Botany*, **52**, 1441-1449.

Bailey, J. P., Bower, A. J. & Lewis, D. H. (1987). Localization of ornithine decarboxylase in barley leaves infected by brown rust (*Puccinia hordei*). *Transactions of the British Mycological Society*, **89**, 83-87.

Bennett, J. & Scott, K. J. (1971). Inorganic polyphosphates in the wheat stem rust fungus and in rust-infected wheat leaves. *Physiological Plant Pathology*, **1**, 185-198.

Brian, P. W. (1967). Obligate parasitism in fungi. *Proceedings of the Royal Society of London, Series B*, **168**, 101-118.

Burrell, M. M. & Lewis, D. H. (1977). Amino acid movement from leaves of *Tussilago farfar* L. to the rust, *Puccinia poarum* Neils. *New Phytologist*, **79**, 327-333.

Bushnell, W. R. (1984). Structural and physiological alterations in susceptible host tissue. In *The Cereal Rusts*, vol. 1, eds. W. R. Bushnell & A. P. Roelfs, pp. 477-507. Academic Press, Orlando.

Bushnell, W. R. & Gay, J. (1978). Accumulations of solutes in relation to the structure and function of haustoria in powdery mildews. In *The Powdery Mildews*, ed. D. M. Spencer, pp. 183-235. Academic Press, London.

Butters, J. A., Burrell, M. M. & Hollomon, D. W. (1985). Purine metabolism in barley powdery mildew and its host. *Physiological Plant Pathology*, **27**, 65-74.

Comhaire, F. (1963). Study of the absorption and repartition of phosphorus in *Triticum vulgare* Link. infected by *Erysiphe graminis*. *Leujeunia*, N. S. **15**, 1-87.

Dekhuijzen, H. M. & Staples, R. C. (1968). Mobilization factors in uredospores and bean leaves infected with bean rust fungus. *Contributions from Boyce Thompson Institute*, **24**, 39-52.

Doodson, J. K., Manners, J. G. & Myers, A. (1965). Some effects of yellow rust (*Puccinia striiformis*) on 14 carbon assimilation and translocation in wheat. *Journal of Experimental Botany*, **16**, 304-317.

Ellingboe, A. H. (1978). A genetic analysis of host-parasite interactions. In *The Powdery Mildews*, ed. D. M. Spencer, pp. 159-181. Academic Press, London.

Farrar, J. F. (1984). Effects of pathogens on plant transport systems. In *Plant Diseases: Infection, Damage and Loss*, eds. R. K. S. Wood & J. G. Jellis, pp. 87-104. Blackwell Scientific Publications, Oxford.

Fric, F. (1978). Absorption and translocation of phosphate in barley plants infected with powdery mildew. *Phytopathologische Zeitschrift*, **91**, 23-32.

Galston, A. W. (1983). Polyamines as modulators of plant development. *Bioscience*, **33**, 382-388.

Gay, J. L. (1984). Mechanisms of biotrophy in fungal pathogens. In *Plant Diseases: Infection, Damage and Loss*, eds. R. K. S. Wood & J. G. Jellis, pp. 49-59. Blackwell Scientific Publications, Oxford.

Gottlieb, D. & Garner, J. M. (1946). Rust and phosphorus distribution in wheat leaves. *Phytopathology*, **36**, 557-564.

Greenland, A. J. & Lewis, D. H. (1984). Amines in barley leaves infected by brown rust and their possible relevance to formation of 'green islands'. *New Phytologist*, **96**, 283-291.

Hancock, J. G. & Huisman, O. C. (1981). Nutrient movement in host-pathogen systems. *Annual Review of Phytopathology*, **19**, 309-331.

Hollomon, D. W. (1979). Evidence that ethirimol may interfere with adenine metabolism during primary infection of barley powdery mildew. *Pesticide Biochemistry and Physiology*, **10**, 181-189.

Hsu, S. C. & Ellingboe, A. H. (1972). Elongation of secondary hyphae and transfer of 35S from barley to *Erysiphe graminis* f.sp. *hordei* during primary infection. *Phytopathology*, **62**, 876-882.

Jäger, K. & Reisener, H. J. (1969). Untersuchungen über Stoffwechselbeziehungen zwischen Wirt und Parasit am Beispiel von *Puccinia graminis* var *tritici* auf Weizen. I. Aufnahme von Aminosäuren aus dem Wirtsgewebe. *Planta*, **85**, 57-72.

Kiraly, Z., Pozsar, B. I. & El-Hammady, M. (1966). Cytokinin activity in rust-infected plants: Juvenility and senescence in diseased leaf tissues. *Acta Phytopathologica Academiae Scientiarum Hungaricae*, **1**, 29-38.

Kiraly, Z., El-Hammady, M. & Pozsar, B. I. (1967). Increased cytokinin activity of rust infected bean and broad bean leaves. *Phytopathology*, **57**, 93-94.

Kneale, J. & Farrar, J. F. (1985). The localization and frequency of haustoria in colonies of brown rust on barley leaves. *New Phytologist*, **101**, 495-505.

Lea, P. J. & Miflin, B. J. (1980). Transport and metabolism of asparagine and other nitrogen compounds within the plant. In *The Biochemistry of Plants*, vol. 5, ed. B. J. Miflin, pp. 569-607. Academic Press, New York.

Livne, A. & Daly, J. M. (1966). Translocation in healthy and rust affected beans. *Phytopathology*, **56**, 170-175.

Manners, J. M. & Gay, J. L. (1983). The host-parasite interface and nutrient transfer in biotrophic parasitism. In *Biochemical Plant Pathology*, ed. J. A. Callow, pp. 163-195. John Wiley & Sons, Chichester.

Martin, T. J. & Ellingboe, A. H. (1978). Genetic control of ^{32}P transfer from wheat to *Erysiphe graminis* f.sp. *tritici* during primary infection. *Physiological Plant Pathology*, **13**, 1-11.

Mendgen, K. (1979). Microautoradiographic studies on host-parasite interactions. II. The exchange of ^{3}H-lysine between *Uromyces phaseoli* and *Phaseolus vulgaris*. *Archives of Microbiology*, **123**, 129-135.

Mendgen, K. (1981). Nutrient uptake in rust fungi. *Phytopathology*, **71**, 983-989.

Mount, M. S. & Ellingboe, A. H. (1969). [32]P and ^{35}S transfer from susceptible wheat to *Erysiphe graminis* f.sp. *tritici* during primary infection. *Phytopathology*, **59**, 235.

Murray, A. J. S. & Ayres, P. G. (1986). Uptake and translocation of nitrogen by mildewed barley seedlings. *New Phytologist*, **104**, 355-365.

Onoe, T., Tani, T. & Haito, N. (1973). The uptake of labelled nucleosides by *Puccinia coronata* grown in susceptible oat leaves. *Report of the Tottori Mycological Institute (Japan)*, **10**, 303-312.

Paul, N. D. & Ayres, P. G. (1988). Nutrient relations of groundsel (*Senecio vulgaris* L.) infected by rust (*Puccinia lagenophorae* Cooke.) at a range of nutrient concentrations. I. Concentrations, contents and distribution of N, P and K. *Annals of Botany*, **61**, 489-498.

Piening, L. J. (1972). Effects of leaf rust on nitrate in rye. *Canadian Journal of Plant Science*, **52**, 842-843.

Pozsar, B. I. & Kiraly, Z. (1966). Phloem transport in rust-infected plants and cytokinin directed long distance movement of nutrients. *Phytopathologische Zeitschrift*, **56**, 297-309.

Raggi, V. (1974). Free and protein amino acids in the pustules and surrounding tissues of rusted bean. *Phytopathologische Zeitschrift*, **81**, 289-300.

Raggi, V. (1976). Amino acids in mycelium of *Sphaerotheca pannosa* var. *persicae* and in the infected and surrounding tissues of peach leaves. *Phytopathologia Mediterranea*, **15**, 110-114.

Rajam, M. V., Weinstein, L. H. & Galston, A. W. (1985). Prevention of a plant disease by specific inhibition of fungal polyamine biosynthesis. *Proceedings of the National Academy of Science, USA*, **82**, 6874-6878.

Roberts, A. M. (1987). Photosynthesis, nitrogen metabolism and shoot: root equilibria in leeks infected with the rust, *Puccinia allii* Rud. Ph.D. Thesis, University of Glasgow.

Roberts, A. M. & Walters, D. R. (1988a). Shoot: root interrelationships in leeks infected with the rust, *Puccinia allii* Rud.: growth and nutrient relations. *New Phytologist*, (in press).

Roberts, A. M. & Walters, D. R. (1988b). Nitrogen assimilation and metabolism in rusted leek leaves. *Physiological Plant Pathology*, **32**, 229-235.

Sadler, R. & Scott, K. J. (1974). Nitrogen assimilation and metabolism in barley leaves infected with the powdery mildew fungus. *Physiological Plant Pathology*, **4**, 235-247.

Scott, K. J. (1972). Obligate parasitism by phytopathogenic fungi. *Biological Reviews*, **47**, 537-572.

Shaw, M. & Samborski, D. J. (1956). The physiology of host-parasite relations. I. The accumulation of radioactive substances at infections of facultative and obligate parasites including tobacco mosaic virus. *Canadian Journal of Botany*, **34**, 389-405.

Shaw, M. & Hawkins, A. R. (1958). The physiology of host-parasite relations. V. A preliminary examination of the level of free endogenous indoleacetic acid in rusted and mildewed cereal leaves and their ability to decarboxylate exogenously supplied radioactive indoleacetic acid. *Canadian Journal of Botany*, **36**, 1-16.

Slesinski, R. S. & Ellingboe, A. H. (1971). Transfer of ^{35}S from wheat to the powdery mildew fungus with compatible and incompatible parasite/host genotypes. *Canadian Journal of Botany*, **49**, 303-310.

Slocum, R. D., Kaur-Sawhney, R. & Galston, A. W. (1984). The physiology and biochemistry of polyamines in plants. *Archives of Biochemistry and Biophysics*, **235**, 283-303.

Spencer-Phillips, P. T. N. & Gay, J. L. (1981). Plasma membrane ATP-ase domains and transport through infected cells. *New Phytologist*, **89**, 393-400.

Thrower, L. B. (1965). Host physiology and obligate fungal parasites. *Phytopathologische Zeitschrift*, **52**, 319-334.

Walters, D. R. (1985). Shoot: root interrelationships: the effects of obligately biotrophic fungal pathogens. *Biological Reviews*, **60**, 47-79.

Walters, D. R. (1986). The effects of a polyamine biosynthesis inhibitor on infection of *Vicia faba* L. by the rust fungus, *Uromyces viciae-fabae* (Pers.) Schroet. *New Phytologist*, **104**, 613-619.

Walters, D. R. (1987). Polyamines: the Cinderellas of Cell Biology. *Biologist*, **33**, 73-76.

Walters, D. R. & Ayres, P. G. (1981). Phosphate uptake and transport by roots of powdery mildew infected barley. *Physiological Plant Pathology*, **18**, 195-205.

Walters, D. R., Wilson, P. W. F. & Shuttleton, M. A. (1985). Relative changes in levels of polyamines and activities of their biosynthetic enzymes in barley infected with the powdery mildew fungus, *Erysiphe graminis* D.C. ex. Merat f.sp. *hordei* Marshal. *New Phytologist*, **101**, 695-705.

Walters, D. R. & Wylie, M. A. (1986). Polyamines in discrete regions of barley leaves infected with the powdery mildew fungus, *Erysiphe graminis*. *Physiologia Plantarum*, **67**, 630-633.

Ward, H. M. (1901). The bromes and their rust fungus (*Puccinia dispersa*). *Annals of Botany*, **15**, 560-562.

Ward, H. M. (1902). Experiments on the effect of mineral starvation on the parasitism of the uredine fungus, *Puccinia dispersa*, on species of *Bromus*. *Proceedings of the Royal Society of London*, **71**, 138-151.

Weinstein, L. H., Osmeloski, J. F., Wettlaufer, S. H. & Galston, A. W. (1987). Protection of wheat against leaf and stem rust and powdery mildew diseases by inhibition of polyamine metabolism. *Plant Science*, **51**, 311-316.

West, H. M. M. & Walters, D. R. (1988). Novel control of fungal plant diseases using inhibitors of polyamine biosynthesis. *Crop Research*, **28**, 97-108.

Woods, A. M. & Gay, J. L. (1983). Evidence for a neckband delimiting structural and physiological regions of the host plasmalemma associated with haustoria of *Albugo candida*. *Physiological Plant Pathology*, **23**, 73-88.

Yarwood, C. E. & Jacobson, L. (1955). Accumulation of chemicals in diseased areas of leaves. *Phytopathology*, **45**, 43-48.

Zaki, A. I. & Durbin, R. D. (1965). The effect of bean rust on the translocation of photosynthate from diseased leaves. *Phytopathology*, **55**, 528-529.

Chapter 8

Effects of fungal pathogens on nitrogen, phosphorus and sulphur relations of individual plants and populations

N. D. Paul

Institute of Environmental and Biological Sciences, University of Lancaster, Bailrigg, Lancaster LA1 4YQ, UK

Introduction

Despite the crucial importance of nitrogen, phosphorus and sulphur to all aspects of plant growth, development and reproduction, our knowledge of the interactions of pathogenic fungi and these nutrients remains extremely limited. Indeed, until recently almost all studies had been concerned with pathogens of the major temperate cereals, wheat and barley. Furthermore, with the notable exception of a few detailed studies of the root pathogens of wheat (e.g. Fitt & Hornby, 1978) data on fungal pathogens other than powdery mildews and rusts were, and still are, virtually absent. Despite this comparative paucity of data, it is clear that fungal infections may have profound consequences for the nutrient relations of their hosts. It is also increasingly evident that many of the observed changes in whole plant nutrient relations following infection, by biotrophic pathogens at least, form part of an integrated response which is, in many ways, no different from responses to other environmental stresses. In considering the effects of pathogens on nitrogen, phosphorus and sulphur, I will make use of our understanding of the responses of uninfected plants to abiotic stress. The use of such data is valuable both in allowing an insight into mechanisms operating in infected plants and in considering the many aspects of nutrient relations where the effects of pathogens remain poorly understood.

In this chapter I will consider the effects of pathogens on the nutrient contents and concentrations of the host (whole plant and both infected and uninfected organs) and on the uptake of nutrients by the roots of infected plants, interactions between populations of hosts and pathogens will also be discussed. The lack of information on necrotrophic and hemibiotrophic infections makes it inevitable that greater emphasis will be placed on biotrophic infections, principally powdery mildews and rusts, and, equally inevitably, I will discuss nitrogen and phosphorus in greater detail than sulphur, which has rarely been studied even in the better worked systems.

Nutrient relations of plants infected with biotrophic pathogens

Changes in contents and concentrations

There are few reports of the contents (i.e. quantity per plant or per organ) and concentrations (i.e. quantity per unit weight) of sulphur in plants infected with powdery mildews. Yarwood & Jacobson (1954) found that exogenously applied ^{35}S accumulated at the site of infection in some mildew infections but not in others. No change in sulphur-containing amino acids was found in peach (*Prunus persicae*) infected with *Sphaerotheca pannosa* (Raggi, 1976). Changes in phosphorus relations are better understood, at least in young barley (*Hordeum vulgare*) plants infected with *Erysiphe graminis* f. sp. *hordei* where phosphorus concentrations have invariably been found to increase in infected leaves (Comhaire, 1963; Fric, 1978; Walters & Ayres, 1981a) and root tissues (Walters & Ayres, 1981a). Powdery mildews appear to have little effect on total nitrogen concentrations in infected barley leaves (Fric, 1978; Walters & Ayres, 1980). However, nitrate concentrations have generally been found to decrease in mildewed leaves (Walters & Ayres, 1980) and reduced nitrogen (free- and protein-amino acids and free ammonium) to increase (Raggi, 1976; Walters & Ayres, 1980; Murray & Ayres, 1986). There is no consistent evidence that free ammonium contributes significantly to elevated concentrations of reduced-nitrogen (see Murray & Ayres, 1986) and it is amino acids which show the most consistent and pronounced increase following infection, although the responses of individual amino acids exhibit substantial variation (see Walters, Chapter 7).

The nutrient relations of rust-infected plants have been investigated in detail in only three systems: brown rust (*Puccinia hordei*) of barley seedlings up to the six leaf stage (Ahmad *et al.*, 1982), *P. lagenophorae* infected groundsel (*Senecio vulgaris*) throughout its development (Paul & Ayres, 1988a & b) and *P. allii* on young leek (*Allium porrum*) plants (Roberts & Walters, 1988a & b). Sulphur was not studied in any of these systems, but in a survey of a number of rusts, Yarwood & Jacobson (1954) found accumulation of exogenously applied ^{35}S in the pustule up to 19 times that in uninfected tissues. Raggi (1974) found accumulation of sulphur-containing amino acids in rust pustules on bean (*Phaseolus vulgaris*) but increases were confined to free amino acids, those incorporated in proteins showed some decline. Tissues around sporulating rust pustules were often found to be depleted of sulphur (Yarwood & Jacobson, 1954), perhaps because the quantity of sulphur-containing amino acids was reduced (Raggi, 1974).

In leek plants, rust infection resulted in increases in the total content of nitrogen and phosphorus, despite the lower dry weight of infected plants (Roberts & Walters, 1988b). In groundsel and barley the total amount of phosphorus per plant was depressed by rust infection and in the latter nitrogen content was also reduced (Ahmad *et al.*, 1982); in groundsel there was no significant change in nitrogen (Paul & Ayres, 1988a). However, reductions in plant dry weight were greater than any decline in nutrient content, with the result that concentrations were increased. In leek and groundsel, elevated phosphate concentration was most pronounced in, but not confined to, infected tissues, while in barley, phosphorus concentration increased more than two-fold in infected leaves but was reduced in uninfected tissues (Ahmad *et al.*, 1982). Indeed, phosphorus accumulation at sites of rust infection is known to occur in a number of rusts, partly as organic phosphate but primarily as polyphosphate. The relative importance of these compounds is discussed by Walters (Chapter 7).

In contrast to phosphorus concentration, the concentration of Kjeldahl-nitrogen in rusted barley leaves was reduced (Ahmad *et al.*, 1982) but in groundsel (Paul & Ayres, 1988a) and leek (Roberts & Walters, 1988b) the concentration of Kjeldahl-nitrogen was increased in infected leaves. More detailed analysis of the form of nitrogen in rusted leek suggested that, since free ammonium-nitrogen decreased, much of the observed increase in reduced nitrogen was due to amino acids (Roberts & Walters, 1988a). In leek, and in the leaves of rye infected with *Puccinia recondita* (Peining, 1972) the concentration of nitrate also increased. Nitrogen concentrations in the uninfected tissues of rusted plants were found to increase in all three species (Ahmad *et al.*, 1982; Paul & Ayres, 1988a; Roberts & Walters, 1988a). The form of nitrogen accumulated in uninfected tissues is not known, and indeed, the expression of nutrient concentrations on the basis of dry weight may be misleading as they may reflect reductions in the percentage dry weight of tissues rather than any change of nutrient concentration in the aqueous phase. It is known that dry weight per unit fresh weight is often reduced in rust-free tissues on infected plants, due, for example, to reduced carbohydrate contents. In groundsel the percentage dry weight content of the roots may be almost halved in heavily rusted plants (e.g. 6·2% at the onset of flower-production as opposed to 11·8% in controls; N. D. Paul, unpublished). If such changes occurred in the experiment in which nutrient relations were studied, the observed increases in nitrogen and phosphorus concentrations per unit dry weight could have occurred even if concentrations per unit fresh weight were substantially reduced by infection. Changes in dry weight content per unit fresh weight may also explain reports of elevated nutrient concentrations per unit dry weight in the uninfected tissues of plants infected with powdery mildew (see above) and have also been invoked to explain increased phosphorus concentration in plants infected with mycorrhizas (Stribley, Tinker & Rayner, 1980).

In biotroph-infected tissues dry weight per unit fresh weight increases and hence increased nitrogen and phosphorus concentrations expressed on a dry weight basis will underestimate changes in concentrations in the aqueous phase. Accumulations in infected tissues reflect concentrations both in host tissues and in the parasite itself and may result from increased import and/or decreased export. These mechanisms and the distribution of nutrients between host and parasite are discussed by Walters (Chapter 7). While such changes are central to increased nutrient concentrations in infected tissues they obviously cannot account for the increased nutrient concentrations of the whole plant which often follows infection (see above): this requires increases in the uptake of nutrients by the roots of rusted or mildewed plants.

Nutrient uptake

In most studies of the effects of plant pathogens, nutrient uptake by the host has been derived, by growth analysis, from differences in nutrient contents at successive harvests. In diseased plants, one limitation of this technique is the unknown loss of nutrients as spores, which is not easily quantified over long periods. Ahmad et al., (1982) found that urediospores of *Puccinia hordei* were rich in phosphorus and nitrogen and while losses to spores may be of little importance in short term studies (e.g. Murray & Ayres, 1986), investigations extending over a number of weeks (e.g. Paul & Ayres, 1988a & b; Roberts & Walters, 1988b) are inevitably liable to underestimate actual uptake. Perhaps the ideal method of studying the nutrient relations of infected plants is the flowing nutrient solution technique (e.g. Clement et al., 1974) which enables long-term uptake to be measured by extraction from the medium, avoiding errors due to loss of spores, as well as by growth analysis. Unfortunately, to date this method has not been applied to plants with foliar infections and in the following discussion of uptake the possibility of underestimation should be borne in mind.

Studies of the effects of powdery mildews on nitrogen uptake have largely been confined to one system, *Erysiphe graminis* on barley. While ammonium uptake was unaffected by mildew infection (Walters & Ayres, 1980), the rate of nitrate uptake by the roots of mildewed barley was inhibited before the onset of sporulation (Walters & Ayres, 1980; Murray & Ayres, 1986) and stimulated after sporulation (Murray & Ayres, 1986). The rates of phosphate uptake by the roots of mildewed plants increased in wheat (Comhaire, 1963) as well as barley (Walters & Ayres, 1981a, 1982) although in the latter Fric (1978) reported decreased phosphorus uptake in heavily infected seedlings. Rust infection in groundsel and leek increased the rates of uptake of both phosphorus and nitrogen when expressed on the basis of root dry weight (Paul & Ayres, 1988a; Roberts & Walters, 1988b). Although no stimulation of phosphate uptake was observed in barley infected with *Puccinia hordei* (Ahmad *et al.*, 1982), the rate of uptake of nitrogen was enhanced.

The range of responses in the uptake of different ions to infection in the various pathosystems studied has discouraged any unified explanation of the underlying mechanisms, especially since the effects may differ at different times within the same system. However, an overall view shows that any decrease in phosphorus or nitrogen (nitrate or ammonium) uptake by plants infected with foliar biotrophs is the exception rather than the rule (Table 8.1), while the uptake of cations such as potassium seems to be invariably increased. A plethora of mechanisms have been proposed for specific ions in particular systems (see Walters, 1985) but, as has often been noted, increased rates of uptake present a dilemma in that they are concurrent with reduced assimilate supply to the roots of infected plants (see Ayres, 1984), a relationship not easily reconciled with the energy requirements of active ion uptake. This anomaly might be explained through a coordinated response of ion uptake to alterations in plant growth regulators (PGRs) in the roots of infected plants (e.g. Ahmad *et al.*, 1982; Walters, 1985). However, our knowledge of the effects of endogenous PGRs on nutrient uptake in healthy plants is rather patchy and superimposed on this is an almost complete absence of reliable data of pathogen-induced changes in these compounds. Thus, while the involvement of PGRs cannot be excluded, any evidence for such a role is, for the present at least, tenuous. The effects of reduced assimilate might be buffered in the roots of infected plants by changes in partitioning between 'growth' and 'maintenance' respiration (Penning de Vries, 1975). Growth respiration has been shown to be more sensitive to reduced assimilate supply than maintenance in healthy plants and shoot infections are known to reduce not only root dry weight accumulation but also cell division in the root tip (Martin & Hendrix, 1974; Walters & Ayres, 1981b). Furthermore, respiration in the roots appears to be largely via the energetically inefficient 'alternative-pathway' (Lambers, 1980) and it is possible that this might be disengaged in infected plants, as it is under conditions of environmental stress, so improving the efficiency with which available assimilate is utilised.

Table 8.1. Summary of the effects of rusts and powdery mildews on the nitrogen and phosphorus relations of the host.

		Host and parasite barley/mildew	barley/rust	leek/rust
Content per plant	nitrogen	increased[1]	decreased[5]	increased[7]
	phosphorus	increased[2]	decreased[5]	increased[7]
Concentration in	nitrogen	unchanged[3]	decreased[5]	increased[7,8]
infected tissues	phosphorus	increased[2]	increased[5,6]	increased[7]
Concentration in	nitrogen	increased[3]	increased[5]	increased[6]
uninfected tissues	phosphorus	increased[2]	decreased[5]	increased[7]
Rate of uptake	nitrogen	incr[1],decr[2,3]	increased[5]	increased[7]
	phosphorus	incr[2],decr[4]	unchanged[5,6]	increased[7]

References: 1, Walters & Ayres, 1980; 2, Murray & Ayres, 1986; 3, Walters & Ayres, 1981a; 4, Fric, 1978; 5, Ahmad et al., 1982; 6, Ahmad et al., 1984; 7, Roberts & Walters, 1987; 8, Roberts & Walters, 1988.

The mechanisms through which changes in nutrient uptake occur in pathogen infected plants may best be considered in the light of recent ideas concerning uptake in intact, healthy plants. That uptake from solutions of low nutrient concentrations can be described in terms equivalent to the Michaelis-Menten equation derived for enzyme kinetics has been known for over 30 years (see Clarkson, 1974). It is also well established that the kinetics of uptake at higher concentrations are more complex but may be described as the sum of two or more terms, each showing Michaelis-Menten type kinetics (Clarkson, 1974). However, despite the widespread acceptance of models based on multiphasic active uptake, there is also a range of evidence which reveals the existence of a linear component to uptake and Borstlap (1981) has shown that many examples of apparently multiphasic uptake are better described as the sum of one (or two) Michaelis-Menten type term(s) and a linear component. Since the saturable uptake mechanism reaches its maximum rate at relatively low external ion concentrations, the linear term assumes increasing importance as concentrations increase but may account for about 30% of uptake even at low concentrations (e.g. Kochian & Lucas, 1982). In considering mechanisms underlying the observed increases in ion uptake in infected plants it is worth noting that all existing

measurements were made on plants grown at high nutrient concentrations, in nutrient solutions, sand culture or, in the case of
Roberts & Walters (1988b), a peat potting compost. In all cases
the ion concentrations were much higher than that at which the
saturable component reaches its maximum and, especially in view
of the growth media used, passive uptake would be expected to
represent a major component of the total. The two uptake mechanisms show very different properties. From the point of view of
the potential effects of pathogens on uptake, perhaps the most important difference is that the mechanism which exhibits Michaelis-
Menten kinetics is active, being critically dependent on the energy
status of the roots whereas the linear, non-saturable component
of uptake is largely, but not entirely independent of energy supply (Kochian & Lucas, 1982). Active uptake may be regulated by
ion concentrations in the external medium or in the root tissue
(see Drew & Saker, 1984). However, in intact plants where both
active and passive uptake is occurring, regulatory mechanisms appear more complex and depend on the nutrient status of the whole
plant not just the roots. Drew & Saker (1984) have suggested that
the limiting step is the control of ion release from the symplast to
the xylem, which may be influenced not only by water flux to the
xylem, and hence by transpirational flux (see Pitman, 1982), but
also by nutrient concentrations in the phloem. Concentrations in
the phloem might in turn be determined by retranslocation from
the shoot and, with transpiration, provides a means by which root
and shoot activity might be integrated.

An insight into the seemingly contradictory changes in ion uptake and assimilate supply in infected plants may come from
studies with healthy plants, particularly those investigating the effects of low root temperatures on nutrient uptake (e.g. MacDuff
et al., 1986; Clarkson, Hopper & Jones, 1986; White, Clarkson &
Earnshaw, 1987), which have highlighted the capacity of the plant
to adapt to inhibition of root metabolism and the potential importance of passive uptake in this response. Active uptake shows a
temperature dependence similar to that of many metabolic systems and in this sense low root temperature is analogous to the
reduced assimilate supply to the roots of infected plants, Furthermore, intact plants grown with the root temperature lower than
that of the shoots show progressive increases in the shoot: root

ratio (S:R), as do plants infected with foliar pathogens (Ayres, 1984). The magnitude of the interaction between the effects of temperature on uptake and on S:R have been elegantly demonstrated by Clarkson *et al.* (1986). During an initial period after its imposition, low root temperature suppressed the rates of uptake of both nitrogen and phosphorus (Fig. 8.1) as expected from metabolic considerations. However, the duration of this phase was short (less than 9 days) and was proscribed by the low-temperature induced increase in S:R. Once adjustments of S:R had oc-

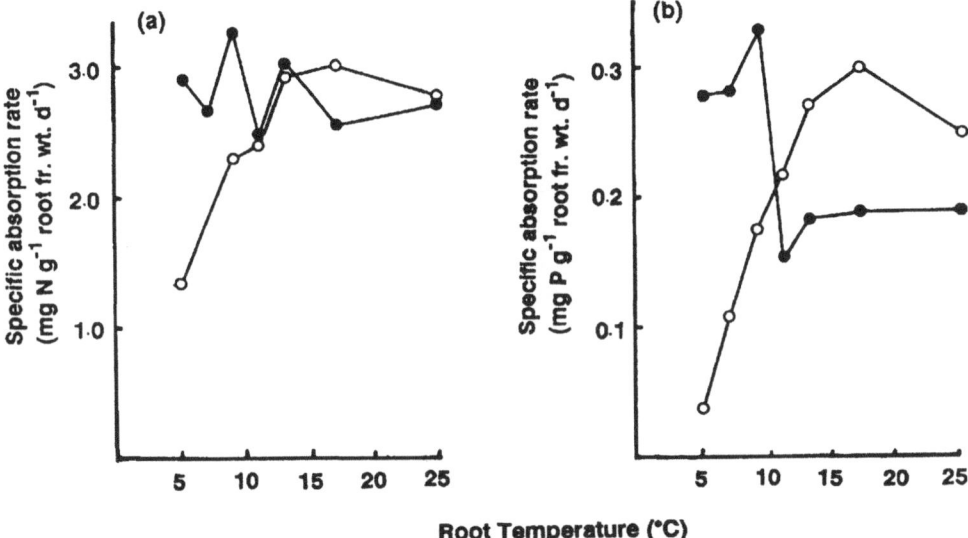

Fig. 8.1. Changes in the rates of uptake of (a) nitrogen and (b) phosphorus in barley grown at a range of root temperatures. In the initial period following the imposition of low root temperatures (open symbols) rates of uptake decline with temperature. However, after longer exposure to low root temperature (closed symbols) the ratio of shoot:root increased and the rates of uptake became insensitive to temperature. Modified from Clarkson *et al.* (1986).

curred, root temperatures down to 5°C caused no reduction in either nitrogen or phosphorus uptake (Fig. 8.1) and the linear, passive component of uptake assumed primary importance. As Clarkson *et al.* (1986) state 'the adjustment of root size ... had a profound effect on the rate of ion absorption per unit root weight'. The increase in S:R results in a requirement for each unit of root to supply relatively more resources to the shoot, i.e. 'shoot demand' on the roots is increased (see Drew & Saker, 1984; Clarkson *et al.*, 1986; White *et al.*, 1987).

The relationship between S:R and uptake per unit root dry weight (SAR_w) was investigated in rust-infected groundsel (Paul & Ayres, 1988b) where a close positive correlation was found to exist for both phosphorus and nitrogen (Fig. 8.2). Such results, together with those from healthy plants, suggest that the frequently observed increases in SAR_ws in infected plants may result from increased shoot demand, and hence that it is the passive component of uptake which is of paramount importance. Thus, increased nutrient uptake per unit root would be the expected outcome whenever foliar pathogens preferentially inhibit root growth. On this basis, it is those cases where uptake is inhibited by shoot infections that raise the most intriguing questions. However, variations in shoot demand may be based on more complex mechanisms than changes in S:R, as is evident, for example, in the deviation of the S:R—SAR_w relationship from linear in the case of nitrogen (Fig. 8.2a) and increased shoot demand as the basis of increased uptake subsumes a number of component mechanisms in addition to S:R. Two such mechanisms are modified recycling of nutrients from shoot to root and altered transpirational fluxes. The role of decreased shoot-root recycling in increased phosphorus and nitrogen uptake in mildewed barley has been acknowledged before (Walters & Ayres, 1981a; Murray & Ayres, 1986). Increased shoot demand for water, especially in rusts where transpiration is stimulated after sporulation, may be especially effective in increasing ion uptake if the hydraulic conductivity of the root system is increased (White *et al.*, 1987), as is indeed the case in mildewed barley (Walters & Ayres, 1982) and rusted beans (Tissera & Ayres, 1988).

Studies of healthy plants emphasise the potential importance of root morphology as well as physiology in responses to increased

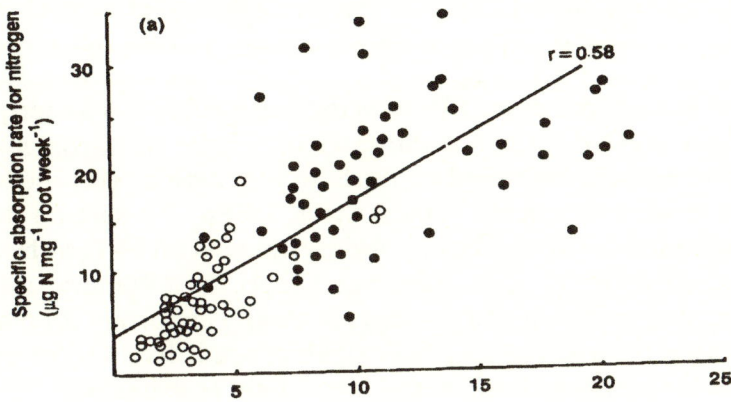

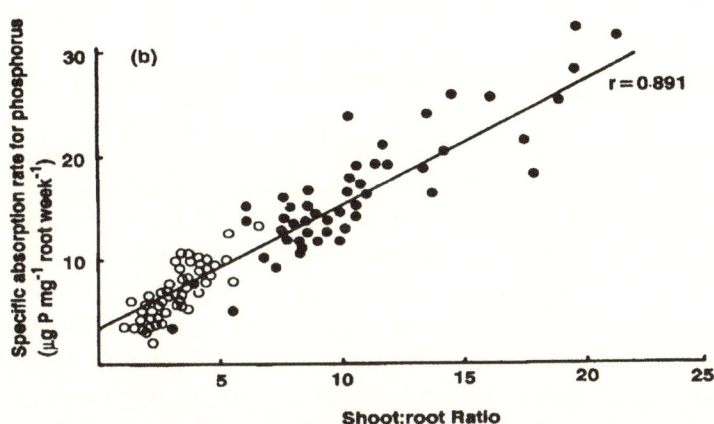

Fig. 8.2. The relationship between shoot:root ratio (S:R) and the rate of uptake of (a) nitrogen and (b) phosphorus on the basis of root dry weight in groundsel infected with *Puccinia lagenophorae* (closed symbols) or uninoculated (open symbols). The wide range of S:R were obtained by growing plants at different dilutions of a nutrient solution. The relationships between S:R and rates of uptake were not changed by infection. Thus, while the rates of uptake of both nitrogen and phosphorus were increased in infected plants, this increase was no greater than expected from their higher shoot:root ratio. Redrawn from Paul & Ayres (1988b).

shoot demand (MacDuff *et al.*, 1986). In rusted wheat (Martin & Hendrix, 1974), bean (Tissera & Ayres, 1988) and groundsel (Fig. 8.3 and see Paul & Ayres, 1986) the length of root per unit dry weight (specific root length, SRL) has been found to increase. In the last system, increased SRL was the key mechanism underlying increases in SAR_w, since if the rates of uptake of nitrogen and phosphorus were expressed on the basis of length no rust-induced increases were apparent (Paul & Ayres, 1988b). To date no study has examined the influence of powdery mildews on SRL, although reductions in root radius have been reported (Walters, 1981). While changes in root morphology are clearly important for plants in solution or sand culture, their significance is likely to be even greater for plants growing in soil in the field (see below).

Clearly, changes in shoot demand as an explanation for the effects of pathogens on nutrient uptake give scope for the complex patterns of change which have been observed, since it encompasses an integrated and dynamic response in both the physiology and morphology of the plant to the effects of infection. It remains to be seen whether increased shoot demand is of general importance in plants infected with foliar pathogens, especially as most of the hosts studied to date share the same nutritional strategy, that of species adapted to nutrient rich habitats (see Chapin, 1980). Leek (Roberts & Walters, 1988b) is an exception as it appears to share much in common with slow-growing wild species adapted to conditions of comparative nutrient poverty (Chapin, 1980). One outstanding feature of leek when infected by rust is the stability of root growth (Roberts & Walters, 1988b). This may reflect the alternative strategy in which S:R varies little in response to environmental stimuli (see Chapin, 1980). However, that changes in nutrient uptake in leek following rust infection were broadly similar to those of the other species studied may indicate that the pattern of responses to infection already reported may be common to a wide range of wild, as well as crop species.

Nutrient relations of plants infected with necrotrophic pathogens

Investigations of changes in nitrogen, phosphorus and sulphur caused by necrotrophs have been confined to pathogens infecting the roots or vascular tissues. For example, root infection by

Phytophthora cinnamomi was found to decrease concentrations of sulphur in the shoots of *Isopogon ceratophylla* (Weste *et al.*, 1980) and *Eucalyptus* spp . (Cahill *et al.*, 1986) and of phosphorus and sulphur in the leaves of Avocado (Whiley *et al.*, 1987). Severe take-all disease of cereals, which is caused by root infection by *Gaeumannomyces graminis*, also causes reductions in shoot phosphate (e.g. Fitt & Hornby, 1978). Such reductions in shoot concentrations may seem the inevitable consequence of root decay. However, in many cases visual assessment of root symptoms are unreliable guides to changes in root function as the cortical tissues of older roots may be lost without impairment of

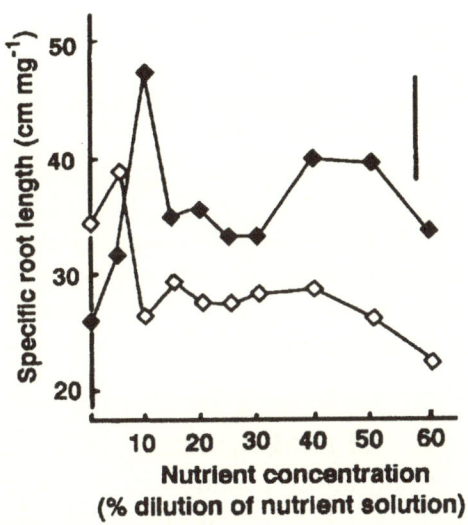

Fig. 8.3. The effect of *Puccinia lagenophorae* on the specific root length (SRL, root length per unit root dry weight) of groundsel grown over a wide range of dilutions of a standard nutrient solution. The SRL of rusted plants (closed symbols) was substantially higher than that of controls (open symbols) at all but extremely low nutrient concentrations. Redrawn from Paul & Ayres (1986a). Points are means of five plants and the bar represents the LSD at $P = 0.05$.

root function (Clarkson & Sanderson, 1971). In wheat a number of pathogens which caused extensive discoloration of the root cortex had no effect on shoot phosphate concentrations (Fitt & Hornby, 1978). Similarly, *G. graminis* causes no reduction in phosphate uptake, despite extensive discoloration of the cortex, until phloem tissues are colonised (Clarkson *et al.*, 1975; Fitt & Hornby, 1978). Once the phloem is invaded by *G. graminis*, regions of the root distal to the site of infection rapidly show decreased phosphorus uptake as a result of the acute assimilate starvation of these tissues (Fitt & Hornby, 1978). The cessation of phloem transport beyond sites of infection by *G. graminis* (or by *Aureobasidium bolleyi*) is very different from the chronic assimilate deprivation found in roots of plants with biotrophic infections of the shoot, and many of the compensatory mechanisms discussed above are unlikely to occur. However, some compensation does take place in regions proximal to *G. graminis* where phosphorus uptake may increase slightly (Clarkson *et al.*, 1975), perhaps because shoot demand on these still functioning tissues is increased. When take-all is mild, compensatory increases in uptake by uninfected roots may buffer the plant from deleterious changes occurring in infected roots (Hornby & Goring, 1972). Compensatory changes in pathogen-free tissues may not occur in all root infections, for example, *Cochliobolus sativus* and *Fusarium culmorum*, caused decreased concentrations of phosphorus in wheat shoots in the absence of phloem colonization (Fitt & Hornby, 1978), possibly because uptake was impaired by phytotoxins.

Nutrient uptake and the growth and development of root systems

In nature it is often limitations on nutrient flux in the soil, rather than in the root, which is the paramount factor limiting uptake, especially of the more immobile ions, such as phosphate (Nye, 1977; Robinson, 1986). Thus, in soil those physiological characteristics of the root discussed above, which stimulated nutrient uptake by rusted or powdery mildew-infected plants in solution or sand culture may be less important than morphological characteristics, both of individual roots and of the complete root system, which allow the plant to explore the maximum volume of soil most efficiently (Fitter, 1987).

Of the morphological features of individual roots, the diameter is perhaps the most important, influencing the efficiency of uptake through effects on both gross exploitation of the soil and its cost, in terms of dry weight investment (Fitter, 1987). Nutrient inflow per unit root length is greater in roots with greater diameters but conversely the volume of soil exploited by a given volume (or weight) of root is greater for finer roots (Fitter, 1987). Thus, especially for immobile ions, the return for a given investment of dry weight is greatest when root radius is low (when SRL is high). The reductions in root diameter, or increase in SRL, found in plants infected with rusts or mildews have already been described (Fig. 8.3). Thus, infected plants may be able to compensate for the limited assimilate available for root growth by enhancing the efficiency with which dry weight is exploited for ion uptake. The effective root diameter for uptake, especially of immobile ions, may be increased by the production of root hairs or by mycorrhizal associations which increase the zone of soil exploited with comparatively small investments in dry weight (Clarkson, 1985). The length of root hairs has been found to increase in plants subjected to low root temperatures (MacDuff et al., 1986) and might also be expected to form part of a complex of adaptions in root morphology in response to foliar biotrophs. However, the effects of foliar pathogens on root hairs remain wholly uninvestigated.

The absence of any information concerning interactions between foliar pathogens and mycorrhizas is another major lacuna in our understanding of nutrient relations of infected plants, especially in view of the increasing awareness of the key role of mycorrhizas in natural systems (see Chapters 9, 10 & 11). Mycorrhizas are known to be sensitive to reduced photosynthesis, for example due to low light intensities, indeed mycorrhizal infection may reduce plant growth when irradiance is low and soil phosphorus content high (e.g. Son & Smith, 1988). Foliar pathogens, like low irradiance, will reduce the supply of photosynthate to the roots and hence it might be expected that shoot infection would reduce or reverse the normal stimulation of plant growth resulting from mycorrhizal infection. Indeed, the effects of foliar parasites may be greater than those of reducing light intensity, since infection sites and mycorrhizas will represent competing sinks for the limited supply of photoassimilate.

While it is the morphology of individual roots which has the greatest influence on immobile ions, it is the architecture of whole root systems which is paramount for mobile nutrients such as nitrate (Fitter, 1987). Root system architecture is complex and as yet poorly understood, despite major advances in its study in recent years (Fitter, 1987). Architecture is determined by metric components, such as the frequency and length of branches and branch angle, and non-metric aspects or topology (Fitter, 1987). Analysis of root branching patterns in infected plants are limited to two concerning powdery mildew of barley and one with *Puccinia recondita tritici* on wheat and, in all three, only metric components of root architecture were considered. In the former, Visarova & Minarcic (1974) found that the seminal roots were shorter in infected plants than in controls and Walters & Ayres (1981b) found that the lengths of root axes and primary laterals were reduced in both the nodal and seminal roots of mildewed seedlings but that the length of secondary laterals was substantially increased. While the total production of root branches was found to be reduced in both studies, it was evident from the results of Walters & Ayres (1981b) that the number of branches per centimetre was greater in infected than control plants. This more compact and relatively more branched root system may compromise the ability of mildewed plants to absorb efficiently mobile nutrients, such as nitrate, in the field. By contrast, stripe rust reduced branching in wheat roots both in seedlings and mature plants, especially in proximal parts of the root system (Martin & Hendrix, 1974). Thus, it would seem that in rusted wheat the limited resources available for root growth are directed towards maximising the volume of soil explored. The contrasting changes in the root branching patterns of plants infected with rusts and powdery mildews is a paradox but may reflect a greater demand of rusted plants for mobile soil resources, water as well as mobile nutrients. Clarification of such changes awaits the application of newly available techniques of analysis to a wider range of host-parasite combinations.

Finally, it must be noted that superimposed on the nutrient foraging strategy of the whole root system is the capacity of roots to proliferate locally in areas of nutrient enrichment (e.g. Robinson & Rorison, 1983). This capacity may be critical in soil, where areas of localised enrichment often occur against a low background nutrient supply, and may represent an alternative strategy to decreased S:R (Crick & Grime, 1987). Such mechanisms remain poorly understood even in healthy plants but preliminary studies have shown that, when grown with the root system divided between separate containers of nutrient-rich and nutrient-poor media, rusted groundsel showed preferential root growth in the former so that root dry weight was 3·7 times that in the latter (N. D. Paul, unpublished). Control groundsel showed less differential root growth (2·6:1) but a lower S:R (3·4 as opposed to 4·2). Thus, rusted groundsel was able to compensate for an increased S:R by flexible partitioning within the root system. The capacity of infected plants to restrict reduced root growth to areas of resource depletion has been shown for water in pea (*Pisum sativum*) infected with powdery mildew (Ayres, 1981) and rusted bean (Tissera, 1986). Thus, increases in S:R in infected plants may be a poor guide of the functional capacity of the root system in natural soils.

Nitrogen, Phosphorus and Sulphur and populations of host and pathogen

It is evident, from numerous reports in the literature, that the severity of crop infection varies substantially in response to applications of fertilizers, especially nitrogen. However, as stated in the review by Zentmyer & Bald (1977), 'the reactions to nutrition are complex and cannot be described by simple rules or principles'. Thus, while the application of nitrogen fertilizers is generally considered to increase the severity of many pathogens in a wide range of crops, the form of nitrogen applied is also critical (Huber & Watson, 1974). In cereal crops, the severity of rusts and powdery mildews is generally increased by the application of nitrate, the preferred nitrogen source of the host, but may be decreased by ammonium. Conversely, of the examples cited by Huber & Watson (1974) there are many where the application of the preferred nitrogen source of the host decreases the severity of diseases caused by necrotrophic fungi. This contrast may reflect differing pathogen responses to host vigour, strong growth often favouring infection and growth of biotrophs (see below) but increasing the capacity of the host to replace tissues destroyed by necrotrophs.

Whether fertilizers alter disease severity by a direct effect of host nutrition on the growth and reproduction of the pathogen or through changes in crop growth and canopy structure, which may affect spore dispersal and the microclimate for infection, has been the subject of debate for at least three decades (e.g. Last, 1962). The relative importance of the two mechanisms appears to vary between pathogens. In the case of *Rhynchosporium secalis* applications of nitrogen in the glasshouse, where changes in canopy characteristics were insignificant, inhibited lesion production (Jenkyn & Griffiths, 1976). Applications of nitrogen fertilizers result in increased leafblotch in the field, where indirect effects of nitrogen on canopy structure thus appear to be paramount (Jenkyn, 1977). By contrast, Bainbridge (1974) showed in the glasshouse that, while conidiospore germination and appressorial formation in *Erysiphe graminis* f. sp. *hordei* were unaffected by nitrogenous fertilizers, the percentage of appressoria which gave rise to infections, the size of colonies and the number of conidia produced per unit area of the pustule were all significantly increased, and the generation interval reduced. Similarly, Tomesh & Structmeyer (1979) found that nitrogen deficiency in cucumber inhibited the area of colonies of *E. cichoracearum*. Thus, stimulation of powdery mildews by nitrogenous fertilizers in the field is probably due to the direct effects of nitrogen on the pathogen. Sulphur stress had no effect on the colony area of *E. cichoracearum* on cucumber but conidiophore production per colony was increased: severe phosphorus deficiency decreased both these measures of pathogen growth (Tomesh & Structmeyer, 1979). The very low level of phosphorus supplied by Tomesh & Structmeyer (1979) may explain the contrast between this report and that from in the field where the application of phosphate fertilizers reduced the severity of mildew infection in barley (Last, 1962). In addition to modifying the growth and spread of the pathogen, nutrient supply may also alter the effects of a given amount of disease on the growth and physiology of the host. *Puccinia lagenophorae* caused relatively smaller reductions in leaf expansion, dry weight growth, relative growth rate and net assimilation rate in groundsel subjected to severe nutrient deficiency than in plants freely supplied with nutrients, even though there were no differences in the severity of infection (Paul & Ayres, 1986b).

Whatever the underlying mechanisms, interactions between diseases and fertilizers, especially nitrogen, have major consequences for crop growth and yields. Powdery mildews in particular have minimal effects on cereal growth, total yield and its components when no nitrogen is applied (Jenkyn, 1977; Darwinkel, 1980). One important consequence of this response is that in the absence of disease control increases in crop yields following fertilizer applications are greatly reduced (Fig. 8.4a). The same principle can be restated to the effect that fungicide application is less necessary when little or no nitrogen is applied. Similarly, Shaner & Finney (1977) observed that slow-mildewing resistance in wheat reduced the severity of infection only when nitrogen was applied at high rates. It seems that the vulnerability of agricultural crops to damaging disease, and the concomitant requirement for efficient control, may be partly the result of the high inputs of fertilizers typical of modern agriculture.

The abundant supply of nutrients available in such crops is obviously remote from that of most natural plant communities, which are often subject to some degree of nutrient limitation (Fitter, 1987). Information on pathogen-nutrient interactions, comparable to that for cereals is completely absent for populations of non-crop species, indeed, our understanding of the effects of pathogens on the population biology of wild plants is extremely limited (Burdon, 1987). However, some idea of the potential interaction between nutrient supply and pathogens comes from the Park Grassland Experiment at Rothamsted, in which areas of grassland have been subjected to defined levels of nitrogen application for almost 100 years. When *Anthoxanthum odoratum* populations adapted to these different nitrogen regimes were grown under uniform conditions in a garden plot (Snaydon & Davies, 1971), there was a close, negative correlation between nitrogen supply in the native habitat and the severity of infection by powdery mildew (Fig. 8.4b). Snaydon & Davies (1971) concluded that ecotypes adapted to abundant nitrogen were more resistant to mildew because they had been subject to more intense selection over many generations by the greater severity of infection under high nitrogen conditions. This result supports the conclusion that the variation in the severity of disease observed in crops will also be apparent in natural systems. If this is indeed the case then the

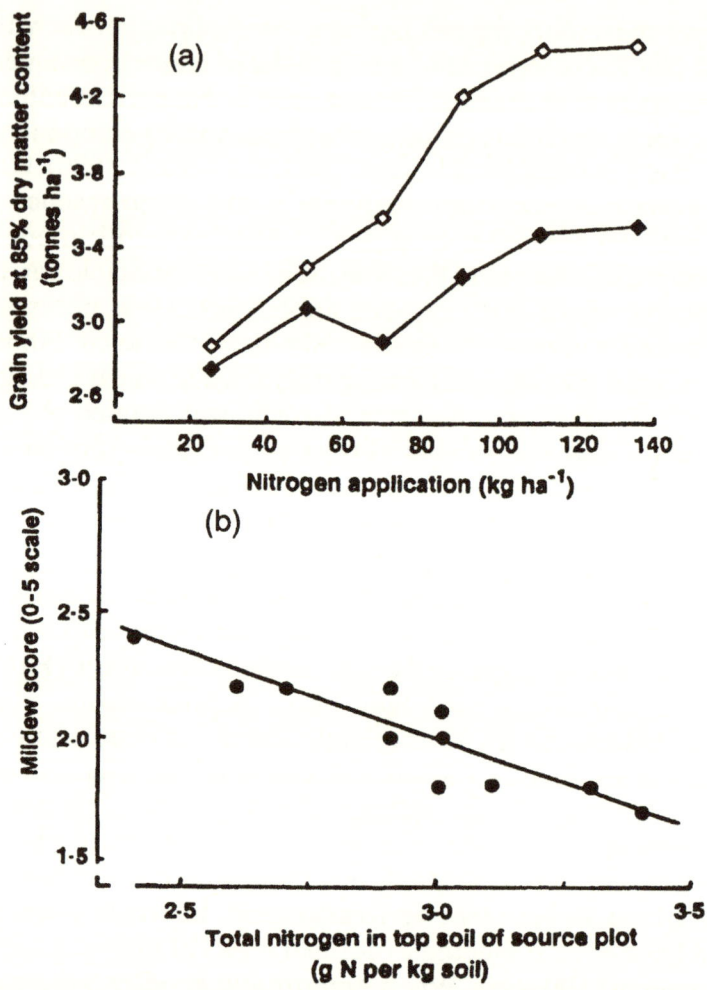

Fig. 8.4. (a) The response of barley grain yield to nitrogen fertilizer in the presence (open symbols) and absence (closed symbols) of powdery mildew control. Without adequate control of mildew, the response of yield to nitrogen application is small. Redrawn from Jenkyn (1977). (b) The severity of powdery mildew infection on forms of *Anthoxanthum odoratum* adapted to differing levels of soil nitrogen. When grown under standard garden conditions, forms adapted to high levels of soil nitrogen were less severely infected than those adapted to a limited nitrogen supply. The apparent greater resistance of forms from nitrogen-rich sites may be attributable to a more intense selection for resistance due to the greater prevalence of mildew. Redrawn from Snaydon & Davies (1971).

lower nutrient supply and tissue concentrations (Chapin, 1980) of wild plants might be expected to limit the prevalence and effect of phytopathogens in natural vegetation, especially in nutrient poor habitats, with major implications not only for the dynamics but also the genetics of populations of wild species.

In natural vegetation in particular, the effects of nutrient supply on competition between the host and uninfected neighbours may be of great importance. For plants in mixtures far more than for spaced individuals any reduction in the size of the root system of infected plants will be profoundly damaging and the relative stability of root length may be critical in maintaining the capacity of the infected host to compete for nutrients. However, since ion uptake seems less sensitive to foliar pathogens than shoot growth and activity, infected plants might be expected to compete comparatively better when nutrients are limiting.

Conclusions

As already noted, our knowledge of the effects of pathogens, especially biotrophs, on host nutrient relations is founded almost solely on studies of plants grown in artificial media in controlled environments. Such methods are not inappropriate for elucidating nutrient accumulation or cycling within the host and the increased nutrient content of infected tissues found under such conditions have also been confirmed in a few field studies. However, our dependence on results obtained under such artificial conditions will inevitably bias our understanding of nutrient uptake. The rates of uptake of nitrogen or phosphorus per unit root dry weight are consistently increased in plants infected with rusts or powdery mildews when grown in solution or sand culture. I have proposed here that such increases are driven by the greater 'shoot demand' (often increased S:R) of infected plants and that it is principally the non-saturable, 'passive' component of ion uptake which is enhanced. The mechanism and regulation of non-saturable uptake is not fully characterised in healthy plants but factors such as transpiration and recycling from the shoot are important and changes in these processes in infected plants are consistent with increased passive uptake. Other, potentially important factors such as counter-ion transport and the ionic balance of the xylem await investigation in infected plants. Regardless of the underlying mechanisms, there remains some doubt as to whether such increases in ion uptake detected under laboratory conditions will also occur under natural conditions in the field. There is an almost complete absence of experimental data concerning the effects of pathogens on many aspects of plant physiology and morphology which determine nutrient uptake in the field. These include aspects of root system growth and architecture, already noted, the effects of infection on nutrient leaching from leaves, on recycling from senescent leaves, and on specific physiological mechanisms which increase nutrient availability at the root surface in soil, all of which may be important to nutrient budgets in the field (see Chapin, 1980). In addition, the increasing evidence of the key role of mycorrhizas, rhizobia and free-living nitrogen-fixing bacteria in the nutrient relations of natural systems, suggests that interactions between mutualistic and plant-parasitic microorganisms may form yet another unknown but important aspect of pathogen-nutrient interactions in wild species.

In view of these many lacunae speculation concerning the overall role of nutrient-pathogen interactions is premature. Nevertheless, results from crops show that interactions not only occur in the field but that they may have major consequences for growth and yield. Furthermore, that pathogen-nutrient interactions in nature may prove to be diverse and far reaching is already evident, for example from nutrient-mediated pathogen-herbivore interactions (Peining, 1974; White, 1984). Clearly, studies of plant nutrition form part of the much wider challenge presented to plant pathology and ecology by our ignorance of the place of pathogens in natural vegetation.

Acknowledgements I am grateful to NERC for financial support and to D. R. Walters, A. J. S. Murray and especially P. G. Ayres for their comments on the manuscript and for numerous valuable discussions.

References

Ahmad, I., Farrar, J. F. Owera, S. A. P. & Whitbread, R. (1982). The distribution of five major nutrients in barley plants infected with brown rust. *Physiological Plant Pathology*, **21**, 335-346.

Ahmad, I., Farrar, J. F. & Whitbread, R. (1984). Fluxes of phosphorus in leaves of barley infected with brown rust. *New Phytologist*, **98**, 361-375.

Ayres, P. G. (1981). Root growth and solute accumulation in pea in response to water stress and powdery mildew. *Physiological Plant Pathology*, **19**, 169-180.

Ayres, P. G. (1984). Effects of infection on root growth and function; consequences for plant nutrient and water relations. In *Plant Diseases: Infection Damage and Loss*, eds. R. K. S. Wood & G. J. Jellis, pp. 105-117. Blackwell, Oxford.

Ayres, P. G. & Paul, N. D. (1986). Foliar pathogens alter the water relations of their hosts with consequences for both host and pathogen. In *Water, Fungi and Plants*, eds, P. G. Ayres & L. Boddy, pp. 267-285. Cambridge University Press, Cambridge.

Bainbridge, A. (1974). Effect of nitrogen nutrition of the host on barley powdery mildew. *Plant Pathology*, **23**, 160-161.

Borstlap, A. C. (1981). Invalidity of the multiphasic concept of ion absorption in plants. *Plant, Cell and Environment*, **4**, 189-195.

Burdon, J. J. (1987). *Diseases and Plant Population Biology*. Cambridge University Press, Cambridge.

Cahill, D., Wookey, C., Weste, G. & Rouse, J. (1986). Changes in the mineral content of *Eucalyptus marginata* and *E. calophylla* grown under controlled conditions and inoculated with *Phytophthora cinnamomi*. *Journal of Phytopathology*, **116**, 18-29.

Chapin, F. S. (1980). The mineral nutrition of wild plants. *Annual Review of Ecology and Systematics*, **11**, 233-260.

Clarkson, D. T. (1974). *Ion Transport and Cell Structure in Plants*. McGraw-Hill, London.

Clarkson, D. T. (1985). Factors affecting mineral nutrient acquisition by plants. *Annual Review of Plant Physiology*, **36**, 77-115.

Clarkson, D. T. & Sanderson, J. S. (1971). Relationship between the anatomy of cereal roots and the absorption of nutrients and water. *Agricultural Research Council Letcombe Laboratory Report, 1970*, 16-25.

Clarkson, D. T., Drew, M. C., Ferguson, I. B. & Sanderson, I. (1975). The effect of take-all fungus *Gaeumannomyces graminis* on the transport of ions by wheat plants. *Physiological Plant Pathology*, **6**, 75-84.

Clarkson, D. T., Hopper, M. J. & Jones, L.H.P. (1986). The effect of root temperature on the uptake of nitrogen and the relative size of the root system in *Lolium perenne*. I. Solutions containing both NH_4^+ and NO_3^-. *Plant, Cell and Environment*, **9**, 535-545.

Clement, C. R., Hopper, M. J., Canaway, R. J. & Jones, L. H. P. (1974). A system for measuring the uptake of ions by plants from flowing solutions of controlled composition. *Journal of Experimental Botany*, **25**, 81-99.

Comhaire, F. (1963). Study of the absorption and repartition of phosphorus in *Triticum vulgare* infected by *Erysiphe graminis*. *Leujeunia N.S.*, **15**, 1-87.

Crick, J. C. & Grime, J. P. (1987). Morphological plasticity of and mineral nutrient capture in two herbaceous species of contrasted ecology. *New Phytologist*, **107**, 403-414.

Darwinkel, A. (1980). Grain production of winter wheat in relation to nitrogen and diseases. II. Relationship between nitrogen dressings and mildew infection. *Journal of Agronomy and Crop Science*, **109**, 309-317.

Drew, M. C. & Saker, L. R. (1984). Uptake and long-distance transport of phosphate, potassium and chloride in relation to internal ion concentrations in barley: evidence for non-allosteric regulation. *Planta*, **160**, 500-507.

Fitt, B. D. L. & Hornby, D. (1978). Effects of root infecting fungi on wheat transport processes and growth. *Physiological Plant Pathology*, **13**, 335-346.

Fitter, A. H. (1987). An architectural approach to the comparative ecology of plant root systems. *New Phytologist*, **106** (Suppl), 61-77.

Fric, F. (1978). Absorption and translocation of phosphate in barley infected with powdery mildew. *Phytopathologische Zeitschrift*, **91**, 23-32.

Hornby, D. & Goring, C. A. I. (1972). Effects of ammonium and nitrate nutrition on take-all disease of wheat in pots. *Annals of Botany*, **70**, 225-231.

Huber, D. M. & Watson, R. D. (1974). Nitrogen form and plant disease. *Annual Review of Phytopathology*, **12**, 139-165.

Jenkyn, J. F. (1977). Nitrogen and leaf diseases of spring barley. In *Fertiliser Use and Plant Health, Proceedings of the 12th Colloquium of the International Potash Institute*, ed. K. Mengel, pp. 119-128. International Potash Institute: Berne, Switzerland.

Jenkyn, J. F. & Griffiths, E. (1976) Some effects of nutrition on *Rhynchosporium secalis*. *Transactions of the British Mycological Society*, **66**, 329-332.

Kochian, L. V. & Lucas, W. J. (1982). Potassium transport in corn roots. I. Resolution of kinetics into a saturable and linear component. *Plant Physiology*, **70**, 1723-1731.

Lambers, H. (1980). The physiological significance of cyanide-resistant respiration in higher plants. *Plant, Cell and Environment*, **3**, 293-302.

Last, F. T. (1962). Effects of nutrition on the incidence of barley powdery mildew. *Plant Pathology*, **11**, 133-135.

MacDuff, J. H., Wild, A., Hopper, M. J. & Dhanoa, M. S. (1986). Effects of temperature on parameters of root growth relevant to nutrient uptake: measurements on oilseed rape and barley grown in flowing nutrient solution. *Plant and Soil*, **94**, 321-332.

Martin, P. J. & Hendrix, J. W. (1974). Anatomical and physiological responses of Baart wheat roots affected by stripe rust. *Washington Agricultural Experimental Station, Technical Bulletin*, No. 77.

Murray, A. J. S. & Ayres, P. G. (1986). Uptake and translocation of nitrogen by mildewed barley seedlings. *New Phytologist*, **104**, 355-365.

Nye, P. H. (1977). The rate limiting step in plant nutrient absorption. *Soil Science*, **123**, 292-297.

Paul, N. D. & Ayres, P. G. (1986a). The effects of nutrient deficiency and rust infection on the relationship between root dry weight and length in groundsel (*Senecio vulgaris* L.). *Annals of Botany*, **57**, 353-360.

Paul, N. D. & Ayres, P. G. (1986b). The effects of rust (*P. lagenophorae*) on the growth of groundsel (*Senecio vulgaris* L.) cultivated under a range of nutrient concentrations. *Annals of Botany*, **58**, 321-331.

Paul, N. D. & Ayres, P. G. (1988a). Nutrient relations of groundsel (*Senecio vulgaris* L.) infected by rust (*Puccinia lagenophorae* Cooke) at a range of nutrient concentrations. I. Concentrations, contents and distribution of N, P and K. *Annals of Botany*, **61**, 489-498.

Paul, N. D. & Ayres, P. G. (1988b). Nutrient relations of groundsel (*Senecio vulgaris* L.) infected by rust (*Puccinia lagenophorae* Cooke) at a range of nutrient concentrations. II. Uptake of N, P and K and shoot-root interactions. *Annals of Botany*, **61**, 499-506.

Penning de Vries, F. W. T. (1975). The cost of maintenance processes in plant cells. *Annals of Botany*, **39**, 77-92.

Peining, L. G. (1972). Effects of leaf rust on nitrate in rye. *Canadian Journal of Plant Science*, **52**, 842-843.

Pitman, M. G. (1982). Transport across plant roots. *Quarterly Review of Biophysics*, **15**, 481-554.

Raggi, V. (1974). Free and protein amino acids in rusted bean leaves. *Phytopathologische Zeitschrift*, **81**, 289-300.

Raggi, V. (1976). Amino acids in mycelium of *Sphaerotheca pannosa* var *persicae* and in infected and surrounding tissues of peach leaves. *Phytopathologia Mediterranea*, **15**, 110-114.

Roberts, A. M. & Walters, D. R. (1988a). Nitrogen assimilation and metabolism in rusted leek leaves. *Physiological and Molecular Plant Pathology*, **32**, 229-235.

Roberts, A. M. & Walters, D. R. (1988b). Root:shoot interrelationships in leeks infected by the rust, *Puccinia allii* Rud.: growth and nutrient relations. *New Phytologist*, in press.

Robinson, D. (1986). Limits to nutrient inflow in roots and root systems. *Physiologia Plantarum*, **68**, 551-559.

Robinson, D. & Rorison, I. H. (1983). A comparison of the responses of *Lolium perenne* L., *Holcus lanatus* L. and *Deschampsia flexuosa* (L.) Trin. to a localised supply of nitrogen. *New Phytologist*, **94**, 263-273.

Shaner, G. & Finney, R. E. (1977). The effect of nitrogen fertilisation on the expression of slow-mildewing resistance in Knox wheat. *Phytopathology*, **67**, 1051-1056.

Snaydon, R. W. & Davies, M. S. (1971). Rapid population differentiation in a mosaic environment. II. Morphological variation in *Anthoxanthum odoratum*. *Evolution*, **26**, 390-405.

Son, C. L. & Smith, S. E. (1988). Mycorrhizal growth responses: interactions between photon irradiance and phosphorus nutrition. *New Phytologist*, **108**, 305-314.

Stribley, D. P., Tinker, P. B. & Rayner, J. H. (1980). Relation of internal phosphorus concentration and plant weight in plants infected with vesicular-arbuscular mycorrhizas. *New Phytologist*, **86**, 261-266.

Tissera, P. (1986). Physiological interactions between rust (*Uromyces viciae-fabae*) and water-stress in faba bean (*Vicia faba*). Ph.D. Thesis, University of Lancaster.

Tissera, P. & Ayres, P. G. (1988). Hydraulic conductivity and anatomy of roots of *Vicia faba* L. infected by *Uromyces viceae-fabae* (Pers.) Shroet. *Physiological and Molecular Plant Pathology*, **32**, 199-207.

Tomesh, R. J. & Structmeyer, B. E. (1979). Effects of nutrient stress on elemental concentrations and powdery mildew growth in cucumber. *Journal of the American Society for Horticultural Science*, **104**, 70-74.

Visarova, G. & Minarcic, P. (1974). The influence of powdery mildew upon the cytokinins and root morphology of barley roots. *Phytopathologische Zeitschrift*, **81**, 49-55.

Walters, D. R. (1981). Root function in mildewed barley. Ph.D. Thesis, University of Lancaster.

Walters, D. R. (1985). Shoot:root interrelationships: the effects of obligately biotrophic fungal pathogens. *Biological Reviews*, **60**, 47-79.

Walters, D. R. & Ayres, P. G. (1980). Effects of powdery mildew disease on the up-take and metabolism of nitrogen by roots of infected barley. *Physiological Plant Pathology*, **17**, 369-379.

Walters, D. R. & Ayres, P. G. (1981a). Phosphate uptake and transport by roots of powdery mildew infected barley. *Physiological Plant Pathology*, **18**, 195-205.

Walters, D. R. & Ayres, P. G. (1981b). Growth and branching patterns of roots of powdery mildew infected barley. *Annals of Botany*, **47**, 159-163.

Walters, D. R. & Ayres, P. G. (1982). Water movement through root systems excised from healthy and mildewed barley: relationship to phosphate transport. *Physiological Plant Pathology*, **20**, 275-284.

Weste, G., Chaudri, M. A., Haukka, M. & Vithanage, K. (1980). The influence of root infection by *Phytophthora cinnamomi* on mineral content of certain native species. *Phytopathologische Zeitschrift*, **99**, 205-214.

Whiley, A. W., Pegg, K. G., Saranah, I. B. & Lagdon, P. W. (1987). Influence of *Phytophthora* root rot on mineral nutrient concentrations in avocado leaves. *Australian Journal of Experimental Agriculture*, **27**, 173-177.

White, P. J., Clarkson, D. T. & Earnshaw, M. J. (1987). Acclimation of potassium influx in rye (*Secale cereale*) to low root temperatures. *Planta*, **171**, 377-385.

White, T. C. R. (1984). The abundance of invertebrate herbivores in relation to the availability of nitrogen in stressed food plants. *Oecologia*, **63**, 90-103.

Yarwood, C. E. & Jacobson, L. (1954) Accumulation of chemicals in diseased areas of leaves. *Phytopathology*, **45**, 43-48.

Zentmyer, G. A. & Bald, J. G. (1977). Management of the environment. In *Plant Disease. I. How Disease is Managed*, eds. I. G. Horsfall & E. B. Cowling, pp. 121-144. Academic Press, New York.

Chapter 9

The nitrogen nutrition of mycorrhizal fungi and their host plants

D. J. Read, J. R. Leake & A. R. Langdale

Department of Plant Sciences, University of Sheffield,
Sheffield S10 2TN, UK

Introduction

A number of reviewers have dealt with mycorrhizas and nitrogen nutrition in recent years (Bowen & Smith, 1981; Alexander, 1983; France & Reid, 1983; Harley & Smith, 1983; Martin *et al.*, 1987). That being the case, justification for a further examination of the subject can be made only if recent advances in some part either of the topic itself, or of a related field, have been of sufficient interest to necessitate its re-appraisal. Some justification for such a re-appraisal now exists on both counts. Within the topic there is increasing evidence of the importance of ericoid and ectomycorrhizal fungi in providing access to organic sources of nitrogen. Since the host plants themselves have little or no access to these resources their fungal associates are seen potentially to play an even more crucial role in the mutualism than has hitherto been accepted (Read, 1987). In the related field of plant physiological ecology it has become widely acknowledged that availability of nitrogen is the key factor determining the photosynthetic capabilities and hence productivities of natural plant communities (Field & Mooney, 1986). This is because the maximum photosynthetic rates (P_{max}) of C_3 plants, under field conditions with saturating light intensities and at normal CO_2 levels, are determined by the concentration of nitrogen in the leaf. Summarising data from studies of a wide range of natural plant communities Field & Mooney (1986) have shown that practically all the data points relating P_{max} as a function of nitrogen cluster around a single line of regression. This suggests a fundamental relationship that is relatively insensitive to differences among species or growth conditions and demonstrates that over a wide diversity of C_3 plants photosynthetic capacity is strongly regulated by leaf nitrogen content. Physiologically, the dependance of photosynthetic rate upon nitrogen content arises directly from the use of the element in the carbon fixing enzymes which contain a large proportion of total leaf nitrogen. It is this interdependence of nitrogen and P_{max} which, on a world scale, makes nitrogen the element which most limits plant growth.

In all cases, except those in which prokaryotic organisms fix dinitrogen gas, the element becomes available to plants through the activities of heterotrophic populations which recycle organic matter. In view of the importance of nitrogen for autotrophs it can be predicted that any environment in which the turnover of nitrogen is restricted will generate powerful pressures leading to the selection of associations between autotrophs and those heterotrophs which are sufficiently specialised to maintain access to nitrogen supplies. In an earlier review (Read, 1985) a relationship was recognised between increasing latitude or altitude, accumulation of organic matter and the occurrence of distinct mycorrhizal types. It was suggested that these types have been selected primarily on the basis of their ability to provide the autotroph with access to the particular type of resource which is most likely to be limiting the growth in each part of the environmental gradient. In the warmer parts of the gradient mineralisation of organic matter takes place rapidly and the relatively high pH characteristic of soils in such systems enables nitrification to proceed with the release of mobile nitrate ions. Under these circumstances the relative immobility of the phosphate ion leads to the development of phosphorus depletion zones around roots, and all the evidence suggests that the vesicular-arbuscular (VA) type of infection found most frequently in this type of environment is primarily involved with provision of increased access to PO_4 ions (see Chapters 10 & 11). While there are reports of enhancement of plant nitrogen content (Possingham & Groot Obbink, 1971), of nitrogen inflow rates (Smith et al., 1986) in VA infected plants as well as of nitrogen transfer through VA mycorrhizal mycelium to the plant (Ames et al., 1983) such functions are justifiably considered to be of secondary importance in this mycorrhizal type and are not considered further here.

In cooler climates, in contrast, where organic matter is turned over more slowly, nitrogen mobilization becomes progressively more restricted. Thus, while low phosphorus availability may still be a factor limiting plant growth, access to nitrogen becomes increasingly important as greater proportions of the element are sequestered in organic combinations. In quantitative terms the plant requirement for nitrogen is approximately ten times greater than that for phosphorus so that any restriction of access to the former very rapidly begins to have adverse effects upon growth and physiology. Evidence will be presented here which confirms that many of those mycorrhizal fungi which are characteristic of habitats in which organic matter accumulates have at least some of the attributes which would enable them to gain access to the nitrogenous components of such materials. Prior to a consideration of these aspects, however, our knowledge of the uptake and assimilation of mineral nitrogen is briefly surveyed.

Utilization and assimilation of inorganic nitrogen

Utilization of inorganic nitrogen by mycorrhizal fungi in pure culture

As indicated above, soils in which the most extensive development of ericoid and ectomycorrhizal systems occur have high organic content, and, partly as a result of the accumulation of this organic material, a predominance of carboxyl groups gives rise to an acid pH. Under these circumstances nitrification is inhibited and ammonium is the main form of mineral nitrogen. Most ectomycorrhizal fungi readily use ammonium as a nitrogen source. In the less extreme parts of the environmental gradient, however, nitrification does take place and it is perhaps not surprising, therefore, that the ericoid endophyte and many ectomycorrhizal fungi can utilise nitrate as well as ammonium.

In a comparative analysis of four ecto-fungi France & Reid (1984) obtained higher dry weight yield of *Pisolithus tinctorius* on nitrate than on ammonium, the reverse situation in *Cenococcum geophilum* and *Thelephora terrestris*, and an equal facility for use of both ions in *Suillus granulatus*. This pattern reflects to some extent the distribution of the fungi in the natural environment, *P. tinctorius* being predominantly a fungus of the arid part of the environmental gradient where nitrification becomes an important process, while *C. geophilum* and *T. terrestris* occur characteristically on organic soils and *S. granulatus* occurs right across the gradient. In some cases, for example *Lactarius rufus*, it has been shown that the fungus completely failed to use nitrate (Bigg & Alexander, 1981). There is, of course, considerable genetic variability amongst individual strains of ectomycorrhizal fungi. Thus, Laiho (1970) demonstrated clearly that whereas some strains of *Paxillus involutus* produced much higher yields on ammonium than on nitrate the reverse was true with other strains. This pattern was largely confirmed for a range of fungi by Lundeberg (1970). Such observations highlight the genetic plasticity of ectomycorrhizal fungi and make clear the need for careful description of the races and origins of organisms being used in assays of nutrient utilization. Indeed, the plasticity in terms of NH_4 and NO_3 utilization appears to be so great as to make generalization about a single species of doubtful validity.

All strains of the ericoid endophyte *Hymenoscyphus ericae* that have so far been examined are able to utilise both NH_4 and NO_3 (Pearson & Read, 1975; Bajwa & Read, 1986) and while there is an initial lag in growth on nitrate, which probably represents a period during which nitrate reductase is induced, final yields on this nitrogen source can be as high as those on ammonium.

Utilization of mineral nitrogen by the mycorrhizal root

In some of the first experimental studies of nitrogen transfer in mycorrhizal systems, Melin & Nilsson (1952) fed ^{15}N-labelled ammonium to mycorrhizal mycelium and detected its transfer to infected pine seedlings. Carrodus (1966) demonstrated that excised beech mycorrhizas absorbed ammonium more rapidly than nitrate. A largely similar situation has since been shown to prevail in a number of important coniferous species (France & Reid, 1983; Rygiewitz *et al.*, 1984 a & b). The latter workers observed that mycorrhizal infection enhanced the uptake of nitrate in only one of the species examined.

Whereas in ericoid mycorrhizas ammonium is a better source of nitrogen than nitrate, infection can significantly improve the levels of nitrate utilization relative to that seen in uninfected plants (Bajwa & Read, 1986).

Assimilation of absorbed inorganic nitrogen

In his studies of beech mycorrhizas Carrodus (1967) showed that absorbed ammonium was rapidly assimilated into glutamine. While glutamate was the major product of nitrate assimilation, glutamine and aspartate were also formed. The principle route of ammonium assimilation in beech ecto-mycorrhizas has since been shown by Martin et al. (1986) to be that involving glutamine synthetase and glutamate synthase. This study, using $^{15}NH_4$, confirmed that glutamine was the major product of ammonium assimilation and suggested that alanine and aspartate were labelled secondarily as a result of amino transferase activity. The evidence suggests that most of these assimilation processes take place in the fungal component of the mycorrhiza, and much interest surrounds the nature of the compounds subsequently transferred to the host plant.

Studies of Reid & Lewis and of France (see France & Reid, 1983) have indicated that glutamine is the major transfer product following assimilation of ammonium by excised beech mycorrhizas. Finlay et al. (1988) have recently studied the assimilation of ammonium in intact mycorrhizal systems of pine infected with Rhizopogon roseolus, Suillus bovinus, Pisolithus tinctorius and Paxillus involutus in natural unsterile peat. Ammonium labelled with ^{15}N was supplied to mycelial fans of the fungi which had grown across a barrier into a root-free portion of a transparent chamber, and incorporation of label into amino-compounds of the mycelia, mycorrhizal roots and shoots of pine was determined after 72 hours. By this time large proportions of the glutamate/glutamine, aspartate/asparagine and alanine in the mycelia of R. roseolus and S. bovinus were labelled with ^{15}N (Fig 9.1). A similar situation was found in Pisolithus tinctorius, but Paxillus involutus contained no asparagine/aspartate. Lower proportions of labelled serine, threonine, tyrosine, lysine, ornithine and arginine were also found in the mycelia. As would be expected, the amount of enrichment declined throughout the transport pathway though the percentage contribution of labelled amino-acid remained relatively high in the

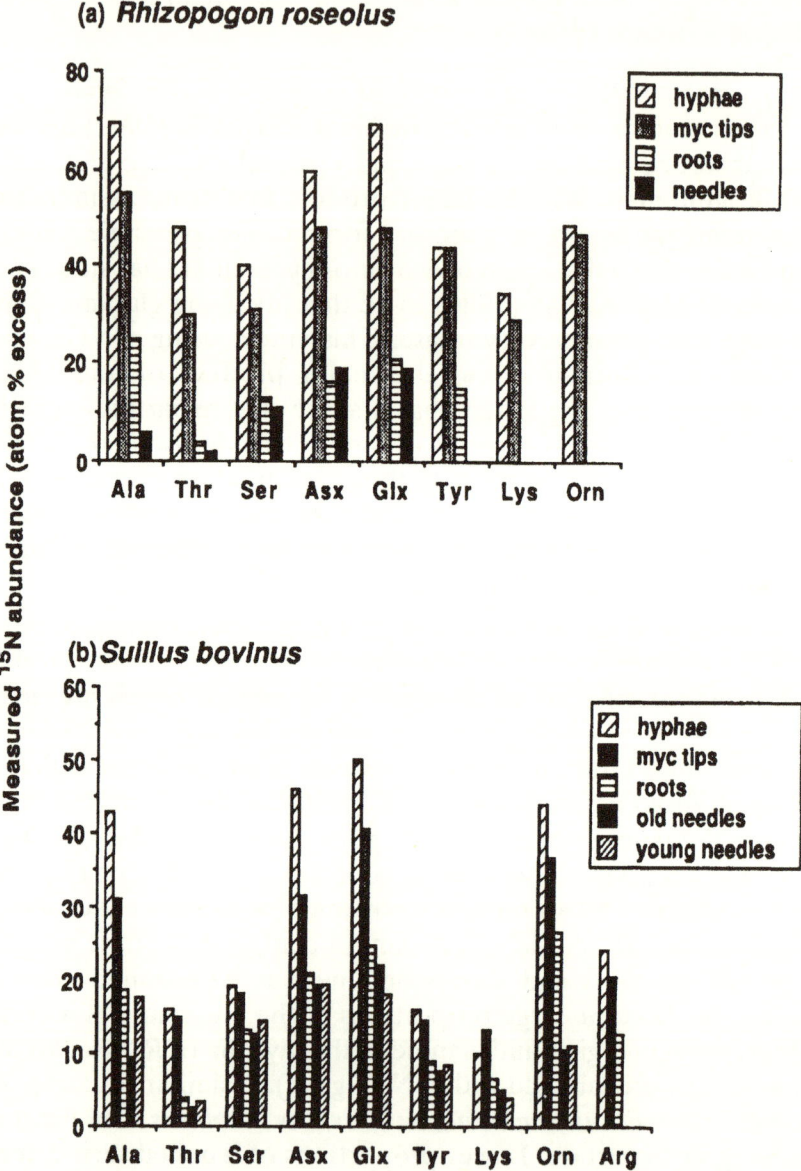

Fig. 9.1. Distribution of ^{15}N amongst amino-acids in different regions of mycorrhizal systems 72 hours after feeding $^{15}NH_4Cl$ to mycelium of two ectomycorrhizal fungi growing from seedlings of *Pinus* in a root chamber; myc = mycorrhizal. From Finlay *et al*., 1988.

mycorrhizal root tips (30-60%), and main root axes (20-40%). Values of between 5 & 30% were detected in the needles.

The free amino acid pools of the fungal mycelia were similar in composition to those reported by Booker (1980), Genetet *et al.* (1984), and Martin *et al.* (1986), as were those of the shoots when compared with data from coniferous needles (Mackenzie & Holme, 1984). Studies such as this of intact systems have the great advantage that normal source-sink relations are maintained so enabling realistic estimates of nutrient fluxes to be obtained. Further studies of this type with sequential harvests being taken at regular short intervals from the time of ^{15}N application should provide definition both of the primary products of ammonium assimilation and of the earliest transferase activities.

Utilization of organic nitrogen
Ericoid systems

Growing at the extreme end of the environmental gradient which leads to the accumulation of recalcitrant mor humus, plants with ericoid mycorrhizas dominate ecosystems in which, by definition, mineralisation processes are slow. There is increasing evidence that their success in these conditions is dependent upon the activities of their mycorrhizal fungi. While the early studies of the function of ericoid mycorrhizas revealed that infection could enhance uptake of ammonium when the mineral ion was the only nitrogen source present (Stribley & Read, 1976), it was also shown (Stribley & Read, 1974) that when ^{15}N labelled ammonium was added to organic heathland soil the amount of label in mycorrhizal plants was less than that in their non-mycorrhizal counterparts even though the mycorrhizal plants contained more nitrogen. This study suggested that infected plants had access to an alternative source of nitrogen, which in the absence of nitrate, was assumed to be organic. In an initial attempt to determine whether infection did, in fact, provide plants with access to such sources, the pattern of uptake of amino-acids by mycorrhizal and non-mycorrhizal plants was examined (Stribley & Read, 1980). This work demonstrated conclusively that whereas uninfected plants could make little use of amino compounds, their infected counterparts assimilated most of the compounds as readily as they did ammonium. Detailed analyses of the nitrogen fractions of heathland soil

(Abuarghub & Read 1988a & b) have since revealed that 'free' amino-acids constitute a significant proportion of the available nitrogen pool, a particularly large concentration occurring in autumn (Fig. 9.2). Mycorrhizal infection will thus provide access

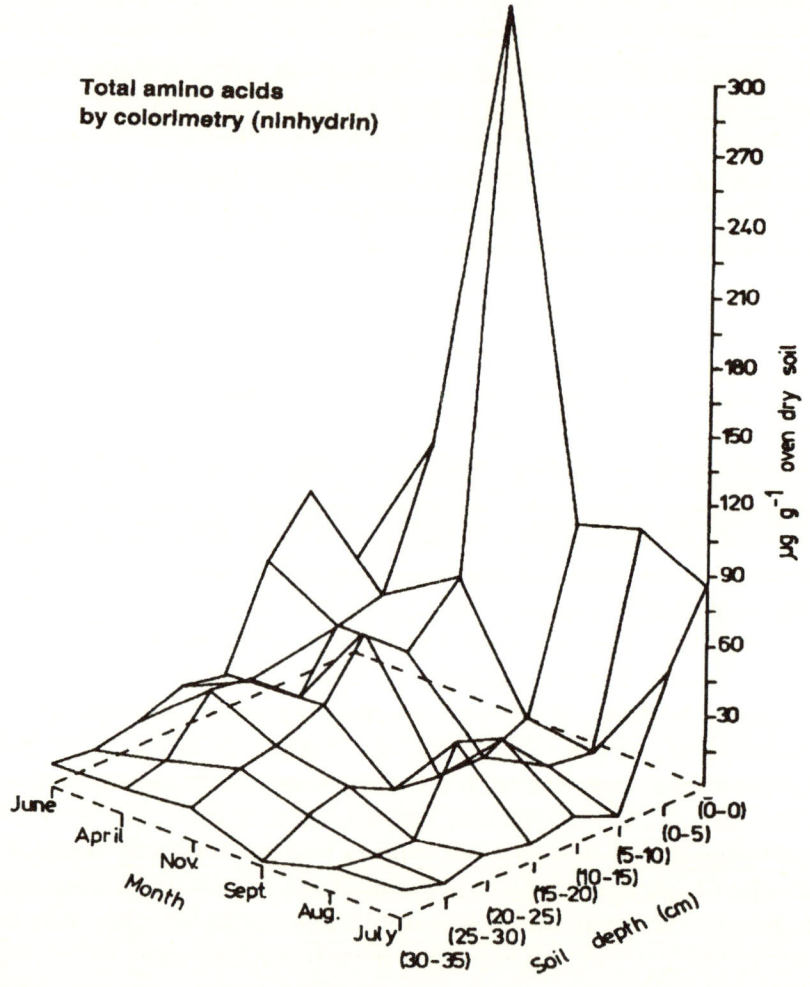

Fig. 9.2. Three dimensional presentation of the 'free' amino acids extracted by dilute ammonium acetate at different depths and times of year under a pure *Calluna* sward from an upland peaty-gley. From Abuarghub & Read, 1988.

to a major component of soil nitrogen which is otherwise unavailable to the plant.

Amino-acids may themselves, however, only be breakdown products of polymeric nitrogen sources so it is necessary to consider the question of the ability of the mycorrhizal fungus to provide access to more complex organic nitrogen compounds. It has now been established that the ericoid mycorrhizal endophyte will utilise not only amino-acids but also peptides (Bajwa & Read, 1985) and proteins (Bajwa, Abuarghub & Read, 1985) as nitrogen sources, and that the assimilated nitrogen is rapidly transferred to infected plants. Production of an extra-cellular proteinase by *Hymenoscyphus ericae* was reported by Read & Bajwa (1985) and Spinner & Hasewandter (1985). Subsequent studies have confirmed and elaborated upon these observations. A more detailed characterization of the factors influencing its activity has since been provided by Leake & Read (1987, 1988). The enzyme has a decidedly acid pH optimum (Fig. 9.3) and a K_m of 0·1 mg protein ml^{-1} which is comparable with those of some acid and alkaline microbial proteinases reported previously, for example 0·15 - 0·54 mg ml^{-1} (Nannipieri *et al.*, 1982) and 0·2 - 0·5 mg ml^{-1} (Chaloupka *et al.*, 1975) but lower than the value of 2·0 - 2·5 mg ml^{-1} obtained for an acid proteinase of the protozoan parasite *Plasmodium* (Levy & Chou, 1974).

Sequential analyses of aliquots of culture filtrates taken while *H. ericae* was growing on media containing protein as a sole nitrogen source have enabled the progress of degradation of the macromolecule to be followed. Growth of the endophyte over the first days lead to rapid decline of the substrate and concomitant increase of amino-acid content of the medium (Fig. 9.4). After 24 days most of the protein had been utilised and at this point the size of the amino-acid pool began to decline. No ammonium was detected in the solution. In view of the proven ability of the endophyte to utilise amino-acids as nitrogen sources, the cleavage of protein into these units is, in terms of energy requirement, the most economical metabolic pathway of nitrogen acquisition. The possible ecological and physiological significance of these observations is discussed later.

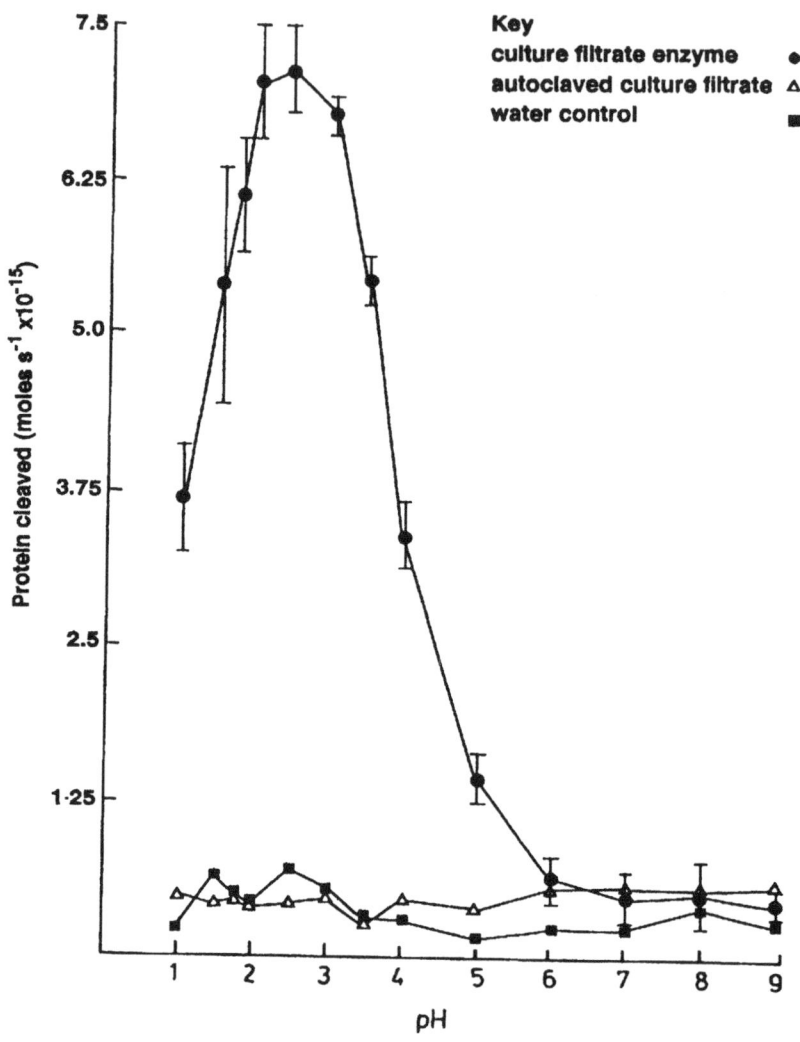

Fig. 9.3. Cleavage of protein by free proteinase enzyme of *Hymenoscyphus ericae* assayed over a pH range with FITC labelled bovine serum albumen (BSA) as substrate. The fungus was first grown on a liquid medium containing unlabelled protein as sole nitrogen source and sterile filtrate was obtained from these cultures at 21 days for enzyme assay.

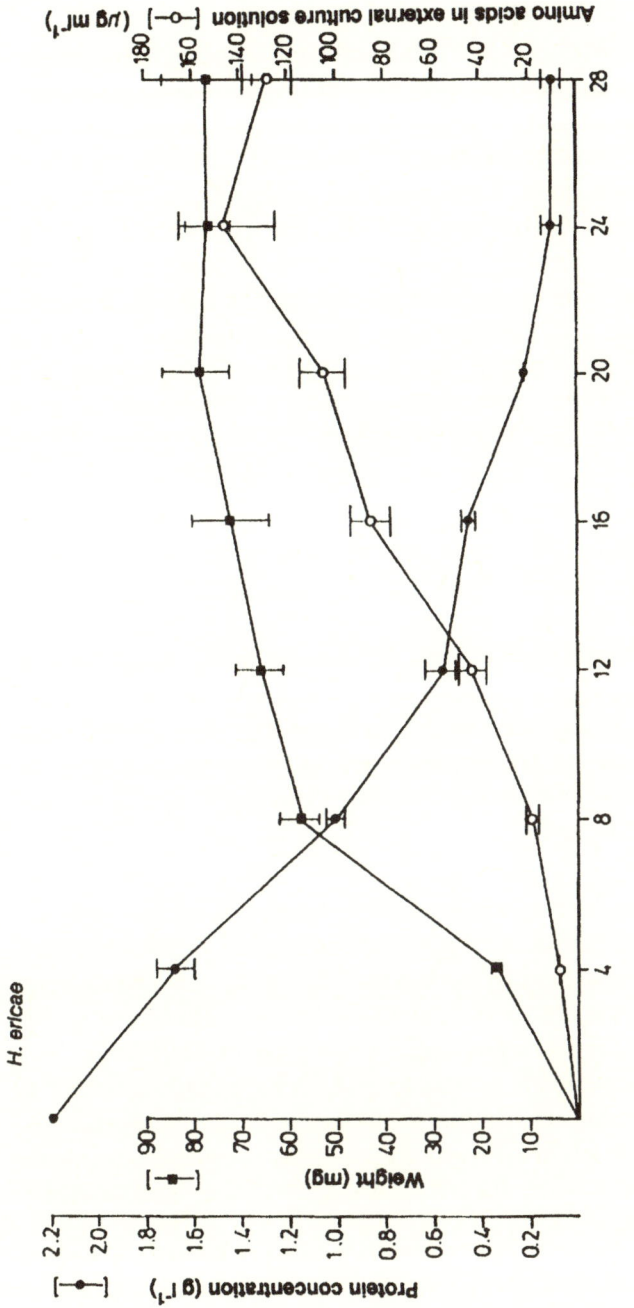

Fig. 9.4. Time course analysis of yield increase, utilization of protein, and release of amino acids into the culture medium when *Hymenoscyphus ericae* was grown over a 28 day period with pure protein as the sole nitrogen source.

Ectomycorrhizal systems

Early studies of Melin (1925) showed that some ectomycorrhizal fungi could utilise amino acids, protein hydrolysate and even, though to a lesser extent, the protein legumin. Later, Melin & Nilsson (1953) demonstrated that ^{15}N labelled glutamate was absorbed by the mycelium of *S. granulatus* and that the nitrogen was transferred to the shoots of pine seedlings infected by the fungi in aseptic culture. The absorption of glutamate, aspartate and their amides by excised beech mycorrhiza was reported by Carrodus (1966) and, using ^{14}C labelled glycine, Sangwanit & Bledsoe (1985) showed that excised mycorrhizas of *Pseudotsuga menziesii* infected with *Hebeloma crustuliniforme* rapidly absorbed the acid which, within 60 minutes, was converted into several other amino compounds.

We have fed ^{14}C-labelled alanine to intact mycorrhizal mycelia growing in association with pine seedlings in unsterile forest soil (Fig. 9.5a) and, by means of autoradiography, followed the distribution of label through the mycelium (Fig. 9.5b). Within 48 hours of commencement of feeding the isotope, label was detected throughout the mycelial network, there being accumulation in the mycorrhizal roots of the interconnected seedlings. Qualitative and quantitative aspects of the transfer of amino-nitrogen and of amino acid carbon in mycorrhizal systems are currently being investigated in more detail.

Despite the early observations of Frank (1894), that mycorrhizal roots were intimately associated with organic material, utilization of such complex sources of nitrogen by ectomycorrhizal fungi has been little studied until recently. The work of Lundeberg (1970) which suggested that ecto fungi were not capable of using humic nitrogen or degrading protein has widely been interpreted to mean that no significant access to polymeric form of nitrogen can be expected in soil. However, it has been shown in numerous studies that ectomycorrhizal roots in forest ecosystems are more intimately associated with the litter than the humified horizon (Harley, 1940; Werlich & Lyr, 1957; Mikola & Laiho, 1962; Meyer 1973; Harvey *et al.*, 1976; Deans, 1979; Alexander & Fairley 1983). On the basis of such observations it has been suggested (Abuzinadah *et al.*, 1986; Read, 1987) that it is ecologically more meaningful to consider the ability of ectomycorrhizal fungi to use the less recalcitrant forms of nitrogen likely to be present in such material. Using the pure protein bovine serum albumin as a model compound, Abuzinadah & Read (1986) showed that a wide range of protein degrading capabilities occurred in ecto fungi. Some, which were termed 'protein fungi' produced higher yield on this substrate than on ammonium, while others termed 'non protein fungi' failed to degrade the macromolecule. When protein fungi such as *Suillus bovinus*, *Amanita muscaria* and *Hebeloma crustuliniforme* were grown in mycorrhizal association with pine, spruce or birch it was shown that nitrogen derived from protein was rapidly transferred to the infected plants. Uninfected plants, in contrast, had

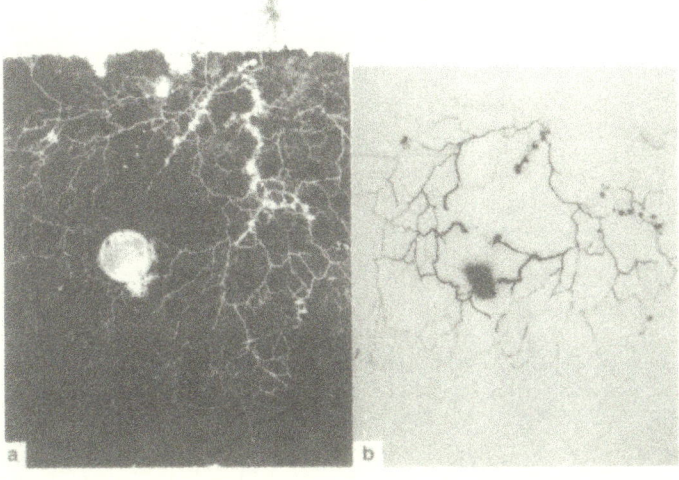

Fig. 9.5. (a) Mycelial network of *Suillus bovinus* interconnecting seedlings of *Pinus sylvestris*, showing feeding dish containing ^{14}C-labelled alanine into which are placed the cut ends of individual mycelial strands. (b) Autoradiograph of 9.5a taken 48 hours after commencement of ^{14}C feeding showing distribution of isotope throughout the network and its accumulation in mycorrhizal roots of the interconnected plants.

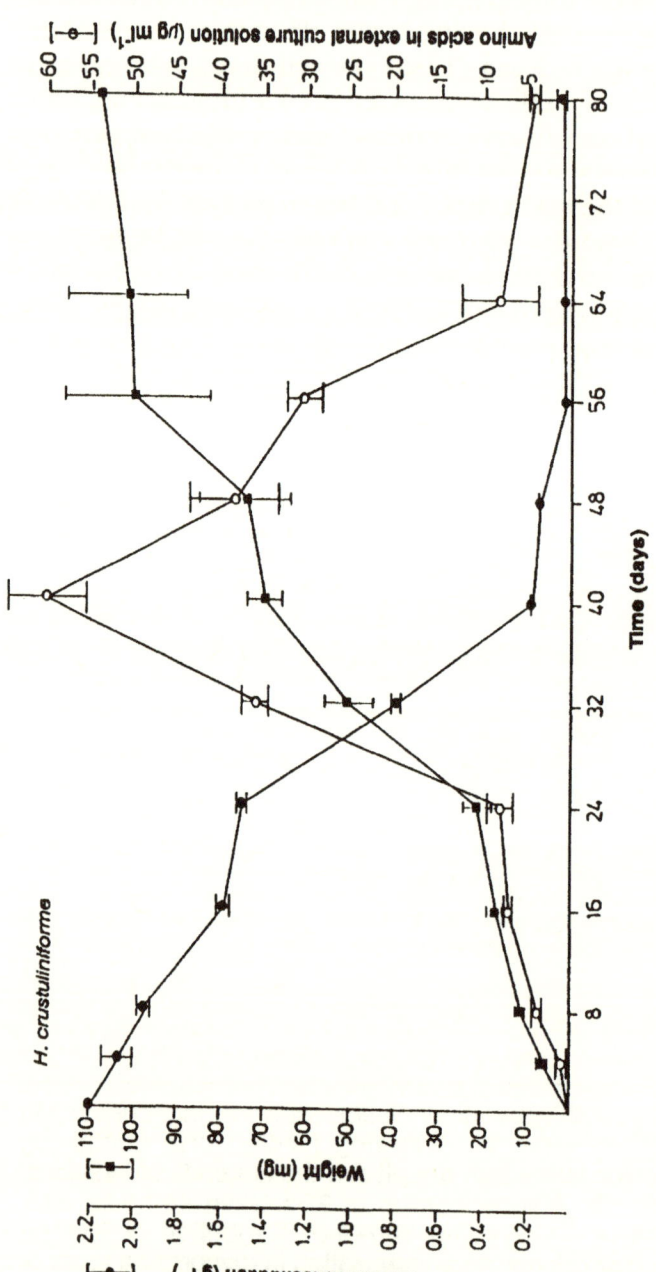

Fig. 9.6. Time course analysis of yield increase, utilization of protein, and release of amino acid into the culture medium when the ectomycorrhizal fungus *Hebeloma crustuliniforme* was grown over an 80 day period with pure protein as the sole nitrogen source.

no access to this nitrogen. It has since been confirmed that, as in the case of the ericoid endophyte, ectomycorrhizal 'protein fungi' produce a free acid proteinase when grown in solution with protein as a sole nitrogen source and that the major products of protein degradation are amino acids (Fig. 9.6), many of which are known from a parallel study of Abuzinadah & Read (1988) to be readily assimilable by the fungi which are releasing them. Again, as in the case of *Hymenoscyphus ericae*, ammonium is not a product of the main phase of protein degradation. In contrast to the situation reported above, Botton *et al.* (1986) could detect only alkaline proteinase activity in *Cenococcum geophilum*. It may be that the strain used was adapted to alkaline soil but details of its origin were not provided.

Discussion

Current evidence indicates that abilities to utilise complex organic sources of nitrogen are far better developed in ericoid and ectomycorrhizal systems than has hitherto been recognised. Such observations justify a re-evaluation both of the function of the mycorrhizal symbiosis and of the overall structure of the nitrogen cycle in heathland and forest ecosystems.

The conventional view of mycorrhizal function sees the fungus as an extension of the root system which, by exploring a volume of soil beyond the depletion zone around roots, increases access to mineral nutrient ions. Simple observations of mycorrhizal structure reveal, however, that in many cases such increases of absorptive capability may not be obtained. The root system of many plants with ericoid mycorrhizas, for example, is itself extremely finely divided into dense wefts of 'hair roots' the net effect being a situation in which all rhizosphere depletion 'shells' overlap. Most of the biomass of the mycorrhizal fungus is located within or close to the root (Read 1983, 1985). In such a situation it is inconceivable that the infection functions as conventionally envisaged. In ectomycorrhizas also, Harley (1978) pointed out that many ensheathed roots are entirely smooth so apparently having few, if any, mycelial connections with soil. It is difficult to see how such a structure can increase absorptive efficiency of a root. If, on the other hand, the fungi were producing diffusible enzymes which alone could provide the plant with access to a growth limiting resource such as nitrogen as it accumulated around the root, this would explain the heavy investment of host carbon in a mycorrhizal structure that did not fulfil a scavenging role. The evidence reported here shows that ericoid and some ectomycorrhizal fungi have the ability to produce such diffusible enzymes and that they readily assimilate the breakdown products of protein degradation.

So far the work has been largely confined to pure culture and there is a need to extend it to soil situations. While many proteins, for example those being released from senescing microbial cells, may be in relatively pure form, some will be rendered inaccessible as a result of chemical binding to polyphenolic compounds, and others will be protected by physical barriers formed by cell walls and lignified structures. A suite of enzymes is likely to be involved in releasing proteinaceous substrates from both chemically bound and physically isolated conditions. Among them polyphenol oxidases, cellulases, peroxidases and lignases are likely to be important. In a preliminary study of mycorrhizal mycelial mats in the decomposition horizon of Douglas fir forests Griffiths *et al.* (1987) have shown a significantly enhanced level of peroxidase and lignase as well as of proteinase activity associated with the fungal component as compared to surrounding soil material. There is obviously an urgent need to examine the field situation in more detail both from the point of view of determining the level of substrate availability and of fungal activity. The revelation of markedly seasonal pattern of amino acid release in heathland soil suggests that there may be temporal as well as spatial heterogeneity in the production and degradation of protein-rich substrate. Whatever the precise pattern of resource capture it is likely that the presence, under the litter layer, of a mat of mycorrhizal roots which have proteolytic capability, will have a profound impact upon nitrogen cycling processes.

In place of the conventional notion that the mycorrhizal fungi are dependent upon release of ammonium by a saprotrophic population, we now contemplate that there will be competition between the two groups for the organic nitrogenous resources. It was suggested by Abuzinadah *et al.* (1986) that the mycorrhizal fungi would be at an advantage in such a competitive situation because they alone were attached to a continuous supply of free carbon with which to synthesise the necessary proteinase enzyme.

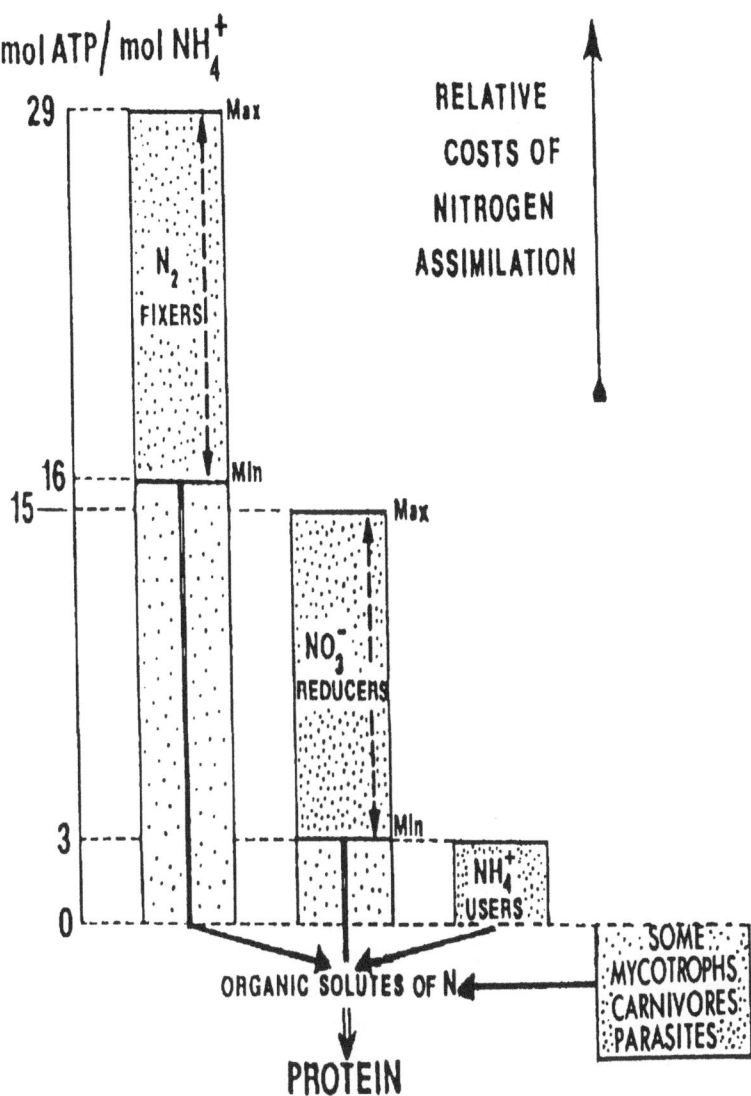

Fig. 9.7. Theoretically based costs of assimilation of different forms of nitrogen by different trophic classes of plants. Ericoid and ectomycorrhizal plants which assimilate nitrogen largely in the ammonium or organic form are seen to suffer the lowest energy costs. Modified from Pate, 1986.

The carbon balance is likely to be of critical importance in determining the outcome of competitive interactions in soils. Dighton *et al.* (1987) investigated the relative rates of breakdown of organic substrates, including hide powder protein, when mycorrhizal seedlings were grown with and without the presence of a saprotrophic fungus in irradiated peat. While their results suggest that the saprotroph reduced the degradative abilities of the mycorrhizal fungus, they must be treated with caution because the interactions were studied in microcosms to which a nutrient solution containing free carbon and mineral salts was added.

In energetic terms the assimilation of simple organic moieties is a highly efficient process. Pate (1986) has calculated the energy costs of assimilation of different types of nitrogen (Fig. 9.7). Assimilation of amino acids is the least energy demanding since they can be absorbed directly and do not require reduction. Ammonium assimilation is, in turn, more energy efficient than is that of nitrate or dinitrogen gas, both of which require several reduction steps. Energetic considerations alone, therefore, suggest that there may be selection for use of amino compounds in situations where organic materials predominate. To be set against this advantage are the costs, as yet unquantified, of synthesis of proteolytic enzymes.

The likely pathways of nitrogen transfer, based upon observations made to date, are shown in Fig. 9.8. The organic nitrogen pathway would be predicted to predominate under circumstances in which mineralisation processes were slow or non existent, whereas pathways of the more traditional kind involving ammonium assimilation would be expected to become more important as turnover of organic matter becomes more rapid. As rates of mineralisation and nutrient release become faster the role of mycorrhizas in nitrogen assimilation probably becomes less critical. In this case, if mycorrhizas provide access to another growth limiting resource, for example phosphorus, infection may be retained. Alternatively, as in most fertile situations, where it is likely to provide the plant with little advantage, infection may be lost altogether.

Emphasis in mycorrhizal research has for a long time been concentrated on the question of the uptake and transfer of the phos-

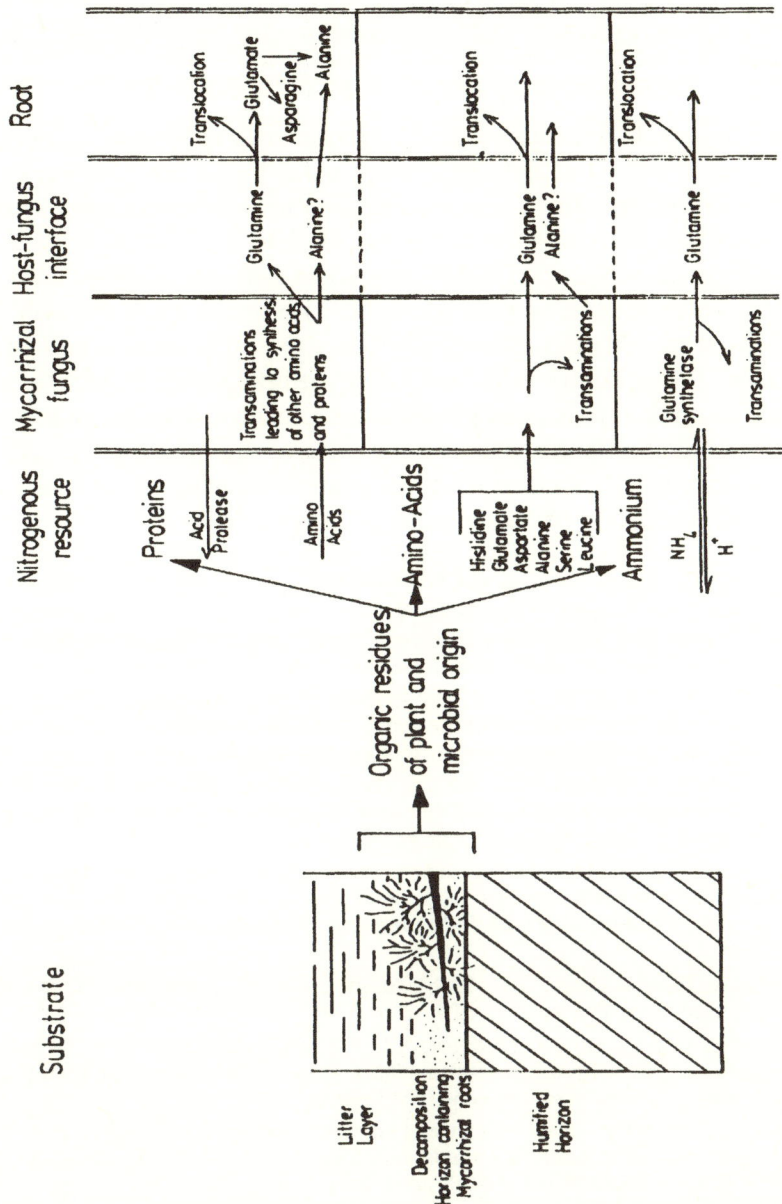

Fig 9.8. The distribution of mycorrhizal roots in relation to the potential nitrogenous substrates in soil and the likely pathways of nitrogen assimilation in ericoid and ectomycorrhizal associations.

phate ion and mycorrhizas have been considered to function simply as extensions of the root system which increase the efficiency
of foraging for mineral nutrients. While this function is well documented and, indeed, may constitute the major role in VA mycorrhizas (see Chapters 10 & 11) there is now sufficient evidence
to support the view, originally expressed by Frank (1894), that
some mycorrhizas have a more fundamental function since their
activities are central to the process of mobilization of nutrients
from organic polymers. Ericoid and some ectomycorrhizal associations are thus seen to play a crucial role in the nutrient cycling
process since through their activities the single most important
element, nitrogen, together probably with phosphorus, can be released from otherwise inaccessible complexes and transported directly to the plants. If we are to understand the nutrient dynamics
of forest and heathland ecosystems a fuller knowledge of these key
nutrient mobilising processes will be essential.

References

Abuarghub, S. M. & Read, D. J. (1988a). The biology of mycorrhiza in the Ericaceae
    XI. The distribution of nitrogen in soil of a typical upland *Callunetum* with special
    reference to the 'free' amino acids. *New Phytologist*, **108**, 425-431.

Abuarghub, S. M. & Read, D. J. (1988b). The biology of mycorrhiza in the Erica-
    ceae XII. Quantitative analysis of individual 'free' amino acids in relation to time
    and depth in the soil profile. *New Phytologist*, **108**, 433-441.

Abuzinadah, R. A. & Read, D. J. (1986a). The role of proteins in the nitrogen nutri-
    tion of ectomycorrhizal plants. I. Utilization of peptides and proteins by ectomy-
    corrhizal fungi. *New Phytologist*, **103**, 481-493.

Abuzinadah, R. A. & Read, D. J. (1986b). The role of proteins in the nitrogen nu-
    trition of ectomycorrhizal plants. III. Protein utilization by *Betula*, *Picea* and *Pinus*
    in mycorrhizal association with *Hebeloma crustuliniforme*. *New Phytologist*, **103**,
    507-514.

Abuzinadah, R. A. & Read, D. J. (1988). Amino acids as nitrogen sources for ecto-
    mycorrhizal fungi. *Transactions of the British Mycological Society*, **91**, 473-479.

Abuzinadah, R. A., Finlay, R. D. & Read, D. J. (1986). The role of proteins in the
    nitrogen nutrition of ectomycorrhizal plants. II. Utilization of protein by mycor-
    rhizal plants of *Pinus contorta*. *New Phytologist*, **103**, 495-506.

Alexander, I. J. (1983). The significance of ectomycorrhizas in the nitrogen cycle. In
    Nitrogen as an Ecological Factor, eds. J. A. Lee, S. McNeill & I. H. Rorison, pp.
    69-93. Blackwell Scientific Publications, Oxford.

Alexander, I. J. & Fairley, R. I. (1983). Effects of N fertilization on populations of
    fine roots and mycorrhizas in spruce humus. *Plant and Soil*, **71**, 49-53.

Ames, R. M., Reid, C. P. P., Porter, L. K. & Cambardella, C. (1983). Hyphal uptake
    and transport of nitrogen from two ^{15}N labelled sources by *Glomus mosseae*, a
    vesicular-arbuscular mycorrhizal fungus. *New Phytologist*, **95**, 381-396.

Bajwa, R. & Read, D. J. (1985). The biology of mycorrhiza in the Ericaceae. IX. Peptides as nitrogen sources for the ericoid endophyte and for mycorrhizal and non-mycorrhizal plants. *New Phytologist*, **101**, 459-467.

Bajwa, R. & Read, D. J. (1986). Utilization of mineral and amino N sources by the ericoid mycorrhizal endophyte *Hymenoscyphus ericae* and by mycorrhizal and non-mycorrhizal seedlings of *Vaccinium*. *Transactions of the British Mycological Society*, **87**, 269-277.

Bajwa, R., Abuarghub, S. & Read, D. J. (1985). The biology of mycorrhiza in the Ericaceae. X. The utilization of proteins and the production of proteolytic enzymes by the mycorrhizal endophyte and by mycorrhizal plants. *New Phytologist*, **101**, 469-486.

Bigg, W. L. & Alexander, I. J. (1981). A culture unit for the study of nutrient uptake by intact mycorrhizal plants under aseptic conditions. *Soil Biology & Biochemistry*, **13**, 77-78.

Booker, C. E. (1980). Free and bound amino acids in the ectomycorrhizal fungus *Pisolithus tinctorius*. *Mycologia*, **72**, 868-881.

Botton, B., El Badaoui, K. & Martin, F. (1985). Induction of extracellular proteases in the ascomycete *Cenococcum geophilum*. In *Proceedings of the 1st European Symposium on Mycorrhizas*, eds. V. Gianinazzi-Pearson & S. Gianinazzi, pp. 403-406. INRA, Paris.

Bowen, G. D. & Smith, S. E. (1981). The effects of mycorrhizas on nitrogen uptake by plants. In *Terrestrial Nitrogen Cycles*, eds. F. E. Clark & T. Rosswall, pp. 237-247. Volume 33 of the Ecological Bulletin, Stockholm.

Carrodus, B. B. (1966). Absorption of nitrogen by mycorrhizal roots of beech. I. Factors affecting the assimilation of nitrogen. *New Phytologist*, **65**, 358-371.

Carrodus, B. B. (1967). Absorption of nitrogen by mycorrhizal roots of beech. II. Ammonium and nitrate as sources of nitrogen. *New Phytologist*, **66**, 1-4.

Chaloupka, J., Obdrzalek, V., Kreckova, P., Nesmeyanova, M. A. & Zalabak, V. (1975). Protease activity in cells of *Bacillus megaterium* during derepression. *Folia Microbiologia*, **20**, 277-288.

Deans, J. D. (1979). Fluctuations of the soil environment and fine root growth in a young sitka spruce plantation. *Plant and Soil*, **52**, 195-205.

Dighton, J., Thomas, E. D. & Latter, P. M. (1987). Interactions between tree roots, mycorrhizas, a saprotrophic fungus and the decomposition of organic substrates in a microcosm. *Biology & Fertility of Soils*, **4**, 145-150.

Field, C. B. & Mooney, H. A. (1986). The photosynthesis-nitrogen relationship in wild plants. In *On the Economy of Plant Form and Function*, ed. T. J. Givnish, pp. 25-55. Cambridge University Press, Cambridge.

Finlay, R. D., Ek, H., Odham, G. & Söderström B. (1988). Mycelial uptake, translocation and assimilation of nitrogen from ^{15}N-labelled ammonium by *Pinus sylvestris* plants infected with four different ectomycorrhizal fungi. *New Phytologist*, **110**, 59-66.

France, R. C. & Reid, C. P. P. (1983). Interactions of nitrogen and carbon in the physiology of ectomycorrhizae. *Canadian Journal of Botany*, **61**, 964-984.

France, R. C. & Reid, C. P. P. (1984). Pure culture growth of ectomycorrhizal fungi on inorganic nitrogen sources. *Microbial Ecology*, **10**, 187-195.

Frank, A. B. (1894). Die Bedeutung der Mykorrhizapilze für die gemeine Kiefer. *Forstwissenschaftliches Centraalblatt*, **16**, 1852-1890.

Genetet. I., Martin. F. & Stewart, G. R. (1984). Nitrogen assimilation in mycorrhizas. Ammonia assimilation in the N-starved ectomycorrhizal fungus *Cenococcum graniforme. Plant Physiology*, **76**, 395-399.

Griffiths. R. P., Caldwell, B. A., Cromack, K., Castellano, M. A., & Morita, R. Y. (1987). A study of chemical and microbial variables ia forest soils colonised with *Hysterangum setchelii* rhizomorphs. In *Proceedings of the 7th North American Conference on Mycorrhizae*, eds. D. M. Sylvia, L. L. Hung & J. H. Graham, p. 196. Institute of Food and Agricultural Sciences, University of Florida, Gainesville, Florida, USA.

Harley, J. L. (1940). A study of the root system of the beech in woodland soils with especial reference to mycorrhizal infection. *Journal of Ecology*, **28**, 107-117.

Harley, J. L. (1978). Ectomycorrhizas as nutrient absorbing organs. *Proceedings of the Royal Society of London, Series B*, **203**, 1-21.

Harley, J. L. & Smith, S. E. (1983). *Mycorrhizal Symbiosis*. Academic Press, London.

Harvey, A. E., Larsen, M. J. & Jurgensen, M. F. (1976). Distribution of ectomycorrhizae in a mature Douglas fir/larch forest system in Western Montana. *Forest Science*, **22**, 393-398.

Laiho, O. (1970). *Paxillus involutus* as a mycorrhizal symbiont of forest trees. *Acta Forestalia Fennica*, **106**, 1-65.

Leake, J. R. & Read, D. J. (1987). Studies on free acid protease of the ericoid endomycorrhizal fungus. In *Proceedings of the 7th North American Conference on Mycorrhizae*, eds. D. M. Sylvia, L. L. Hung & J. H. Graham, pp. 333-334. Institute of Food and Agricultural Sciences, University of Florida, Gainesville, Florida, USA.

Leake, J. R. & Read, D. J. (1988). The biology of mycorrhiza in the Ericaceae. XIII. Some characteristics of the extracellular protease activity of the ericoid endophyte *Hymenoscyphus ericae. New Phytologist*, in press.

Levy, M. R. & Chou, S. C. (1974). Some properties and susceptibility to inhibitors of partially purified acid proteases from *Plasmodium berghei* and from ghosts of mouse red cells. *Biochimica et Biophysica Acta*, **334**, 423-430.

Lundeberg, G. (1970). Utilization of various nitrogen sources, in particular bound soil nitrogen, by mycorrhizal fungi. *Studia Forestalia Sueccica*, **79**, 1-95.

MacKenzie, S. L. & Holme, K. R. (1984). Analysis of conifer leaf free amino acids by gas liquid chromatography. *Journal of Chromatography*, **299**, 387-396.

Martin, F. (1985). ^{15}N-NMR studies of nitrogen assimilation and amino acid biosynthesis in the ectomycorrhizal fungus *Cenococcum graniforme. FEBS Letters*, **182**, 350-354.

Martin, F., Stewart, G. R., Genetet, I. & Le Tacon, F. (1986). Assimilation of $^{15}NH_4{}^+$ by beech (*Fagus sylvatica* L.) ectomycorrhizas. *New Phytologist*, **102**, 85-94.

Martin, F., Ramstedt, M. & Söderhall, K. (1987). Carbon and nitrogen metabolism in ectomycorrhizal fungi and ectomycorrhizas. *Biochemie*, **69**, 569-581.

Melin, E. (1925). *Untersuchungen über die Bedeutung der Baummykorrhize*. G. Fischer, Jena.

Melin. E. & Nilsson, H. (1952). Transport of labelled nitrogen from an ammonium source to pine seedlings through mycorrhizal mycelium. *Svensk Botanisk Tidskrift*, **46**, 281-285.

Melin, E. & Nilsson, H. (1953). Transfer of labelled nitrogen from glutamic acid to pine seedlings through the mycelium of *Boletus variegatus* (Sw.) Fr. *Nature*, **171**, 134.

Meyer, F. H. (1973). Distribution of ectomycorrhizae in native and man-made forests. In *Ectomycorrhizae*, eds. G. C. Marks & T. T. Kozlowski, pp. 79-105. Academic Press, New York.

Mikola, P. & Laiho, O. (1962). Mycorrhizal relations in the raw humus layer of northern spruce forests. *Communicationes Instituti Forestalia Fennicae*, **55**, 1-13.

Nannipieri, P., Ceccanti, B., Cervelli, S. & Conti, C. (1982). Hydrolases extracted from soil, kinetic parameters of several enzymes catalysing the same reaction. *Soil Biology & Biochemistry*, **14**, 429-432.

Pate, J. (1986). Economy of symbiotic nitrogen function. In *On the Economy of Plant Form and Function*, ed. T. J. Givnish, pp. 299-325. Cambridge University Press, Cambridge.

Pearson, V. & Read, D. J. (1975). The physiology of the mycorrhizal endophyte of *Calluna vulgaris*. *Transactions of the British Mycological Society*, **64**, 1-7.

Possingham, J. V. & Groot Obbink, J. (1971). Endotrophic mycorrhiza and the nutrition of grape vines. *Vitis*, **10**, 120-130.

Read, D. J. (1983). The biology of mycorrhiza in the Ericales. *Canadian Journal of Botany*, **61**, 985-1004.

Read, D.J. (1985). The structure and function of the vegetative mycelium of mycorrhizal roots. In *The Ecology and Physiology of the Fungal Mycelium*, eds. D. H. Jennings & A. D. M. Rayner, pp. 215-240. Cambridge University Press. Cambridge.

Read, D. J. (1987). In support of Frank's organic nitrogen theory. *Angewandt Botanik*, **61**. 25-37.

Read, D. J. & Bajwa, R. (1985). Some nutritional aspects of the biology of ericaceous mycorrhizas. *Proceedings of the Royal Society of Edinburgh*, **85B**, 317-332.

Rygiewitz, P. T., Bledsoe, G. S. & Zasoski, R. J. (1984a). Effects of ectomycorrhizae and solution pH on ^{15}N ammonium uptake by coniferous seedlings. *Canadian Journal of Forest Research*, **14**, 885-892.

Rygiewitz, P. T., Bledsoe, G. S. & Zasoski, R. J. (1984b). Effects of ectomycorrhizae and solution pH on ^{15}N nitrate uptake by coniferous seedlings. *Canadian Journal of Forest Research*, **14**, 893-899.

Sangwanit, V. & Bledsoe, C. (1985). Organic nitrogen uptake by axenically-grown mycorrhizal coniferous roots. In *Proceedings of the 6th North American Conference on Mycorrhizas*, ed. R. Molina, p. 346. Forest Research Laboratory, Oregon State University, Corvallis, Oregon, USA.

Smith, F. A., Smith, S. E., St John, B. J., & Nicholas, D. J. D. (1986). Inflow of N & P into roots of mycorrhizal and non-mycorrhizal onions. In *Proceedings of the 1st European Symposium on Mycorrhizas*, eds. V. Gianinazzi-Pearson & S. Gianinazzi, pp. 371-375. INRA, Paris.

Spinner, S. & Haselwandter, K. (1985). Proteins as nitrogen sources for *Hymenoscyphus* (= *Pezizella*) *ericae*. In *Proceedings of the 6th North American Conference on Mycorrhizae*, ed. R. Molina, p. 422. Forest Research Laboratory, Oregon State University, Corvallis, Oregon, USA.

Stribley, D. P. & Read, D. J. (1974). The biology of mycorrhiza in the Ericaceae. IV. The effect of mycorrhizal infection on uptake of ^{15}N from labelled soil by *Vaccinium macrocarpon* Ait. *New Phytologist*, **73**, 1149-1155.

Stribley, D. P. & Read, D. J. (1976). The biology of mycorrhiza in the Ericaceae VI. The effects of mycorrhizal infection and the concentration of ammonium on growth of cranberry (*Vaccinium macrocarpon* Ait.) in sand culture. *New Phytologist*, **77**, 63-72.

Stribley, D. P. & Read, D. J. (1980). The biology of mycorrhiza in the Ericaceae. VIII. The relationship between mycorrhizal infection and the capacity to utilize simple and complex organic nitrogen sources. *New Phytologist*, **86**, 365-371.

Werlich, I. & Lyr, H. (1957). Über die Mykorrhizaausbildung von Kiefer (*Pinus sylvestris* L.) und Buche (*Fagus sylvatica* L.) auf verschiedenen Standorten. *Archiv Forstwiss*, **61**, 1-23.

Chapter 10

Acquisition of phosphorus by VA mycorrhizal fungi and the growth responses of their host plants

B. A. D. Hetrick

Department of Plant Pathology, Kansas State University, Throckmorton Hall, Manhatten, Kansas 665016, USA

Introduction

While phosphorus is an extremely important mineral for plant nutrition, it is only sparingly available in most soils. The majority of phosphorus is present in soil in insoluble inorganic forms. At low pH, iron and aluminium phosphates are common while apatite, hydroxyapatite or possibly tricalcium phosphates occur at higher pH (Barber, 1984). Phosphorus can also be present in organic forms, the availability of which is controlled by mineralization rates in soil. The phosphorus content of soils averages approximately 0.05% by weight, but that of the soil solution is much lower. For example, Barber (1984) compared a wide range of soils and found the average phosphorus content of soil solution to be 0.05 mg P l^{-1}.

As plants remove phosphorus from the soil solution, a depletion zone of 1-2 mm may develop around roots (Bhat & Nye, 1974; Lewis & Quirk, 1967). Following development of this depletion zone, availability of phosphorus to the plant is not so much determined by the concentration of phosphorus in the soil, but rather by the rate of movement of phosphorus through the soil to replenish the soil solution surrounding the root. Phosphorus availability to plants is often considered to be diffusion limited because phosphorus is virtually immobile in soil.

Nye & Tinker (1977) suggested that diffusion of phosphorus in soil is related to volumetric water content, tortuosity (a measure of how difficult a pathway nutrients must take to reach a plant root), the diffusion coefficient of orthophosphate in water, and the buffering capacity of the soil. The buffering capacity of soil is a function of the intensity of phosphorus (or phosphorus in soil solution) and the total quantity of phosphorus in a soil. In soils fertilized with a readily available form of phosphorus, the level of phosphorus in the soil solution increases and phosphorus uptake by plants may be most limited by the rate at which the plant can absorb the element. In this situation, mycorrhizal fungi may offer little benefit to plant phosphorus uptake.

In contrast, soils amended with relatively insoluble fertilizer forms, such as rock phosphate, show little change in solution phosphorus but the overall quantity of phosphorus increases. In these soils and in native, nonfertilized soils which are naturally low in available phosphorus, phosphorus uptake by the plant is strongly diffusion-limited and plant acquisition of the element depends on replenishment of the depletion zone around the roots. It is in these cases that mycorrhizal fungi may make an important contribution to phosphorus acquisition by plants.

Since the pioneering research of Mosse (1957) it has been demonstrated repeatedly, and it is widely accepted, that mycorrhizal plants contain considerably higher concentrations of phosphorus than do nonmycorrhizal plants. While a number of mechanisms have been proposed to explain this phenomenon (Tinker, 1975), the main mechanism is considered to be absorption of phosphorus from soil by hyphae of mycorrhizal fungi and translocation of this nutrient through hyphae into roots colonized by the mycorrhizal fungus. In this way the soil outside the root's depletion zone is explored for nutrient and the ability of mycorrhizal plants to obtain phosphorus is extended far beyond the capability of a nonmycorrhizal plant (Rhodes & Gerdemann, 1975). It has been estimated that the hyphae external to the root may virtually double the absorbing surface of plants (Gianinazzi-Pearson & Gianinazzi, 1983). These external hyphae absorbed and transported phosphorus over 7 cm distance to a root while nonmycorrhizal plant roots could not absorb phosphorus even 1 cm away from the root (Rhodes & Gerdemann, 1975). In addition to this physical absorbing surface provided by mycorrhizal fungi, Cress, Throneberry & Lindsey (1979) have suggested that external hyphae have a greater affinity (lower 'K_m') for phosphorus than do nonmycorrhizal roots. This mechanism was questioned by Karunaratne, Baker & Barker (1986) who observed no such increased affinity of mycorrhizas for phosphorus. They concluded that mycorrhizal infection simply increases the number of uptake sites per unit area of root.

The greater phosphorus uptake of mycorrhizal plants can lead to considerable increases in plant growth (Mosse, 1973; Menge, 1983; Hayman, 1983). For example, biomass increases of 306% for corn, 156% for subterranean clover, 80-100% for white clover, 29-34% for ryegrass, 48-3155% for onions, 250% for strawberries, and 220% for wheat have been observed (Menge, 1983). These levels of growth response are generally maxima achieved by assessing growth response under ideal conditions. Thus, these levels of yield increase may represent the potential rather than the actual benefit from the mycorrhizal symbiosis which can be realized under field conditions. Indeed, responses to mycorrhizal symbiosis in the field are notoriously difficult to predict (Hayman, 1983), perhaps because the bulk of research to date has been conducted with sterilized soils and fails to consider the significant impact of the soil microflora. Also, most experiments have studied single plants in single pots and fail to consider the influence of inter- or intra-specific plant competition.

This chapter examines phosphorus acquisition by mycorrhizal fungi and the resulting growth enhancement for host plants as an interaction of phosphorus availability in soil, the dependence of the plant on the symbiosis for phosphorus acquisition and the ability of the fungus to absorb and translocate phosphate. The modifying effect of the soil microflora and interplant competition on this interaction will also be discussed.

Phosphorus availability in soil
Microbial solubilization of inorganic forms of phosphorus

While phosphorus availability in soil is intimately related to the parent material of which a soil is composed, mineralization and solubilization are microbially related processes which can directly control phosphorus availability (Tiessen, Stewart & Cole, 1984). The relatively recent observations that the rhizosphere microflora of mycorrhizal plants differs from that of nonmycorrhizal plants (Meyer & Linderman, 1986; Ames, Reid & Ingham, 1984) invites speculation about the impact of this difference in microbial composition on phosphorus availability in the rhizosphere of mycorrhizal roots.

It is clear that rock phosphate-solubilizing bacteria can themselves increase plant yield (Bajpai & Rao, 1971; Kabesh, Saber & Young, 1975), and the synergistic interaction between these bacteria and mycorrhizal fungi may further improve plant growth (Azcon, Barea & Hayman, 1976; Raj, Bagyaraj & Manjunath, 1981). However, contradictory results have also been reported by Lee & Bagyaraj (1986) who observed that phosphorus-solubilizing bacteria did not increase shoot dry weight or phosphorus uptake, whether these bacteria were applied alone or in combination with mycorrhizal fungi. Also, Lee & Bagyaraj (1986) observed no benefit from sulphur-oxidizing *Thiobacillus* bacteria which are known to solubilize phosphorus during sulphur oxidation. Based on these results it is impossible to conclude whether or not phosphorus-solubilizing bacteria play any significant role in availability of phosphorus to mycorrhizal fungi. Since most studies to date have examined the reintroduction of phosphorus-solubilizing microorganisms into sterilized soil, little information is available on the role of these microbes in mycorrhizal plant growth under natural conditions. It is also unclear whether phosphorus-solubilizing microorganisms are more common in the rhizosphere of mycorrhizal plants than nonmycorrhizal plants.

The ability of mycorrhizal plants themselves to obtain phosphorus from relatively unavailable (inorganic) phosphates has been sharply contested for many years. It was generally believed that mycorrhizal and nonmycorrhizal plants use the same phosphate sources (Sanders & Tinker, 1971; Mosse, Hayman & Arnold, 1973; Gianinazzi-Pearson *et al.*, 1981; Brechet & LeTacon, 1984). This conclusion, reached by comparing the specific activity of mycorrhizal and nonmycorrhizal plants following ^{32}P addition to soil, assumes that added ^{32}P remains in the labile pool and does not exchange with unavailable forms of phosphorus. However, Bolan *et al.* (1984) demonstrated that precipitation or exchange of ^{32}P with phosphorus sources unavailable to plant growth does occur. Hence, lack of difference in specific activity of mycorrhizal and nonmycorrhizal plants does not preclude use of unavailable phosphorus forms by mycorrhizal plants. Subsequently, Bolan, Robson & Barrow (1987) amended soil with three inorganic phosphorus sources of varied solubility. Mycorrhizal plants benefitted more than nonmycorrhizal plants from these fertilizers but the degree of difference was greatest with the most insoluble phosphorus fertilizer. In other words, the less soluble the phosphorus source, the greater the benefit from mycorrhizal fungus inoculation. This is not surprising since phosphorus availability to plants is still diffusion-limited following fertilization with insoluble forms of phosphorus. It is in precisely these situations that the benefits of mycorrhizal infections are maximised.

The data of Bolan *et al.* (1987) and others (Daft & Nicholson, 1966; Hall, 1975; Powell & Daniel, 1978), suggest that mycorrhizal plants may be able to use relatively unavailable phosphorus sources. However, since mycorrhizal inoculation in most experiments is accomplished with inoculum consisting of rhizosphere sievings, nonaxenic spores, or soil containing spores, infected root fragments and hyphae, it is unclear whether the benefit to mycorrhizal plants arising from application of insoluble phosphorus is a result of solubilization of the phosphorus by the fungi themselves or by their accompanying microflora.

Microbial mineralization of organic forms of phosphorus

Mineralization of phosphorus from organic matter is another means by which the element becomes available for plant growth. It has been suggested that if mineralization rates are high, mycorrhizal colonization could be shed by the plant in response to higher levels of available phosphorus (Baylis, 1967; Daft & Nicholson, 1969; Crush, 1973). Nutrient release by mineralization could certainly explain why microorganisms other than mycorrhizal fungi improve plant growth when added to steamed soil (Hetrick *et al.*, 1988a). Furthermore, if the need for mycorrhizas decreases as a consequence of mineralization, this could partially explain why mycorrhizal growth response and root colonization are reduced in nonsterile soil (Hetrick, *et al.*, 1988a). However, since plant growth in nonsterile soil is still reduced as compared to that in steamed soil, it seems most realistic to assume that both mineralization and microbial competition for available nutrients may occur more or less simultaneously.

Tallgrass prairie soils are highly productive but contain relatively little plant-available phosphorus as predicted by soil tests (Halm, Stewart & Halstead, 1972). Instead, organic phosphorus is abundant and mineralization appears to be rate-limiting (Cole *et al.*, 1978; Blair, Till & Smith, 1977). Plants native to tallgrass prairie are highly dependent on mycorrhizal symbiosis (Hetrick, Kitt & Wilson, 1988c), suggesting that mycorrhizal fungi, or their associated microflora, play a significant role in mineralization of organic matter. That the fungi themselves mineralize organic matter is suggested by the superior growth of mycorrhizal plants in steamed soil from which the soil microflora has been virtually eliminated (Hetrick *et al.*, 1988a). Mycorrhizal fungi are also found in association with organic matter in soils (Dowding, 1959; Mosse, 1959; Nicolson, 1959; Koske, Sutton & Shepard, 1975; Nicolson & Johnston, 1979). St. John, Coleman & Reid (1983) observed that, while mycorrhizal fungal hyphae did not appear to grow toward organic matter deposits, hyphal length was greater around organic matter, probably because of increased hyphal branching in these sites. This suggests that hyphae adopted an absorptive growth pattern when confronted with a nutrient source.

Mineralization of organic matter by mycorrhizal fungi is further suggested by Warner & Mosse (1980), Hepper & Warner (1983), and Warner (1984), who observed that saprotrophic growth by mycorrhizal fungi was related to soil organic matter. More recently, it was demonstrated that acid phosphatase activity is higher in the rhizosphere of mycorrhizal plants than nonmycorrhizal plants (Dodd *et al.*, 1987). It is impossible to conclude from these data whether it is mycorrhizal fungi themselves or the associated microflora which actually effect mineralization of nutrient. It is entirely possible that mineralization is accomplished by the soil microflora and that mycorrhizal hyphae merely take advantage of the released nutrient by positioning themselves and proliferating near organic matter. It is also possible that mycorrhizal plants, by altering their rhizosphere microflora (Barea, Azcon & Hayman, 1975; Azcon *et al.*, 1976; Hayman, 1978; Bagyaraj & Menge, 1978;

Krishna, Balakrishna & Bagyaraj, 1982; Ames *et al.*, 1984; Meyer & Linderman, 1986; Ames, Mihara & Bethlenfalvay, 1987) actually increase mineralization rates in soil. That mycorrhizal fungi exert some direct control on their rhizosphere is implied by the results of Morandi, Bailey & Gianinazzi-Pearson (1984), who found mycorrhizal plants to contain higher levels of isoflavonoids, glyceollin I, coumestrol, and diadzein than nonmycorrhizal plants. Although Morandi *et al.* (1984) suggest that these levels of phenolic compounds explain the greater resistance of mycorrhizal plants to plant pathogens, these compounds could also be expected to impact on nonpathogenic microorganisms if they leak from roots or are released from killed plant cells. If mycorrhizas stimulate mineralization by differentially increasing the population of soil microorganisms involved in mineralization, this would be an extremely important role for these fungi in nutrient cycling. However, the impact of mycorrhizal symbiosis on mineralization has not been studied and is a subject of speculation.

Acquisition of phosphorus and plant response
Role of soil fertility
Phosphorus acquisition by mycorrhizal fungi and subsequent benefit to plant growth, or growth response, is related to the degree to which plants rely on the symbiosis for nutrient absorption. Gerdemann (1975) used the term mycorrhizal dependence (MD) to describe this concept, defining it as 'the degree to which a plant is dependent on the mycorrhizal condition to produce its maximum growth or yield at a given level of soil fertility.' The reliance of a plant on the symbiosis was defined quantitatively by Menge *et al.* (1982) as percentage growth response:

$$\text{MD} = \frac{\text{dry wt of mycorrhizal plant}}{\text{dry wt of nonmycorrhizal plant}} \times 100\%$$

Subsequently, Plenchette, Fortin & Furlan (1983) proposed that mycorrhizal dependence should more accurately be expressed as a percentage value termed relative field mycorrhizal dependency (RFMD):

$$\text{RFMD} = \frac{(\text{dry wt mycorrhizal plant}) - (\text{dry wt nonmycorrhizal plant})}{\text{dry wt mycorrhizal plant}} \times 100\%$$

The latter assessment of dependency recognizes that a plant cannot be more than 100% dependent on the symbiosis.

Differences in the mycorrhizal dependence of plants probably relate largely to innate genetically determined characteristics such as root morphology, growth rate or carbohydrate allocation. However, soil fertility can determine what portion of the potential benefit from mycorrhizal symbiosis is realized. Recognising this, Plenchette *et al.* (1983) proposed that the RFMD value should carry a superscript to designate the soil phosphorus level at which it was derived. However, available soil phosphorus levels are not the sole determinants of mycorrhizal dependence. For example, Menge *et al.* (1982) demonstrated that although dependence of Troyer citrange (citrus rootstock) varied in 26 soils tested, each was positively correlated with pH and inversely correlated with extractable soil P, Zn, Mn, Cu, percentage organic matter, and cation exchange capacity. Using regression analysis, soil P, Mn, and Zn were shown to have the greatest predictive value for mycorrhizal dependence. In contrast, in a similar study using big bluestem (*Andropogon gerardii* Vitman) in 15 native prairie and agricultural soils, mycorrhizal dependence was not strongly correlated with any individual soil parameter, but a combination of parameters (K, Ca, Cu, and Fe) allowed soils resulting in high or low dependence to be differentiated (Kitt, Hetrick & Wilson, 1988). Thus, soil fertility is an important determinant of mycorrhizal phosphorus uptake and of a plant's dependence on mycorrhizas.

Role of root morphology

A relationship between root morphology and mycorrhizal dependence was first elucidated by Baylis (1970). He suggested that root hairs and mycorrhizas were alternative means for obtaining phosphorus with obligatory mycotrophy occurring in plants with poorly developed roothairs. Subsequently, Baylis (1974) observed that there is a 'sharply defined level of available phosphorus, related to root morphology, above which growth without mycorrhizas proceeds steadily and below which the species is mycotrophic.' Since this hypothesis was originally proposed, the relationship between root morphology, available phosphorus level, and mycorrhizal dependence has been confirmed for a wide range of plant species (St. John, 1980; Chilvers & Daft, 1981; Saif, 1987).

In support of this theory, it was suggested that the root hair density of certain 'nonhost' plants (plants which normally do not form mycorrhizas) is actually a physical deterrent to root colonization (Ocampo, Martin & Hayman, 1980) and that root hairs and mycorrhizas may, in fact, be mutually exclusive (Chilvers & Daft, 1981). In contrast, Pope *et al.* (1983) suggested that root length and fineness of the root system might better predict mycorrhizal dependence than root hair density. However, in tropical forage species, root production, fibrosity, and geometry were more closely related to dependence than was root hair length (Saif, 1987). Throughout these studies, the shorter and coarser (less finely branched) the root system, the greater was the mycorrhizal dependence of the host plants. In five species of orange, root fineness (length per unit weight of dry root) was also an important predictor of mycorrhizal dependence, perhaps because a fine root system contained greater surface area for absorption of water (Graham & Syvertsen, 1985). Since ability to absorb water is related to ability to take up phosphorus, they suggested an inverse relationship between hydraulic conductivity and mycorrhizal dependence.

These studies demonstrate a significant relationship between mycorrhizal dependence and root morphology. While the growth-altering impact of ectomycorrhizal fungi on host root morphology is well documented (Harley & Smith, 1983), similar changes in root morphology are not generally attributed to vesicular-arbuscular mycorrhizal fungi (Gerdemann, 1971). Only recently have any differences in the geometry of mycorrhizal and nonmycorrhizal plant roots been observed. Fitter (1985a), using graph theory to model root topology, found that both mycorrhizal and nonmycorrhizal plants grew initially in an ordered pattern, presumably facilitating efficient exploration of soil. However, nonmycorrhizal plants began to branch and change to a more random growth pattern sooner than mycorrhizal plants. Presumably, the latter growth pattern facilitates more effective absorption of nutrients from soil. Mycorrhizal plants stayed in the less-branched, exploratory growth pattern longer, perhaps because mycorrhizal hyphae performed absorption for the plants and branching of roots was less necessary. Whether such changes in growth pattern reflected nutritional differences between mycorrhizal and nonmycorrhizal plants was unclear.

Since VAM fungi produce hormones (Allen, Moore & Christensen, 1980) or may stimulate hormone production by the host plant (Edress, Davis & Burger, 1984), mycorrhizal fungi may alter rooting strategy in ways not equalled by phosphorus fertilization, although mycorrhizal change in root architecture might still be a response to host nutrition. In support of this hypothesis, Hetrick *et al.* (1988b) have demonstrated that the mycorrhizal condition significantly reduces root branching in big bluestem grass. By maintaining a more elongate, exploratory root growth pattern, mycorrhizal plants exploit a greater soil volume for nutrients and have access to larger nutrient pools. They also limit energy loss due to unnecessary and redundant root maintenance if mycorrhizal hyphae perform absorption functions for the plant. Since plants with coarser root systems are generally more dependent on mycorrhizal symbiosis, Hetrick *et al.* (1988b) have further suggested that, by reducing root branching in host plants, mycorrhizal fungi can actually increase mycorrhizal dependence. The changes in root branching were not explained by differences in phosphorus availability but seemed to be directly controlled by the fungi. In the presence of the soil microflora, however, root branching was further decreased, suggesting that mycorrhizal plants can explore an even greater soil volume for nutrients when microbial competition further limits nutrient availability. Thus, while phosphorus availability alone could not explain the observed differences in root topology of mycorrhizal plants, Hetrick *et al.* (1988b) suggest that the mycorrhizal fungi control root topology in response to soil conditions.

Role of host plant physiology

While root morphology is an important determinant of mycorrhizal dependence, host plant physiology also contributes. Variation in plant growth rate affects mycorrhizal dependence, with slow growing plants having generally lower dependence (Hall, 1975; Graham & Syvertson, 1985). Root exudation also affects mycorrhizal dependence. When mycotrophic and nonmycotrophic plant species were compared, plants which formed mycorrhizal associations had higher rates of root exudation, but no specific qualitative differences in the exudate of these plants were detected (Schwab, Leonard & Menge, 1984). This relationship between root exudation and mycorrhizal symbiosis was also demonstrated by Schwab, Johnson & Menge (1982) who found that a normally nonmycorrhizal plant species (*Chenopodium quinona*) became weakly mycorrhizal following application of sublethal dosages of simazine, a herbicide which increased root exudation. Apparently, some minimum rate of exudation is necessary to initiate and maintain mycorrhizal symbiosis. The greater exudation rate of mycorrhizal plants may result from selective pressure to supply substrate for the fungus, while, alternatively, the lower exudation rate of nonmycorrhizal or weakly mycorrhizal plants could result from selection pressure to conserve photosynthate (Schwab *et al.*, 1984). Certainly, loss of photosynthate to a fungal symbiont when that symbiont is not necessary to plant growth and survival would be counterproductive.

A relationship between root exudation and mycorrhizal dependence was also examined by Azcon & Ocampo (1981), who observed that wheat cultivars which displayed higher levels of mycorrhizal dependence also had relatively high rates of root exudation. While the degree to which plants exude nutrients is, in part, genetically determined, it is obvious from comparisons of mycotrophic and nonmycotrophic plant species that root exudation may reflect variation in the ability of a plant's root system to absorb nutrients adequately without mycorrhizal symbiosis. However, subtle, more temporal differences in root exudation also occur and have been related to the phosphorus content of host plant tissues (Ratnayake, Leonard & Menge, 1978; Graham, Leonard & Menge, 1981). Presumably, phosphorus mediates the membrane permeability of roots and therefore stimulates or inhibits the mycorrhizal symbiosis. In this way the host plant can control loss of photosynthate and decrease mycorrhizal dependence when nutrient is not limited and the symbiosis is unnecessary, i.e. when adequate amounts of nutrients can be supplied by the root system itself. From this perspective, root morphology and root exudation may be interrelated and in concert define a plant's mycorrhizal dependence and growth response.

This interrelationship of root exudation and root morphology led Hetrick *et al.* (1988b) to propose a model in which plant and fungus possess interrelated controls on the symbiosis. According to this model, in response to the nutritional status of the plant, root exudation rate changes and thereby determines whether or not the fungus will colonize and function in the plant roots (Ratnayake *et al.*, 1978; Graham *et al.*, 1981). Once the symbiosis is established, in response to fertility and rhizosphere conditions encountered in the soil, the mycorrhizal fungus alters the host plant root branching pattern to control the volume of soil that will be explored for nutrient (Hetrick *et al.*, 1988b).

Ability of mycorrhizal fungi to acquire phosphorus
Impact of the soil microflora

Acquisition of phosphorus by mycorrhizal fungi is directly related to the development and survival of mycorrhizal fungal hyphae in soil. External hyphae of mycorrhizal fungi are subject to a wide range of synergistic and antagonistic interactions with the soil microflora which can contribute to or reduce mycorrhizal phosphorus acquisition. The potential impact of these interactions is considerable and led Daniels & Menge (1980) to suggest that nonpathogenic soil microorganisms could be viewed as plant pathogens if they limit mycorrhizal nutrient uptake and growth response.

Certainly, hyperparasites of mycorrhizal fungi may limit the potential contribution of mycorrhizas to plant growth by reducing inoculum levels in soil (Daniels & Menge, 1980). Similarly, collembola and microarthropods which graze on mycorrhizal fungal spores and hyphae probably reduce mycorrhizal phosphorus uptake. The impact of these grazers is potentially considerable, since they could destroy hyphal bridges between plants and limit hyphal growth in soil. Since hyphal extension in soil is critical to nutrient uptake by the fungi, Fitter (1985b) has suggested that, under field conditions, mycorrhizal symbiosis may be of more limited benefit to plant growth than is implied by pot experiments.

Fitter's (1985b) contention is supported by the fact that growth response to mycorrhizas and root colonization by the fungi in non-sterile soil frequently does not equal that achieved in sterilized soil to which only mycorrhizal fungi are returned (Mosse, 1977; Mosse, Hayman & Ide, 1969; Habte & Aziz, 1985; Hetrick, Kitt & Wilson,

1986). This suppression of mycorrhizal growth response and root colonization in nonsterile soil is attributable to the actions of the soil microorganisms rather than to any chemical or physical differences because addition of 1 or 10% nonsterile soil to pasteurized soil results in suppression similar to that occurring in 100% nonsterile soil (Hetrick *et al.*, 1986). While this evidence for microbial involvement in the suppression phenomenon is indirect, more direct evidence was found when bacteria and fungi, isolated from nonsterile soil by dilution plating, were returned to sterilized soil (Hetrick *et al.*, 1988a). By this method, bacteria which directly suppress mycorrhizal phosphorus uptake and subsequent growth response were detected. Although Hetrick *et al.* (1988a) observed no deleterious effects of soil fungi on growth response, at least some fungi probably also limit growth response and root colonization, since application of the fungicide metalaxyl to soil increases mycorrhizal root colonization (Groth & Martinson, 1983).

While hyperparasites or VAM-grazers could explain some of these observations, another mechanism has also been proposed. The soil microflora may compete with mycorrhizal hyphae for available nutrient. This seems likely, since addition of phosphorus to infertile soils can eliminate the plant growth depression observed in the presence of the soil microflora (Hetrick *et al.*, 1988b). Also, higher phosphorus levels are tolerated, without eliminating mycorrhizal growth response, when the soil microflora is present (Hetrick *et al.*, 1988b). Wilson, Hetrick & Kitt (1988) observed that mycorrhizal fungal spore germination was lower in nonsterile phosphorus-deficient soil than in steamed soil. Spore germination in nonsterile soil was stimulated by addition of phosphorus while a similar amendment to steamed soil inhibited germination. Whether nutrient is directly necessary for spore germination or indirectly stimulates spore germination by encouraging growth of the soil microflora adjacent to the spore is unknown. However, reduced spore germination as a result of microbial competition for nutrient could explain the lower root colonization levels and mycorrhizal growth response observed in the presence of the soil microflora.

The relationship of soil nutrient levels to suppression of plant growth in nonsterile soil was further studied by Kitt *et al.* (1988)

who observed plant growth in 15 soils and determined that the magnitude of suppression could be described by regression equations using soil Na, Zn, Mn, organic matter, Mg, and NH4. High levels of suppression occurred when levels of Na, Mg, and organic matter were low and Zn, Mn, and NH4 were abundant. Thus, while competition for nutrients may limit growth response and root colonization, nutrients other than phosphorus may also be important in determining the degree of suppression which occurs in a particular soil.

Impact of interplant competition

As previously mentioned, mycorrhizal fungal acquisition of phosphorus has been studied almost exclusively in greenhouse pot experiments which ignore intra- and inter-specific competitive interactions of plants. Plant density and spatial pattern effects, potential influences of the above and below-ground microclimate of a natural plant stand, and the impact of these interactions on growth response have also been ignored. As plant roots and mycorrhizal fungi compete for soil nutrients, it seems likely that phosphorus availability will be modified in ways not predicted by chemical soil tests of nutrient availability. Above-ground competition between plants could also influence below-ground interactions and growth response (Harper, 1977).

Most studies involving mycorrhizal host plant effects on other host plants have been conducted with the facultative mycotroph, ryegrass (*Lolium perenne*). For example, Fitter (1977) observed that ryegrass dry weight and root length decreased when grown in competition with Yorkshire fog (*Holcus lanatus*). Root colonization by the mycorrhizal fungus increased in ryegrass, however, in response to that competition. Presumably, *H. lanatus* was better able than *L. perenne* to take advantage of the symbiosis, and mycorrhizal colonization was ultimately deleterious to *L. perenne*. Thus, even in low phosphate soil, mycorrhizal symbiosis may be deleterious if the plant is a poor competitor for resources (Fitter, 1977).

Similar results were obtained when clover (*Trifolium repens*) and ryegrass were cropped together. Hall (1978) observed that, while clover grown alone benefitted from mycorrhizal symbiosis only at lower phosphorus levels, clover grown with ryegrass benefitted from the symbiosis at higher phosphorus levels. However, clover dry weight and phosphorus content were still reduced in the presence of ryegrass. Mycorrhizal symbiosis improved the ability of clover to compete with ryegrass, and this competitive advantage was attributed to its superior ability to compete for phosphorus. Subsequently, Buwalda (1980), also using clover and ryegrass, observed that the competitive advantage of clover was reduced at high nitrogen and phosphorus levels, confirming that clover's competitive advantage in low fertility soils is related to its superior nutrient uptake ability. Mycorrhizal symbiosis, which at low fertility functions more efficiently in clover than in ryegrass, is apparently a major factor influencing competition between these plants.

The potential role of mycorrhizal symbiosis in influencing outcomes of interspecific competition has also been confirmed with two tallgrass prairie plant species, big bluestem (*Andropogon gerardii*) (obligately mycorrhizal) and Junegrass (*Koeleria pyranidata*) (facultatively mycorrhizal) (B. A. D. Hetrick *et al.*, unpublished). In greenhouse studies, dry weight of mycorrhizal big bluestem was not influenced by the presence of Junegrass but dry weight of nonmycorrhizal big bluestem was significantly reduced when grown with Junegrass, despite phosphorus fertilization. In contrast, dry weight of mycorrhizal Junegrass was significantly lower in the presence of mycorrhizal big bluestem. In the absence of the symbiosis, Junegrass was not affected by the presence of big bluestem. Hetrick, Kitt & Wilson (1988c) have demonstrated that cool-season (C_3) grasses like Junegrass are significantly less dependent on mycorrhizal symbiosis than are warm-season (C_4) grasses like big bluestem. They have also confirmed a relationship between root morphology and mycorrhizal dependence for these plants. The detrimental effect of the symbiosis on the competitive ability of facultative mycotrophs like Junegrass may have resulted in selection for temporal niche divergence (shifting growth phenology to earlier in the season), thus reducing competition and allowing the stable coexistence of these two groups.

Janos (1980) has suggested that obligate mycotrophs dominate in plant communities established in nutrient-deficient soils because the absorption of nutrients via hyphae affords the greatest exploration of the soil for nutrient. Janos (1980) further suggests that, in these nutrient deficient soils, obligate mycotrophs would have a competitive advantage over nonmycotrophs or facultative mycotrophs because development of abundant root hairs or fine root systems, adaptations for nutrient acquisition without the symbiosis, are a costly expenditure for the plant, particularly if they are ineffective. Root hairs are less effective than hyphae in absorbing phosphorus from nutrient-poor soils because phosphorus depletion zones around root hairs overlap and subsequently, root hairs may compete with each other for nutrients (Janos, 1980).

Conclusions

Mycorrhizal symbiosis may be most important to plant growth when phosphorus availability in soil is diffusion-limited. In these situations absorption of nutrients by mycorrhizal fungal hyphae and translocation of these nutrient through hyphae to plants can significantly improve plant growth. Phosphorus acquisition by mycorrhizal fungi and the resulting growth response in plants is envisioned as an interaction of phosphorus availability in soil, the plant's reliance on the symbiosis and the ability of the fungus to absorb and translocate phosphorus.

Phosphorus availability in soil is microbially mediated. A growing body of circumstantial evidence suggests that mycorrhizal fungi or their accompanying microflora may improve soil phosphorus availability by solubilizing inorganic forms of phosphorus or by mineralization of organic phosphorus.

The dependency of plants on mycorrhizal fungi for phosphorus acquisition is influenced by soil fertility, root morphology and host plant physiology, Apparently, the plant controls the presence or absence of the symbiosis in response to internal phosphorus concentrations. The fungus then controls plant root branching patterns and, in effect, the volume of soil explored for nutrient.

The ability of mycorrhizal fungi to acquire phosphorus is influenced by the soil microflora. Hyperparasites and grazers may directly limit mycorrhizal fungus inoculum level and hyphal development in soil, but reduced phosphorus uptake by mycorrhizal fungi may also result from microbial competition for phosphorus. Competition between plants of differing mycorrhizal dependencies also affects mycorrhizal phosphorus acquisition. Mycorrhizal symbiosis may be detrimental to plants with lower mycorrhizal dependency, although the mechanism controlling these competitive interactions needs further research.

References

Allen, M. F., Moore, T. S., Jr. & Christensen, M. (1980). Phytohormone changes in *Bouteloua gracilis* infected by vesicular-arbuscular mycorrhizae. I. Cytokinin increases in the host plant. *Canadian Journal of Botany*, **58**, 371-374.

Ames, R. N., Mihara, K. L. & Bethlenfalvay, G. J. (1987). The establishment of microorganisms in vesicular-arbuscular mycorrhizal and control treatments. *Biology and Fertility of Soils*, **3**, 217-223.

Ames, R. N., Reid, C. P. P. & Ingham, E. R. (1984). Rhizosphere bacterial population responses to root colonization by a vesicular-arbuscular mycorrhizal fungus. *New Phytologist*, **96**, 555-563.

Azcon, R., Barea, J. M. & Hayman, D. C. (1976). Utilization of rock phosphate in alkaline soils by plants inoculated with mycorrhizal fungi and phosphate-solubilizing bacteria. *Soil Biology and Biochemistry*, **8**, 135-138.

Azcon, R. & Ocampo, J. A. (1981). Factors affecting the vesicular-arbuscular infection and mycorrhizal dependency of thirteen wheat cultivars. *New Phytologist*, **87**, 677-685.

Bagyaraj, D. C. & Menge, J. A. (1978). Interaction between a VA mycorrhizal and *Azotobacter* and their effects on rhizosphere microflora and plant growth. *New Phytologist*, **80**, 567-573.

Bajpai, P. D. & Rao, W. V. B. S. (1971). Phosphate solubilising bacteria. Part III. Soil inoculation with phosphorus solubilising bacteria. *Soil Science and Plant Nutrition*, **17**, 46-53.

Barber, S. A. (1984). *Soil nutrient bioavailability: A mechanistic approach*. John Wiley & Sons, New York.

Barea, J. M., Azcon, R. & Hayman, D. S. (1975). Possible synergistic interaction between *Endogone* and phosphate-solubilizing bacteria in low phosphate soils. In *Endomycorrhizas*, eds. F. E. Sanders, B. Mosse & P. B. Tinker, pp. 409-417. Academic Press, London.

Baylis, G. T. S. (1967). Experiments on the ecological significance of phycomycetous mycorrhizas. *New Phytologist*, **66**, 231-243.

Baylis, G. T. S. (1970). Root hairs and phycomycetous mycorrhizas in phosphorus deficient soil. *Plant and Soil*, **33**, 713-716.

Baylis, G. T. S. (1974). The evolutionary significance of phycomycetous mycorrhizas. In *Mechanisms of Regulation of Plant Growth*, bulletin 12, pp. 191-193. The Royal Society of New Zealand.

Bhat, K. K. S. & Nye, P. H. (1974). Diffusion of phosphate to plant roots in soil. II. Uptake along the roots at different times and effect of different levels of phosphorus. *Plant Soil*, **41**, 365-382.

Blair, G. J., Till, A. R. & Smith, R. C. G. (1977). The phosphorus cycle — What are the sensitive areas? In *Reviews in Rural Science III*, Proceedings of a Symposium held at the University of New England, ed. G. J. Blair, pp. 9-19. The University of New England, Armidale, NSW, Australia.

Bolan, N. S., Robson, A. D., Barrow, N. J. & Alymore, L. A. G. (1984). Specific activity of phosphorus in mycorrhizal and non-mycorrhizal plants in relation to the availability of phosphorus to plants. *Soil Biology and Biochemistry*, **16**, 299-304.

Bolan, N. S., Robson, A. D. & Barrow, N. J. (1987). Effects of vesicular-arbuscular mycorrhiza on the availability of iron phosphates to plants. *Plant and Soil*, **99**, 401-410.

Brechet, C. & LeTacon, F. (1984). Responses of endomycorrhizal and non-mycorrhizal plants of *Acer pseudoplatanus* to different levels of soluble and non-soluble phosphorus. *European Journal of Forestry Pathology*, **14**, 68-77.

Buwalda, J. G. (1980), Growth of a clover-ryegrass association with VA mycorrhizas. *New Zealand Journal of Agricultural Research*, **23**, 379-383.

Chilvers, M. T. & Daft, M. J. (1981). Mycorrhizas of the Lilii florae. II. Mycorrhiza formation and incidence of root hairs in field grown *Narcissus* L., *Tulipa* L. and *Crocus* L. cultivars. *New Phytologist*, **89**, 247-261.

Cole, C. V., Elliott, E. T., Hunt, H. W. & Coleman, D. C. (1978). Trophic interactions in soils as they affect energy and nutrient dynamics. V. Phosphorus transformations. *Microbial Ecology*, **4**, 381-387.

Cress, W. A., Throneberry, G. O. & Lindsey, D. L. (1979). Kinetics of phosphorus absorption by mycorrhizal and nonmycorrhizal tomato roots. *Plant Physiology*, **64**, 484-487.

Crush, J. R. (1973). Significance of endomycorrhizas in tussock grassland in Otago, New Zealand. *New Zealand Journal of Botany*, **11**, 645-660.

Daft, M. J. & Nicholson, T. H. (1966). Effect of *Endogone* mycorrhiza on plant growth. *New Phytologist*, **65**, 343-350.

Daft, M. J. & Nicholson, T. H. (1969). Effect of *Endogone* mycorrhiza on plant growth. II. Influence of soluble phosphate on endophyte and host in maize. *New Phytologist*, **68**, 945-952.

Daniels, B. A. & Menge, J. A. (1980). Hyperparasitization of vesicular-arbuscular mycorrhizal fungi. *Phytopathology*, **70**, 584-588.

Dodd, J. C., Burton, C. C., Burns, R. G. & Jefferies, P. (1987). Phosphatase activity associated with the roots and the rhizosphere of plants infected with vesicular-arbuscular mycorrhizal fungi. *New Phytologist*, **107**, 163-172.

Dowding, E. S. (1959). Ecology of *Endogone*. *Transactions of the British Mycological Society*, **42**, 449-457.

Edris, M. H., Davis, R. M. & Burger, D. W. (1984). Influence of cytokinin production in sour orange. *Journal of the American Society of Horticultural Science*, **109**, 587-590.

Fitter, A. H. (1977). Influence of mycorrhizal infection on competition for phosphorus and potassium by two grasses. *New Phytologist*, **79**, 119-125.

Fitter, A. H. (1985a). Functional significance of root morphology and root system architecture. In *Ecological Interactions in Soil*, eds. A. H. Fitter, D. Atkinson, D. J. Read & M. B. Usher, pp. 87-106. Blackwell Scientific Publications, Oxford.

Fitter, A. H. (1985b). Functioning of vesicular-arbuscular mycorrhizas under field conditions. *New Phytologist*, **99**, 257-265.

Gerdemann, J. W. (1971). Fungi that form the vesicular-arbuscular type of endomycorrhizae. In *Mycorrhizae*, ed. E. Hacskaylo, pp. 9-18. Publication number 1189 of the USDA Forest Service, Washington D. C., USA.

Gerdmann, J. W. (1975). Vesicular-arbuscular mycorrhizae. In *The Organization and Structure of Roots*, eds. J. G. Torrey & D. T. Clarkson, pp. 575. Academic Press, London, New York, San Francisco.

Gianinazzi-Pearson, V., Fardeau, J., Asimi, S. & Gianinazzi, S. (1981). Source of additional phosphorus absorbed from soil by vesicular-arbuscular mycorrhizal soybeans. *Physiologie Végétale*, **19**, 33-43

Gianinazzi-Pearson, V. & Gianinazzi, S. (1983). The physiology of vesicular-arbuscular mycorrhizal roots. *Plant Soil*, **71**, 197-209.

Graham, J. H., Leonard, R. T. & Menge, J. A. (1981). Membrane-mediated decrease in root exudation responsible for phosphorus inhibition of vesicular-arbuscular mycorrhiza formation. *Plant Physiology*, **68**, 548-552.

Graham, J. H. & Syvertsen, J. P. (1985). Host determinants of mycorrhizal dependency of citrus rootstock seedlings. *New Phytologist*, **101**, 667-676.

Groth, D. E. & Martinson, C. A. (1983). Increased endomycorrhizal infection of maize and soybeans after soil treatment and metalaxyl. *Plant Disease*, **67**, 1377-1378.

Habte, M. & Aziz, T. (1985). Response of *Sesbania grandiflora* to inoculation of soil with vesicular-arbuscular mycorrhizal fungi. *Applied and Environmental Microbiology*, **50**, 701-703.

Hall, I. R. (1975). Endomycorrhizas of *Metrosideros umbellata* and *Weinmannia racemosa*. *New Zealand Journal of Botany*, **13**, 463-472.

Hall, I. R. (1978). Effects of endomycorrhizas on the competitive ability of white clover. *New Zealand Journal of Agricultural Research*, **21**, 509-515.

Halm, B. J., Stewart, J. W. B. & Halstead, R. H. (1972). The phosphorus cycle in a native grassland ecosystem. In *Proceedings of a Symposium on Isotopes and Radiation in Soil-Plant Relationships including Forestry*, pp. 571-589. International Atomic Energy Agency, Vienna.

Harley, J. L. & Smith, S. E. (1983). *Mycorrhizal Symbiosis*. Academic Press, London.

Harper, J. L. (1977). *Population Biology of Plants*. Academic Press, New York.

Hayman, D. S. (1978). Endomycorrhizae. In *Interactions between Nonpathogenic Soil Microorganisms and Plants*, eds. Y. R. Dommergues & S. V. Krupa, pp. 401-442. Elsevier, Amsterdam, Oxford & New York.

Hayman, D. S. (1983). The physiology of vesicular-arbuscular endomycorrhizal symbiosis. *Canadian Journal of Botany*, **61**, 944-963.

Hepper, C. & Warner, A. (1983). Role of organic matter in growth of a vesicular-arbuscular mycorrhizal fungus in soil. *Transactions of the British Mycological Society*, **81**, 155-156.

Hetrick, B. A. D., Kitt, D. G. & Wilson, G. T. (1986). The influence of phosphorus fertilization, drought, fungal species and soil microorganisms on mycorrhizal growth response in tallgrass prairie plants. *Canadian Journal of Botany*, **64**, 1199-1203.

Hetrick, B. A. D., Wilson, G. W. T., Kitt, D. G. & Schwab, A. P. (1988a). Effects of soil microorganisms on mycorrhizal contribution to growth of big bluestem grass in nonsterile soil. *Soil Biology and Biochemistry*, **20**, 501-507.

Hetrick, B. A. D., Leslie, J. F., Wilson, G. W. T. & Kitt, D. G. (1988b). Physical and topological assessment of effects of a vesicular-arbuscular mycorrhizal fungus on root architecture of big bluestem. *New Phytologist*, **110**, 85-96.

Hetrick, B. A. D., Kitt, D. G. & Wilson, G. W. T. (1988c). Mycorrhizal dependence and growth habit of warm-season and cool-season tallgrass prairie plants. *Canadian Journal of Botany*, **66**, 1376-1380.

Janos, D. P. (1980). Mycorrhizae influence tropical succession. *Biotropica*, **12**, 56-64.

Kabesh, M. O., Saber, M. S. & Young, M. (1975). Effect of phosphate dissolving bacteria and phosphate fertilization on the P-uptake by pea plants cultivated in a calcareous soil. *Zeitschrift für Pflanzenernahrung und Bodekunde*, **6**, 506-511.

Karunaratne, R. S., Baker, J. H. & Barker, A. V. (1986). Phosphorus uptake by mycorrhizal and nonmycorrhizal roots of soybean. *Journal of Plant Nutrition*, **9**, 1303-1313.

Kitt, D. G., Hetrick, B. A. D. & Wilson, G. W. T. (1988). Relationship of soil fertility to suppression of the growth response of mycorrhizal big bluestem in nonsterile soil. *New Phytologist*, **109**, 473-481.

Koske, R. E., Sutton, J. C. & Sheppard, B. R. (1975). Ecology of *Endogone* in Lake Huron sand dunes. *Canadian Journal of Botany*, **53**, 87-93.

Krishna, K. R., Balakrishna, A. N. & Bagyaraj, D. J. (1982). Interaction between a vesicular-arbuscular mycorrhizal fungus and *Streptomyces cinnamomeous* and their effects on finger millet. *New Phytologist*, **92**, 401-405.

Lee, A. & Bagyaraj, D. J. (1986), Effect of soil inoculation with vesicular-arbuscular mycorrhizal fungi and either phosphate rock dissolving bacteria or thiobacilli on dry matter production and uptake of phosphorus by tomato plants. *New Zealand Journal of Agricultural Research*, **29**, 525-531.

Lewis, D. G. & Quirk, J. P. (1967). Phosphate diffusion in soil and uptake by plants. III. P32 movement and uptake by plants as indicated by P32 autoradiography. *Plant Soil*, **26**, 445-453.

Menge, J. A. (1983). Utilization of vesicular-arbuscular mycorrhizal fungi in agriculture. *Canadian Journal of Botany*, **61**, 1015-1024.

Menge, J. A., Jarrell, W. M., Labanauskas, C. K., Ojala, J. C., Haszar, C., Johnson, E. L. V. & Sibert, D. (1982). Predicting mycorrhizal dependency of Troyer citrange on *Glomus fasciculatus* in California citrus soils and nursery mixes. *Soil Science Society of America Journal*, **46**, 762-768.

Meyer, J. R. & Linderman, R. G. (1986). Selective influence on populations of rhizosphere or rhizoplane bacteria and actinomycetes by mycorrhizas formed by *Glomus fasciculatus*. *Soil Biology and Biochemistry*, **18**, 191-196.

Morandi, D., Bailey, J. A. & Gianinazzi-Pearson, V. (1984). Isoflavonoid accumulation in soybean roots infected with vesicular-arbuscular mycorrhizal fungi. *Physiological Plant Pathology*, **24**, 357-364.

Mosse, B. (1957). Growth and chemical composition of mycorrhizal and nonmycorrhizal apples. *Nature*, **179**, 922-924.

Mosse, B. (1959). Observations on the extramatrical mycelium of a vesicular-arbuscular endophyte. *Transactions of the British Mycological Society*, **42**, 439-448.

Mosse, B. (1973). Advances in the study of vesicular-arbuscular mycorrhiza. *Annual Review of Phytopathology*, **11**, 171-195.

Mosse, B. (1977). Plant growth responses to vesicular-arbuscular mycorrhiza. X. Responses of *Stylosanthes* and maize to inoculation in unsterile soils. *New Phytologist*, **78**, 277-288.

Mosse, B., Hayman, D. S. & Arnold, D. (1973). Plant growth responses to vesicular-arbuscular mycorrhiza. V. Phosphate uptake from 32P labelled soil solution by three plant species. *New Phytologist*, **72**, 809-815.

Mosse, B., Hayman, D. S. & Ide, G. J. (1969). Growth responses of plants in unsterilized soil to inoculation with vesicular-arbuscular mycorrhiza. *Nature*, **224**, 1031-1032.

Nicolson, T. H. (1959). Mycorrhiza in the Gramineae. I. Vesicular-arbuscular endophytes, with special reference to the external phase. *Transactions of the British Mycological Society*, **42**, 421-438.

Nicolson, T. H. & Johnston, C. (1979). Mycorrhiza in the Gramineae. III. *Glomus fasciculatus* as the endophyte of pioneer grasses in a maritime sand dune. *Transactions of the British Mycological Society*, **72**, 261-268.

Nye, P. H. & Tinker, P. B. (1977). *Solute Movement in the Soil-Root System*. Blackwell Scientific Publishers, Oxford.

Ocampo, J. A., Martin, J. & Hayman, D. S. (1980). Influence of plant interactions on vesicular-arbuscular mycorrhizal infections. I. Host and non-host plants grown together. *New Phytologist*, **84**, 27-35.

Plenchette, C., Fortin, J. A. & Furlan, V. (1983). Growth responses of several plant species to mycorrhizae in a soil of moderate phosphorus fertility. I. Mycorrhizal dependency under field conditions. *Plant and Soil*, **70**, 199-209.

Pope, P. E., Chaney, W. R., Rhodes, J. D. & Woodhead, S. H. (1983). The mycorrhizal dependency of four hardwood tree species. *Canadian Journal of Botany*, **61**, 412-417.

Powell, C. L. & Daniel, J. (1978). Mycorrhizal fungi stimulate uptake of soluble and insoluble phosphate fertilizer from a phosphate-deficient soil. *New Phytologist*, **80**, 351-358.

Raj, J., Bagyaraj, D. J. & Manjurath, A. (1981). Influence of soil inoculation with vesicular-arbuscular mycorrhiza and a phosphate-dissolving bacterium on plant growth and 32P uptake. *Soil Biology and Biochemistry*, **13**, 105-108.

Ratnayake, M., Leonard, R. T. & Menge, J. A. (1978). Root exudation in relation to supply of phosphorus and its possible relevance to mycorrhizal formation. *New Phytologist*, **81**, 543-552.

Rhodes, L. H. & Gerdemann, J. W. (1975). Phosphate uptake zones of mycorrhizal and nonmycorrhizal onions. *New Phytologist*, **75**, 555-561.

Saif, S. R. (1987). Growth responses of tropical forage plant species to vesicular-arbuscular mycorrhizae. *Plant and Soil*, **97**, 25-35.

Sanders, F. E. & Tinker, P. B. (1971). Mechanism of absorption of phosphate from soil by *Endogone* mycorrhizas. *Nature*, **233**, 278-279.

Schwab, S. M., Johnson, E. L, V. & Menge, J. A. (1982). Influence of simazine on formation of vesicular-arbuscular mycorrhizae in *Chenopodium quinona* Willd. *Plant and Soil*, **64**, 283-287.

Schwab, S. M., Leonard, R. T. & Menge, J. A. (1984). Quantitative and qualitative comparison of root exudates of mycorrhizal and nonmycorrhizal plant species. *Canadian Journal of Botany*, **62**, 1227-1231.

St. John, T. V. (1980). Root size, root hairs and mycorrhizal infection: a reexamination of Baylis's hypothesis with tropical trees. *New Phytologist*, **84**, 483-487.

St. John, T. V., Coleman, D. C. & Reid, C. P. P. (1983). Association of vesicular-arbuscular mycorrhizal hyphae with soil organic particles. *Ecology*, **64**, 957-959.

Tiessen, H., Stewart, J. W. B. & Cole, C. V. (1984). Pathways of phosphorus transformations in soils of different pedogenesis. *Soil Science Society of America Journal*, **48**, 853-858.

Tinker, P. B. H. (1975). Effects of vesicular-arbuscular mycorrhizas on higher plants.
Symposium of the Society for Experimental Biology, **29**, 325-349.

Warner, A. (1984). Colonization of organic matter by vesicular-arbuscular mycorrhizal fungi. *Transactions of the British Mycological Society*, **82**, 252-354.

Warner, A. & Mosse, B. (1980). Independent spread of vesicular-arbuscular mycorrhizal fungi in soil. *Transactions of the British Mycological Society*, **74**, 407-446.

Wilson, G. W. T., Hetrick, B. A. D. & Kitt, D. G. (1988). Suppression of VA mycorrhizal fungus spore germination by nonsterile soil. *Canadian Journal of Botany*, in
press.

Chapter 11

Phosphorus metabolism in mycorrhizas

V. Gianinazzi-Pearson & S. Gianinazzi

*Laboratoire de Phytoparasitologie, Station d'Amélioration des Plantes,
INRA, BV 1540, 21034 DIJON CEDEX, France*

Introduction

Because of the fundamental role of phosphorus in biological systems (Westheimer, 1987), organisms have developed specialised mechanisms which enable them to obtain this element from their environment. For many land plants mycorrhizal associations are one way of guaranteeing adequate phosphorus absorption from reserves in the soil.

The central role of the fungal symbiont in the provision of greater efficiency of phosphate absorption has been shown both *in vitro* and *in vivo* (see Harley & Smith, 1983, for references; Hale & Sanders, 1982; Jakobsen, 1986). Modifications of phosphate uptake properties of root systems arising from mycorrhizal infection will, consequently, depend largely upon (a) development of extramatrical hyphae in soil and their absorption of phosphorus (see also Hetrick, Chapter 10), (b) translocation of phosphorus through hyphae, and (c) mycelial development within root tissues and transfer of phosphorus from fungus to root cells (Smith & Gianinazzi-Pearson, 1988).

Phosphorus sources exploited by mycorrhizal systems

There is now no doubt that mycorrhizas readily absorb soluble phosphate from the labile pool of phosphorus in soils (see Harley & Smith, 1983, for references; Thomas *et al.*, 1982; Gianinazzi-Pearson, 1986), but their role in the removal or breakdown of insoluble, condensed or complexed forms of phosphate has still not been clearly elucidated.

Mycorrhizal plants sometimes respond more than nonmycorrhizal plants to application of insoluble or poorly soluble phosphate fertilizers (Pairunan, Robson & Abbott, 1980; Cabala-Rosand & Wild, 1982; Harley & Smith, 1983)(Table 11.1), and we have recently been investigating the mechanisms involved in the growth enhancement of vesicular-arbuscular (VA) infected plants. In two tropical acid soils amended with a poorly soluble rock phosphate and labelled with ^{32}P in the soil solution, VA-mycorrhizal palms grew better and used more fertilizer than those which were nonmycorrhizal (Table 11.1). However, similar values for specific activity and percentage phosphorus derived from fertilizer were observed in all the plants, indicating that they had adsorbed from the same pool of phosphorus in the soil. It is highly probable, therefore, that the response of the mycorrhizal plants to the fertilizer was due to their more efficient absorption of those phosphate ions drawn into solution following natural dissolution of the rock phosphate by the low soil pH. This is confirmed by the fact that application of soluble triple superphosphate fertilizer gave identical results to those obtained with rock phosphate (Table 11.1). In experiments using ^{32}P-labelled tricalcium phosphate, Azcon-Aguilar et al. (1986) showed that in a neutral soil, where this fertilizer remains insoluble, VA mycorrhiza formation did not, in fact, improve plant utilization of this form of phosphate. Ecto- and ericoid mycorrhizal fungi can cause acidification of culture media (see Harley & Smith, 1983; Bajwa & Read, 1985; St John et al., 1985), but whether this is of significance for phosphate dissolution in soils by corresponding mycorrhizal root systems has not been determined.

Mycorrhizal infection may, however, increase the ability of root systems to break down other condensed or complexed forms of phosphate since it is known that the activity of root surface phytases and acid phosphatases can be enhanced by mycorrhiza formation (Williamson & Alexander, 1975; Antibus et al., 1981; Gianinazzi-Pearson & Gianinazzi, 1981, 1986; Doumas et al., 1986; Dodd et al., 1987). Ectomycorrhizas can also affect the activity of surface phosphatases which break down inorganic pyrophosphate and tripolyphosphate (Doumas et al., 1986). Although stimulation of enzyme activity can vary with the fungal species in the mycorrhizal complex (Antibus et al., 1981; Gianinazzi-Pear-

Table 11.1. Utilization of two phosphate fertilizers (66 ppm P) by oil palms in a ^{32}P-labelled sandy clay acid soil* from Soubré (Ivory Coast).

Plant treatment	Fertiliser	Dry weight (g shoot^{-1})	Fertiliser utilisation coefficient (%)	% P derived from fertiliser	Specific activity (^{32}P/^{31}P)
Non-Mycorrhizal	none	0·6	-	-	-
	rock phos-phate	1·5	4·3	65	1·7
	triple super-phosphate	1·6	5·0	75	1·9
Mycorrhizal	none	2·4	-	-	-
	rock phos-phate	3·3	16·8	76	1·9
	triple super-phosphate	3·1	13·8	74	2·0

* Soil water pH = 5·2; ppm P (Olsen) = 3·9. Unpublished data of Blal, Morel, Gianinazzi-Pearson, Fardeau & Gianinazzi.

---

son & Gianinazzi, 1986; Dodd *et al.*, 1987), it is not possible to say whether the effects are host-mediated and/or due to fungal enzymes. In the case of ecto- and ericoid mycorrhizas, the fungal symbionts are known to produce and excrete acid phosphatases and phytases in pure culture (Pearson & Read, 1975; Ho & Zak, 1979; Calleja *et al.*, 1980; Dighton, 1983; Bousquet, Mousain & Salsac, 1986; Doumas *et al.*, 1986; Mousain & Salsac, 1986; Straker & Mitchell, 1986), and electron microscope studies have revealed that hyphae of the mycorrhizal fungi growing on the surface of host roots have intense acid phosphatase activity associated with their walls (Dexheimer *et al,*, 1986; Gianinazzi-Pearson, Bon-fante-Fasolo & Dexheimer, 1986).

Straker & Mitchell (1986) purified and characterised acid phos-phatases of an ericoid mycorrhizal fungus from *Erica hispidula*. They found that a large proportion of the enzyme activity was as-sociated with a high molecular weight wall-bound phosphatase which could use a wide range of phosphate esters as substrates, in-cluding, as with ectomycorrhizas, pyrophosphate, This phospha-tase has now been purified further and used to raise polyclonal antibodies (Straker et al., 1989). Ultrastructural localization of acid phosphatase using cytochemical techniques and indirect immunogold labelling with the anti-phosphatase antiserum confirmed that the enzyme is associated with the hyphal wall of the ericoid fungus in pure culture (Fig. 11.1 a & b). Using the same antiserum it has been possible to detect the enzyme on hyphae associated with the host root (Fig. 11.1d). Preliminary light and electron microscope studies (V. Gianinazzi-Pearson, D. Mousain, J. Dexheimer, J. C. Cleyet-Marel, J. Lei & S. Gianinazzi, unpublished data) using polyclonal antibodies against phosphatase of *Pisolithus tinctorius* (Bousquet, 1988) indicate that, as for ericoid mycorrhizas, acid phosphatase associated with the fungal hyphae in ectomycorrhizas is also the same as that produced by the fungus in pure culture.

It is highly probable, therefore, that ericoid and ectomycorrhizal fungi can play a role in the mobilization of phosphorus from certain condensed or complexed forms of organic and inorganic phosphate in the soil. Such a role for VA fungi has yet to be demonstrated and more research in this area is clearly necessary as phosphate mobilising enzymes could be of considerable ecological importance in natural soils (Woolhouse, 1969; MacLachlan, 1976).

Fig. 11.1. *(Facing page)* Electron (a-c) and light (d & e) micrographs of cytochemical localisation of acid phosphatase in an ericoid mycorrhizal fungus from *Erica hispidula.*

(a) β-glycerophosphatase activity (pH 5·0) and (b) immunogold labelling (anti-phosphatase antiserum 1/10000) of wall phosphatase (arrows) in hyphae growing in pure culture;

(d) vizualization of antigenic sites of the enzyme (arrows) using the silver enhancement technique (Vanderbosch, 1986), on hyphae associated with a host root;

(c) and (e) controls prepared without anti-phosphatase antiserum with no labelling (open arrows) of fungus.

Unpublished data of V. Gianinazzi-Pearson, C. J. Straker & S. Gianinazzi.

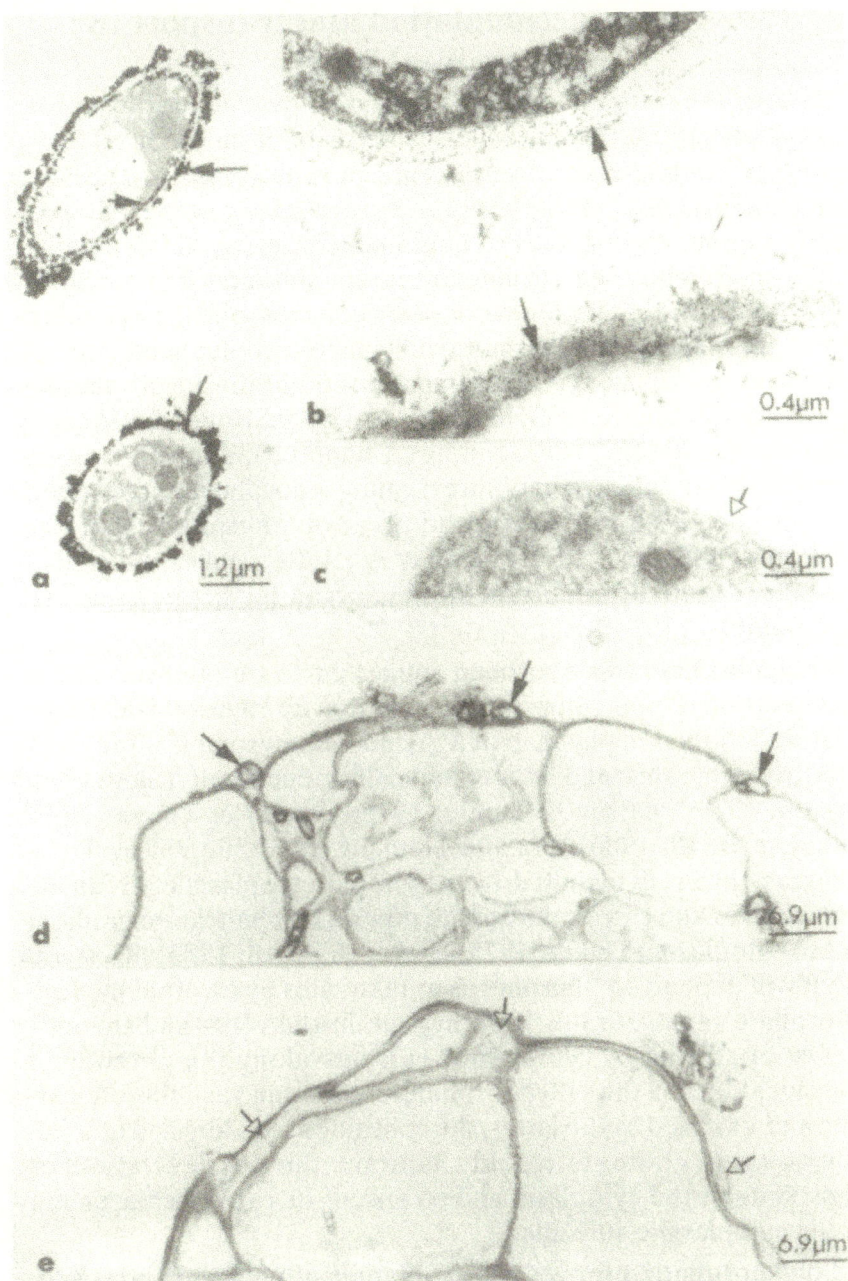

Phosphorus accumulation and transport by mycorrhizal fungi

Whatever the source of phosphate used and whatever the mycorrhizal association, the extramatrical hyphae of mycorrhizal fungi absorb, accumulate and translocate phosphate ions to the host root (see Harley & Smith, 1983). Absorption is an active, metabolically dependent process (Gianinazzi-Pearson & Gianinazzi, 1986) and hyphae can accumulate large amounts of phosphorus which would otherwise interfere with cell metabolism, by converting it into condensed, osmotically inactive polyphosphate, of which at least a part is compartmentalised in the fungal vacuoles (see Harley & Smith, 1983; Martin *et al.*, 1985; Straker & Mitchell, 1985; Ashford *et al.*, 1986; Jennings, Chapter 1). The ability of ectomycorrhizal fungi to mobilise rapidly phosphorus from their polyphosphate store under conditions of phosphate shortage (Harley & Brierley, 1954; Martin *et al.*, 1985) makes the sheath tissue a potentially important storage organ for ectomycorrhizal host plants.

Polyphosphate has also been suggested to be involved in the mechanism of phosphorus translocation by mycorrhizal fungi. Translocation along hyphae of VA and ectomycorrhizal fungi occurs at high rates and is metabolically dependent (Skinner & Bowen, 1974; Pearson & Tinker, 1975; Cooper & Tinker, 1978; Finlay & Read, 1986). For the moment, experimental evidence suggests that it is mainly driven by rapid cytoplasmic streaming which mixes and carries phosphate down a concentration gradient in the cytoplasm (Tinker, 1975; Harley & Smith, 1983). This gradient will depend on phosphorus uptake rates by external hyphae, phosphate removal rates from hyphae in the roots and the efficiency of stirring by cytoplasmic currents along the pathway of translocation. As the polyphosphate-containing vacuoles are carried and/or mixed by the latter, the continuous loading and unloading of their contents would maintain the concentration of phosphate in the cytoplasm and so ensure its rapid displacement in the cytoplasmic currents.

Polyphosphate turnover and rapid rates of phosphorus translocation are not, however, distinguishing features of mycorrhizal fungi and they can be found in saprotrophic and parasitic fungal groups (Beever & Burns, 1980). What does appear to be specific to mycorrhizal fungi is the persistent functional compatibility they establish with host tissues (Gianinazzi-Pearson, 1984), and their continuous transfer of phosphorus to the phytosymbiont.

Phosphorus transfer from myco- to phytosymbiont

Large areas of surface contact between fungus and host cells are features of all mycorrhizas but processes occurring at the fungus-host interface are poorly understood. It is not known what triggers phosphate loss from hyphae in the presence of host root cells. The value for phosphorus flux of 780 nmol $m^{-2} min^{-1}$ calculated, by Cox & Tinker (1976), to occur across the fungus-plant interface in arbuscules of VA mycorrhizas is similar to those for phosphorus influx into fungal hyphae and giant algal cells (Beever & Burns, 1980). Cells do not normally have high rates of phosphorus efflux and so phosphorus loss to the host in mycorrhizas must result from modifications in fungal efflux, possibly brought about through host-induced increase in hyphal membrane permeability. However, no mechanisms for such effects are known. Efflux of phosphorus from fungal hyphae can be greatly increased by treatments, like high external sugar concentrations, which affect plasmalemma integrity (Beever & Burns, 1980), and it has been previously suggested that phosphate transfer from fungus to host in mycorrhizas may somehow be coupled to carbohydrate uptake from the host pool in which they bathe (Harley & Smith, 1983). Phosphate release would lead to reduced cytoplasmic concentrations of phosphorus in hyphae and consequently mobilization of polyphosphate stores. Vacuolar polyphosphate granules disappear from the terminal hyphae developing in host cells in VA mycorrhizas (Cox *et al.*, 1975; Strullu *et al.*, 1981) (Fig. 11.2) and polyphosphatases are characteristically localised within the internal mycelium (Capaccio & Callow, 1982).

Structural modifications occurring in host and fungus in the region of cellular contact between the mycorrhizal symbionts lead to the formation of a simplified interface which is highly favourable for the exchange of metabolites between them (Scannerini & Bonfante-Fasolo, 1983; Gianinazzi-Pearson, 1984; Duddridge & Read, 1984). Furthermore, when physiologically active cells of the symbionts come into contact with each other, ATPase activity is

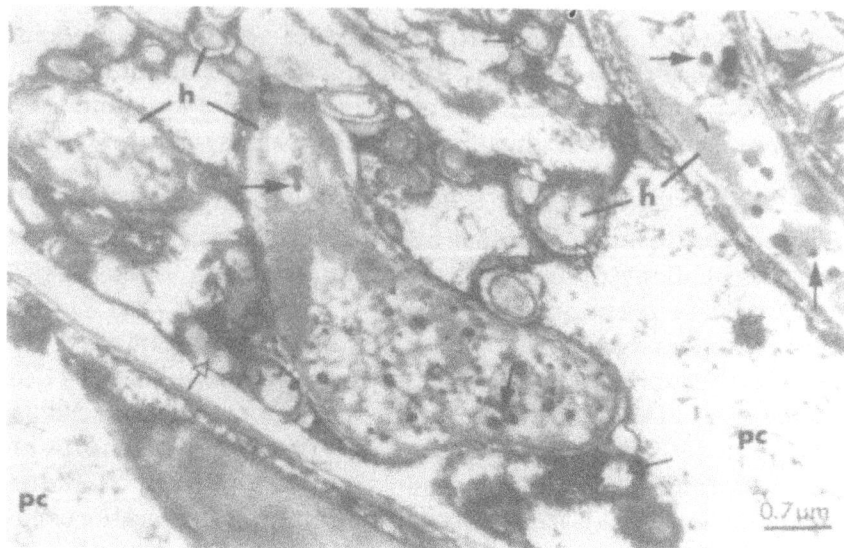

Fig. 11.2. Electron micrograph of arbuscular hyphae (h) of *Glomus fasciculatum* in a parenchymal cell (pc) of the root cortex of *Rubus idaeus*; electron dense polyphosphate granules are present (arrows) in vacuoles of larger hyphae and absent (open arrows) from fine arbuscule branches (illustration courtesy of D. Morandi).

associated with the plasmalemma of both (Marx *et al.*, 1982; Gianinazzi *et al.*, 1983; Lei & Dexheimer, 1988), indicating the existence of an energy generating system for transmembrane transport of metabolites by either or both partners (Smith & Smith, 1986). In the case of VA mycorrhizas, localization of host plasmalemma-bound ATPase activity around the fungus only occurs in the parenchyma cells of the root cortex and in the presence of living branches of the arbuscules (Gianinazzi-Pearson & Gianinazzi, 1988; Gianinazzi-Pearson *et al.* 1988). It is not known whether this specific localization involves activation of the enzyme, or results from induced changes in genome expression of the host cell with *de novo* synthesis of ATPase; these possibilities are presently being investigated using immunocytochemical techniques. The overall result of VA mycorrhiza formation is, consequently, that the parenchyma cortical cells acquire a new function of phosphate absorption, generally restricted to the outer root cells, and which is directly linked to the activity of the fungus in channelling phosphate to these cells.

It is the persistence of this cellular and functional compatibility between host and fungal cells as mycorrhizal infections develop which is a particular distinguishing feature of this type of mutualism. Mutualistic phases have been observed in the very early stages of pathogen infections (Lucas, 1977; O'Connell, 1987) but these are transient, short lived features which rapidly degenerate into pathogenic, fungus dominated infections or resistance.

Effects of external phosphate on mycorrhiza development and phosphorus metabolism

Mycorrhizas are most effective in phosphate nutrition at low to moderate levels of phosphorus in the growth medium and, in general, mycorrhizal effects decrease with increasing levels of available phosphate (Harley & Smith, 1983; Gianinazzi-Pearson, 1986; Hetrick, Chapter 10). Although this is mainly explained by the better phosphorus nutrition of nonmycorrhizal plants, increasing phosphate can also affect development and physiology of the mycorrhizal infection.

Abbott, Robson & De Boer (1984) have shown that increases in soluble phosphate reduce the development of extramatrical hyphae in VA mycorrhizal systems well before significantly affecting percentage infection levels within roots. Arbuscule development in tissues is also reduced before overall infection levels drop (Trouvelot, Kough & Gianinazzi-Pearson, 1986) so that the essential phases for absorption and transfer of phosphorus are the first to be affected by increasing soil phosphate concentrations. It has been suggested that such phosphate effects may be due to membrane-mediated changes in root exudation or in metabolite concentrations, which would reduce the availability of metabolites to the fungus (Graham, Leonard & Menge, 1982; Schwab, Menge & Leonard, 1983), but this remains to be confirmed. In recent experiments by Son & Smith (1988) on effects of phosphate and light supplies on phosphate metabolism in mycorrhizal onions (*Allium cepa*), it was shown that in plants growing at high light intensities phosphate fertilizer reduced overall infection levels but phosphorus inflow rates remained considerably higher into mycorrhizal than non mycorrhizal plants. However, phosphorus inflow rates

decreased in mycorrhizal plants with similar infection development at the highest phosphorus level when irradiance was reduced, whilst there was no effect on inflow into non mycorrhizal
plants. Reductions of inflow were, therefore, probably due to reduced activity of the fungal pathway of phosphate transfer to the
plant, rather than to reduced infection or to direct effects on host
phosphorus uptake. This reduced fungal activity at low levels of
light supply and high phosphate concentrations could be due to
reduced growth of external mycelium and/or arbuscule formation
(which were not measured), or due to modifications in the nutrient
transport and/or transfer processes in the mycorrhizal system.

Very little is known about the effect that high phosphate concentrations may have on the phosphate physiology of mycorrhizal
fungi. In VA fungi, phosphate can affect rates of cytoplasmic
streaming in hyphae developing from germinating spores (Pons &
Gianinazzi-Pearson, 1984) and alkaline phosphatase activity in
VA fungal mycelium developing within roots (Gianinazzi, Gianinazzi-Pearson & Dexheimer, 1979; Gianinazzi-Pearson & Gianinazzi, 1986) is considerably reduced by applications of phosphate
fertilizer to soil (Gianinazzi-Pearson & Gianinazzi, 1983).

Conclusions

The distinguishing feature of mycorrhizal fungi is not so much
their ability to accumulate and move large quantities of phosphate
along their hyphae since certain pathogenic and saprotrophic
fungi also do this, but rather their ability to sustain functional connections between soil and plant tissues which is coupled with the
establishment of a durable phase of compatibility with host cells,
enabling the transfer of phosphate to take place. The uptake and
translocation of phosphate by the fungi, together with transport
processes at the host-fungus interface, must be rate-limiting steps
which determine the supply of phosphorus to mycorrhizal plants.
Much more effort is needed to further our understanding of the
molecular basis of compatibility at the host-fungus interface in mycorrhizas, and of the mechanisms whereby fungal and root cells interact to control nutrient uptake and transfer between the
symbionts.

References

Abbott, L. K., Robson, A. D. & De Boer, G. (1984). The effect of phosphorus on the formation of hyphae in soil by the vesicular-arbuscular mycorrhizal fungus *Glomus fasciculatum*. *New Phytologist*, **97**, 437-446.

Antibus, R. K., Croxdale, J. G., Miller, O. K. & Linkins, A. E. (1981). Ectomycorrhizal fungi of Salix rotundifolia. III. Resynthesized mycorrhizal complexes and their surface phosphatase activities. *Canadian Journal of Botany*, **59**, 2458-2465.

Ashford, A. E., Peterson, R. L., Dwarte, D. & Chilvers, G. A. (1986). Polyphosphate granules in eucalypt mycorrhizas: determination by energy dispersive X-ray microanalysis. *Canadian Journal of Botany*, **64**, 677-687.

Azcon-Aguilar, C., Gianinazzi-Pearson, V., Pardeau, J. C. & Gianinazzi, S. (1986). Effect of vesicular-arbuscular mycorrhizal fungi and phosphate solubilising bacteria on growth and nutrition of soybean in a neutral-calcareous soil amended with ^{32}P-^{45}Ca tricalcium phosphate. *Plant and Soil*, **96**, 3-15.

Bajwa, R. & Read, D. J. (1985). The biology of mycorrhiza in the Ericaceae. IX. Peptides as nitrogen sources for the ericoid endophyte and for mycorrhizal and nonmycorrhizal plants. *New Phytologist*, **101**, 459-467.

Beever, R. E. & Burns, D. J. W. (1980). Phosphorus uptake storage and utilisation by fungi. *Advances in Botanical Research*, **8**, 128-219.

Bousquet, N. (1988). Étude immunoenzymatique de phosphatases acides de champignons ectomycorhiziens. Doctorat Thesis, Université des Sciences et Techniques de Languedoc, Montpellier, France.

Bousquet, N., Mousain, D. & Salsac, L. (1986). Influence de l'orthophosphate sur les activités phosphatases de *Suillus granulatus* en culture *in vitro*. *Physiologie Végétale*, **24**, 153-162.

Cabala-Rosand, P. & Wild, A. (1982). Direct use of low grade phosphate rock from Brazil as fertilizer. II. Effects of mycorrhizal inoculation and nitrogen source. *Plant and Soil*, **65**, 363-373.

Calleja, M., Mousain, D., Lecouvreur, B. & d'Auzac, J. (1980). Influence de la carence phosphatée sur les activités phosphatases acides de trois champignons mycorhiziens: *Hebeloma edurum* Metrod., *Suillus granulatus* (L. ex Fr.) O. Kuntze et *Pisolithus tinctorius* (Pers.) Cooker et Couch. *Physiologie Végétale*, **18**, 489-504.

Capaccio, L. C. M. & Callow, J. A. (1982). The enzymes of polyphosphate metabolism in vesicular-arbuscular mycorrhizas. *New Phytologist*, **91**, 81-91.

Cooper, K. M. & Tinker, P. B. (1978). Translocation and transfer of nutrients in vesicular-arbuscular mycorrhizas. II. Uptake and translocation of phosphorus, zinc and sulphur. *New Phytologist*, **81**, 43-52.

Cox, G. & Tinker, P. B. (1976). Translocation and transfer of nutrients in vesicular-arbuscular mycorrhizas. I. The arbuscule and phosphorus transfer: a quantitative ultrastructural study. *New Phytologist*, **77**, 371-378.

Cox, G., Sanders, F. E., Tinker, P. B. & Wild, J. A. (1975). Ultrastructural evidence relating to host-endophyte transfer in a vesicular-arbuscular mycorrhiza. In *Endomycorrhizas*, eds. F. E. Sanders, B. Mosse & P. B. Tinker, pp. 297-312. Academic Press: London & New York.

Dexheimer, J., Aubert-Dufresne, M. P., Gérard, J., Letacon, F. & Mousain, D. (1986). Étude de la localisation ultrastructurale des activités phosphatasiques acides dans deux types d'ectomycorhizes: *Pinus nigra nigricans/Hebeloma crustu-*

liniforme et *Pinus pinaster/Pisolithus tinctorius. Bulletin de la Société Botanique Française,* **133,** 343-352.

Dighton, J. (1983). Phosphatase production by mycorrhizal fungi. *Plant and Soil,* **71,** 455-462.

Dodd, J. C., Burton, C. C., Burns, R. G. & Jeffries, P. (1987). Phosphatase activity associated with the roots and the rhizosphere of plants infected with vesicular-arbuscular mycorrhizal fungi. *New Phytologist,* **107,** 163-172.

Doumas, P., Berjaud, C., Calléja, M., Coupé, M., Espiau, C. & d'Auzac, J. (1986). Phosphatases extracellulaires et nutrition phosphatases chez les champignons ectomycorhiziens et les plantes-hôtes. *Physiologie Végétale,* **24,** 173-184.

Duddridge, J. A. & Read, D. J. (1984). The development and ultrastructure of ectomycorrhizas. II. Ectomycorrhizal development on pine *in vitro. New Phytologist,* **96,** 575-582.

Finlay, R. D. & Read, D. J. (1986). The structure and function of the vegetative mycelium of ectomycorrhizal plants. II. The uptake and distribution of phosphorus by mycelial strands of interconnecting plants. *New Phytologist,* **103,** 157-165.

Gianinazzi, S., Gianinazzi-Pearson, V. & Dexheimer, J. (1979). Enzymatic studies on the metabolism of vesicular-arbuscular mycorrhiza. III. Ultrastructural localisation of acid and alkaline phosphatase in onion roots infected by *Glomus mosseae* (Nicol. Gerd.). *New Phytologist,* **82,** 127-132.

Gianinazzi, S., Dexheimer, J., Gianinazzi-Pearson, V. & Marx, C. (1983). Role of the host-arbuscule interface in the symbiotic nature of VA mycorrhizal associations: ultracytological studies of processes involved in phosphate and carbohydrate exchange. *Plant and Soil,* **71,** 211-215.

Gianinazzi-Pearson, V. (1984). Host-fungus specificity, recognition and compatibility in mycorrhizae. In *Genes Involved in Microbe Plant Interactions. Advances in Plant Gene Research, Basic Knowledge and Application,* eds. E. S. Dennis, B. Hohn, Th. Hohn, P. King, I. Schell & D. P. S. Verma, pp. 225-253. Springer-Verlag: Vienna & New York.

Gianinazzi-Pearson, V. (1986). Mycorrhizae: a potential for better use of phosphate fertilizer. *Fertilizers and Agriculture,* **92,** 3-12.

Gianinazzi-Pearson, V. & Gianinazzi, S. (1981). Role of endomycorrhizal fungi in phosphorus cycling in the ecosystem. In *The Fungal Community; its Organization and Role in the Ecosystem,* eds. D. T. Wicklow & G. C. Carrol, pp. 637-652, Marcel Dekker Inc., New York.

Gianinazzi-Pearson, V. & Gianinazzi, S. (1983). The physiology of vesicular-arbuscular mycorrhizal roots. *Plant and Soil,* **71,** 197-209.

Gianinazzi-Pearson, V. & Gianinazzi, S. (1986). The physiology of improved phosphate nutrition in mycorrhizal plants. In Physiological and Genetical Aspects of Mycorrhizae, eds. V. Gianinazzi-Pearson & S. Gianinazzi, pp. 101-109. INRA Publications, Paris.

Gianinazzi-Pearson, V. & Gianinazzi, S. (1988). Morphological integration and functional compatibility between symbionts in vesicular-arbuscular endomycorrhizal associations. In *Cell to Cell Signals in Plant, Animal and Microbial Symbiosis,* NATO ASI series vol. 17, eds. S. Scannerini, D. C. Smith, P. Bonfante-Fasolo & V. Gianinazzi-Pearson, pp. 73-84. Springer-Verlag, Berlin.

Gianinazzi-Pearson, V., Bonfante-Fasolo, P. & Dexheimer, J. (1986). Ultrastructural studies of surface interactions during adhesion and infection by ericoid endomycorrhizal fungi. In *Recognition in Microbe-Plant Symbiotic and Pathogenic Interactions*, NATO ASI series vol. 4, ed. B. Lugtenberg, pp. 273-282. Springer-Verlag, Berlin.

Gianinazzi-Pearson, V., Gianinazzi, S., Dexheimer, J., Morandi, D., Trouvelot, A. & Dumas, E. (1988). Recherche sur les mécanismes intervenant dans les interactions symbiotiques plant-champignons endomycorhizogènes VA. *Cryptogamie Mycologie*, **9**, 201-209.

Graham, J. H., Leonard, R. T. & Menge, J. A. (1982). Interactions of light intensity and soil temperature with phosphorus inhibition of vesicular-arbuscular mycorrhiza formation. *New Phytologist*, **91**, 683-690.

Hale, K. A. & Sanders, F. E. (1982). Effects of benomyl on vesicular-arbuscular mycorrhizal infection of red clover (*Trifolium pratense* L.) and consequences for phosphorus inflow. *Journal of Plant Nutrition*, **5**, 1355-1367.

Harley, J. L. & Brierley, J. K. (1954). The uptake of phosphate by excised mycorrhizal roots of the beech. VI. Active transport of phosphate from the fungal sheath into the host tissue. *New Phytologist*, **53**, 240-252.

Harley, J. L. & Smith, S. E. (1983). *Mycorrhizal Symbiosis*. Academic Press: London & New York.

Ho, I. & Zak, B. (1979). Acid phosphatase activity of six ectomycorrhizal fungi. *Canadian Journal of Botany*, **57**, 1203-1205.

Jakobsen, I. (1986). Vesicular-arbuscular mycorrhiza in field-grown crops. III. Mycorrhizal infection and rates of phosphorus inflow in pea plants. *New Phytologist*, **104**, 573-581.

Lei, J. & Dexheimer, J. (1988). Ultrastructural localisation of ATPase activity in the *Pinus silvestris/Laccaria laccata* ectomycorrhizal association. *New Phytologist*, **108**, 329-334.

Lucas, R. L. (1977). The movement of nutrients through fungal mycelium. *Transactions of the British Mycological Society*, **69**, 1-9.

MacLachlan, K. D. (1976). Comparative responses in plants to a wide range of phosphorus situations. *Australian Journal of Agriculture Research*, **27**, 323-341.

Martin, F., Marchal, J. P., Timinska, A. & Canet, D. (1985), The metabolism and physical state of polyphosphate in ectomycorrhizal fungi. A ^{31}P nuclear magnetic resonance study. *New Phytologist*, **101**, 275-290.

Marx, C., Dexheimer, J., Gianinazzi-Pearson, V. & Gianinazzi, S. (1982). Enzymatic studies on the metabolism of vesicular-arbuscular mycorrhiza. IV. Ultra-cytoenzymological evidence (ATPase) for active transfer processes in the host-arbuscule interface. *New Phytologist*, **90**, 37-43.

Mousain, D. & Salsac, L. (1986). Utilisation du phytate et activités phosphatases acides chez *Pisolithus tinctorius*, basidiomycète mycorhizien. *Physiologie Végétale*, **24**, 193-200.

O'Connell, R. J. (1987). Absence of a specialised interface between intracellular hyphae of *Colletotrichum lindemuthianum* and cells of *Phaseolus vulgaris*. *New Phytologist*, **107**, 725-734.

Pairunan, A. K., Robson, A. D. & Abbott, L. K. (1980). The effectiveness of vesicular-arbuscular mycorrhizas in increasing growth and phosphorus uptake of subterranean clover from phosphorus sources of different solubilities. *New Phytologist*, **84**, 327-338.

Pearson, V. & Read, D. J. (1975). The physiology of the mycorrhizal endophyte of *Calluna vulgaris*. *Transactions of the British Mycological Society*, **64**, 1-7.

Pearson, V. & Tinker, P. B. (1975). Measurement of phosphorus fluxes in the external hyphae of endomycorrhizas. In *Endomycorrhizas*, eds. F. E. Sanders, B. Mosse & P. B. Tinker, pp. 277-287. Academic Press: London.

Pons, F. & Gianinazzi-Pearson, V. (1984). Influence du phosphore, du potassium, de l'azote et du pH sur le comportement *in vitro* de champignons endomycorhizogenes à vésicules et arbuscules. *Cryptogamie Mycologie*, **5**, 87-100.

Scannerini, S. & Bonfante-Fasolo, P. (1983). Comparative ultrastructural analysis of mycorrhizal associations. *Canadian Journal of Botany*, **61**, 917-943.

Schwab, S. M., Menge, J. A. & Leonard, R. T. (1983). Quantitative and qualitative effects of phosphorus on extracts and exudates of sudangrass in relation to vesicular-arbuscular mycorrhiza formation. *Plant Physiology*, **73**, 761-765.

Skinner, M. F. & Bowen, G. D. (1974). The uptake and translocation of phosphate by mycelial strands of pine mycorrhizae. *Soil Biology and Biochemistry*, **6**, 53-56.

Smith, F. A. & Smith, S. E. (1986). Movement across membranes: physiology and biochemistry. In *Physiological and Genetical Aspects of Mycorrhizae*, eds. V. Gianinazzi-Pearson & S. Gianinazzi, pp. 75-84. INRA Publications, Paris.

Smith, S. E. & Gianinazzi-Pearson, V. (1988). Physiological interactions between symbionts in vesicular-arbuscular mycorrhizal plants. *Annual Review of Plant Physiology and Plant Molecular Biology*, **39**, 221-244.

Son, C. L. & Smith, S. E. (1988). Mycorrhizal growth responses: interactions between photon irradiance and phosphorus nutrition. *New Phytologist*, **108**, 305-314.

St. John, B. J., Smith, S. E., Nicholas, D. J. D. & Smith, F. A. (1985). Enzymes of ammonium assimilation in the mycorrhizal fungus *Pezizella ericae* (Read). *New Phytologist*, **100**, 579-584.

Straker, C. J. & Mitchell, D. T. (1985). The characterization and estimation of polyphosphates in endomycorrhizas of the Ericaceae. *New Phytologist*, **99**, 431-440.

Straker, C. J. & Mitchell, D. T. (1986). The activity and characterization of acid phosphatases in endomycorrhizal fungi of the Ericaceae. *New Phytologist*, **104**, 243-256.

Straker, C. J., Gianinazzi-Pearson, V., Gianinazzi, S., Cleyet-Marel, J. C. & Bousquet, N. (1989). Electrophoretic and immunological studies on acid phosphatase from a mycorrhizal fungus of *Erica hispidula* L. *New Phytologist*, **111**, in press.

Strullu, D. G., Gourret, J. P., Garrec, J. P. & Fourcy, A. (1981). Ultrastructure and electron-probe microanalysis of the metachromatic vacuolar granules occurring in *Taxus* mycorrhizas. *New Phytologist*, **87**, 537-545.

Thomas, G. W., Clarke, C. A., Mosse, B. & Jackson, R. M. (1982). Source of phosphate taken up from two soils by mycorrhizal (*Thelephora terrestris*) and non-mycorrhizal *Picea sitchensis* seedlings. *Soil Biology and Biochemistry*, **14**, 73-75.

Tinker, P. B. (1975). Effects of vesicular-arbuscular mycorrhizas on higher plants. *Symposium of the Society for Experimental Biology*, **29**, 325-329.

Trouvelot, A., Kough, J. & Gianinazzi-Pearson, V. (1986). Mesure du taux de mycorhization VA d'un système radiculaire. Recherche de méthodes d'estimation ayant une signification fonctionnelle. In *Physiological and Genetical Aspects of Mycorrhizae*, eds. V. Gianinazzi-Pearson & S. Gianinazzi, pp. 217-221. INRA Publications, Paris.

Vanderbosch, K. A. (1986). Light and electron microscopic visualization of uricase by immunogold labelling of sections of resin-embedded soybean nodules. *Journal of Microscopy*, **143**, 187-197.

Westheimer, F. H. (1987). Why nature chose phosphates. *Science*, **235**, 1173-1178.

Williamson, B. W. & Alexander, I. J. (1975). Acid phosphatase localised in the sheath of beech mycorrhiza. *Soil Biology and Biochemistry*, **7**, 195-198.

Woolhouse, H. W. (1969). Differences in the properties of the acid phosphatases of plant roots and their significance in the evolution of edaphic ecotypes. In *Ecological Aspects of the Mineral Nutrition of Plants*, ed. I. H. Rorison, pp. 357-380. Blackwell, Oxford.

Chapter 12

Nitrogen relations of mat-forming lichens

P. D. Crittenden

Department of Botany, University of Nottingham,
Nottingham NG7 2RD, UK

Introduction

Of the three principal types of mutualistic symbiosis found in terrestrial environments – lichens, mycorrhizas and legumes/actinorhizas, the lichens are the least well understood. Interactions between the symbionts in mycorrhizas and legumes are known to involve bilateral exchanges of nutrients: photosynthate moving from photobiont to microsymbiont and mineral nutrients from microsymbiont to photobiont. These nutrient fluxes have been studied intensively, mainly in attempts to determine costs and benefits to the vascular plant hosts. In contrast, our view of the interplay between symbionts in lichens has been coloured almost entirely by the results of studies on carbon nutrition (cf. Smith & Douglas, 1987). With the exception of nitrogen metabolism in cyanobacterial lichens, about which there is now a growing body of information, knowledge of nitrogen, phosphorus, potassium and sulphur relationships in lichens is comparatively rudimentary.

This situation has arisen partly because lichenology has not benefited from an economic impetus comparable to that driving research on the legume and mycorrhizal symbioses, but it also reflects difficulties posed by lichens as experimental systems. For example, experiments in which photobionts are grown either in the presence or absence of a heterotrophic symbiont have been of paramount importance in elucidating interactions in higher plant symbioses. In the case of lichens this approach is blocked because, among other things, lichens do not readily grow under laboratory conditions. Moreover, frequent reports of 'negative' results, such as absence of clear seasonal changes in thallus nutrient concentrations (Lewis Smith, 1978; Puckett, 1985) and the apparent insensitivity of growth and photosynthetic rates to inorganic nutrient additions (Armstrong, 1977; Carstairs & Oechel, 1978; Bailey & Larson, 1982; Matthes-Sears, Nash & Larson, 1987), may have done much to create an impression that lichen mineral nutrient relations is an unrewarding field of enquiry. These problems make the subject all the more challenging.

There are several crucial questions on which quantitative information is needed: (i) at what rates and from what sources do mineral elements arrive at lichen thalli; (ii) how efficiently do lichens capture nutrients from these supplies; (iii) does uptake satisfy potential growth requirement; (iv) by which pathways in the thallus do mineral nutrients reach the photobiont? These questions are discussed in this chapter with particular reference to nitrogen and to the results of recent work on the ecology of mat-forming lichens in subarctic woodlands. Ecological and physiological parallels between lichens and mycorrhizas are briefly discussed.

Aspects of the ecology of subarctic lichen woodland

Lichen woodlands occur extensively on well-drained terrain throughout the northern boreal and forest-tundra zones. They consist of open stands of trees with lichen-dominated ground cover and often develop as seral communities on sites of forest fires (Ahti, 1977; Kershaw, 1977; Auclair, 1983; Foster, 1985) but may also arise in the absence of fire to form mature and stable vegetation (Auclair, 1983; Rowe, 1984; Payette *et al.*,1985). Lichen woodland soils are acid, oligotrophic podzols or brunisols (Kershaw, Rouse & Bunting, 1975; Ahti, 1977; Moore, 1980; Ugolini *et al.*, 1981; Sepponen, 1985; Rowe & Acton, 1985). The most important mat-forming lichens in the ground cover are species of *Cladonia, Cetraria, Stereocaulon* and *Alectoria* but three species, *Cladonia stellaris, C. mitis* and *S. paschale*, are especially abundant and each frequently forms continuous carpets. The abundance of *S. paschale* is of particular interest because it is a tripartite lichen containing the cyanobacterium *Stigonema* in cephalodia and achieves appreciable rates of nitrogen fixation (Kallio & Kallio, 1975; Huss-Danell, 1977, 1978; Crittenden & Kershaw, 1978, 1979).

Fig. 12.1. Birch-lichen woodland near Ai'bmejav'ri, Finnmarksvidda, Norway (69° 05'N 24° 50'E): open stands of *Betula pubescens* ssp. *tortuosa* with interjacent lichen mats composed almost exclusively of *Cladonia stellaris*. The birch trees, each with an understorey of dwarf shrubs (principally *Empetrum hermaphroditum*, *Vaccinium vitis-idaea* and *B. nana*), appear as green islands in aerial view (b). *Betula nana* is seen in the foreground of (a).

In Europe, lichen woodland is best represented in the pine (*Pinus sylvestris*) and subarctic mountain birch (*Betula pubescens* ssp. *tortuosa*) forests of northernmost Fennoscandia (Hämet-Ahti, 1963; Oksanen & Ahti, 1982). In Finnmarksvidda, in particular, there are large tracts of open birch-*Cladonia stellaris* woodlands (Fig. 12.1). Beneath the canopy of each tree, where litter fall has resulted in comparatively deep accumulation of fibrous organic matter (cf. F horizon) (Fig. 12.2a), communities of dwarf shrubs occur comprised in the main of *Empetrum hermaphroditum*, *Vaccinium vitis-idaea* and *Betula nana* (Fig. 12.1). The reasons for the limited invasion of the inter-tree areas by these shrubs are not fully understood but probably include low soil nutrient status, periodic summer drought and phytotoxicity of lichen products (cf. Rowe, 1984; Bakaeva & Galanin, 1985). The suggestion that soil nutrient availability may restrict dwarf shrub distribution in these habitats is supported by the results of a bioassay in which growth of mycorrhizal *V. vitis-idaea* in a controlled environment was found to be poor in surface organic soil collected beneath lichen mats (cf. H horizon) and the periphery of birch canopies compared to that in organic topsoil taken closer to tree bases (Fig. 12.2b). There is also evidence that lichen secondary products may inhibit seed germination and plant growth (Lawrey, 1984, 1986; Kershaw, 1985) but most work on this topic has involved the use of ecologically improbable concentrations of metabolites and hence the ecological relevance of many of the reported effects remains to be critically assessed. Although it is indeed likely that lichen metabolites are washed into soil during rainfall (Malicki, 1965; García-Junceda & Xavier Filho, 1986; P. D. Crittenden & R. P. Beckett, unpublished data) they decompose comparatively rapidly and do not accumulate substantially (Dawson, Hrutfiord & Ugolini, 1984; Vainstein & Ravinskaya, 1984). At first sight the view of Brown & Mikola (1974) and Fisher (1979) that lichen ground cover can inhibit the formation of mycorrhizas does not appear to be borne out here since a cursory examination revealed both birch trees and dwarf shrubs to have well developed mycorrhizal associations even in the case of roots that extend out under lichen mats.

Read (1983, 1984) has suggested that vesicular arbuscular, ecto- and ericoid mycorrhizas are each associated with a specific set of

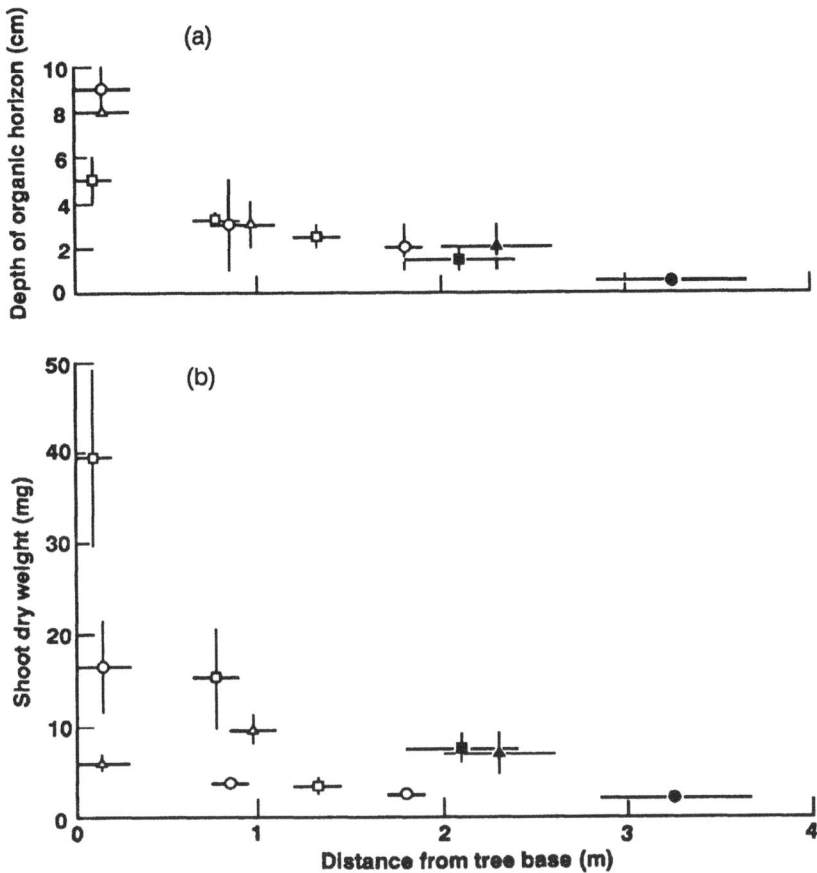

Fig. 12.2. (a) Relationship between distance from tree base and approximate depth of the organic soil horizon. Data for three birch trees each at different sites in Finnmarksvidda, Norway: *Cladonia stellaris*-rich woodland at Ai'bmejav'ri (circles, 69° 05'N 24° 50'E) and Av'zi (triangles, 69° 01'N 23° 15'E); *Stereocaulon paschale*-rich woodland, Kautokeino (squares, 69° 01'N 23° 10'E). Horizontal bars indicate areas in which measurements were made and samples of surface soil organic matter collected (see below) beneath both birch canopies (open symbols) and lichen mats (closed symbols). Vertical bars give ranges of values. (b) Shoot dry weight of *Vaccinium vitis-idaea* grown for 15 weeks in a controlled environment growth room in organic soils identified in (a). Surface sterilized seed of *V. vitis-idaea* was germinated on distilled water agar. When the cotyledons were expanded fully (45 days after sowing) seedlings were each transfered to 60 ml of soil contained in nylon mesh (150 μm aperture) bags embedded in washed silica sand in 150 ml capacity free-draining pots; the sand and soil surfaces were covered in opaque plastic beads to inhibit growth of algae and bryophytes. Sufficient deionized water was added to the sand at 1 or 2 day intervals to maintain the soil/sand system at field capacity with minimum leaching losses. Growth-room conditions: 330-365 μmol m^{-2} s^{-1} photosynthetically active radiation at soil level; 16/8h, 16/10°C light/dark cycle; c. 78% relative humidity. Mean values (n = 20) are plotted together with 95% confidence intervals. All plants were found to be heavily infected with mycorrhizal endophyte at harvest.

soil circumstances typical of a broad latitudinal and altitudinal zone. He considers ericoid mycorrhizas to be characteristic of marginal ecological situations at higher latitudes and altitudes where the capacity of the endophytes to exploit organic forms of nitrogen and phosphorus in nutrient poor heathland soils underlies their success (Read, Leake & Langdale, Chapter 9). This scheme might perhaps be extended further to include large areas of drier terrain in the subarctic and arctic where soil conditions appear to be even more extreme than those to which ericoid mycorrhizas are well adapted. In these habitats the potential for nutrient uptake from soil is so small and the vigour of vascular plants so reduced that yet another distinct type of mutualistic symbiosis between fungi and photoautotrophs becomes predominant. This involves the association of microalgae (and/or cyanobacteria) with highly differentiated aerial fungal mycelia to form lichen thalli. Since the fungal partner invests the cells of the photobiont in a sheath-like manner, there are certain structural similarities between lichen thalli and ectomycorrhizal roots. However, unlike mycorrhizas these mat-forming lichens probably derive relatively few nutrients from soil, the strategy developed instead is one of exploiting nutrient deposition from the atmosphere.

Scavenging nitrogen from the atmosphere

The widely held belief that lichens derive nutrients from the atmosphere is founded on two main lines of evidence. First, there are often good correlations between element concentrations in lichen thalli and those measured or expected in atmospheric deposits. Thus, when exposed to air pollution lichens may accumulate, among other things, sulphur (Gilbert, 1969; Pyatt, 1973; Tomassini et al., 1976; Pakarinen, 1981b; Zakshek, Puckett & Percy, 1986), heavy metals (e.g. Nieboer et al., 1972; Tomassini et al., 1976; Pilegaard, 1978, 1979; Addison & Puckett, 1980; Garty, Galun & Hochberg, 1986; Zakshek et al., 1986) and radionuclides (e.g. Tuominen & Jaakkola, 1973; Hutchison-Benson, Svoboda & Taylor, 1984; Sheard, 1986; Ellis & Smith, 1987), and at coastal sites they may accumulate chloride (Brown & Di Meo, 1972; Larson, Matthes-Sears & Nash, 1985; Matthes-Sears et al., 1987). Secondly, lichens rapidly and actively absorb key ions, such as ammonium, nitrate and phosphate from solutions similar in concentration to rainwater (Smith, 1962; Farrar, 1976; Lang, Reiners & Heier, 1976; Fletcher, 1976; Hällbom & Bergman, 1983; Reiners & Olson, 1984; Nieboer et al., 1984). On the basis of kinetic studies on phosphorus uptake in *Hypogymnia physodes*, Farrar (1976) predicted that input in rainfall would be sufficient to satisfy the growth requirement for phosphorus in this species. There is also unequivocal evidence of accumulation of minerals from substrates, notable examples being crustaceous lichens growing on calcareous rocks amassing calcium (Syers, Birnie & Mitchell, 1967) and those on metal-rich substrata amassing elements such as iron, copper, cadmium and zinc (Brodo, 1973; Tuominen & Jaakkola, 1973; Purvis, 1984). However, since the substrata on which lichens grow are characteristically severely deficient in available nitrogen and phosphorus it is unlikely that lichens can derive significant quantities of these elements from this source. It should be noted that correlations between the chemical composition of lichens and that of their substrata can also result from atmospheric processes viz. deposition of aeolian particles derived from local soils or bedrock which then become entrapped in the thallus (Nieboer, Richardson & Tomassini, 1978; Nieboer & Richardson, 1981; Puckett, 1985).

Data concerning accumulation of atmospheric deposits usually reveal little about rates of deposition and retention and, while there is much literature relating to elements such as cadmium, zinc and lead, there is a dearth of information on the essential macronutrients. Data on ion uptake from solutions are usually obtained using procedures the ecological relevance of which is questionable; immersing or spraying lichens with nutrient solutions may not adequately simulate rainfall and nutrient delivery under field conditions. For example, abrupt rewetting of air-dry lichen thalli can result in an efflux of metabolites into the bathing solution (Farrar & Smith, 1976; Lang et al., 1976; Millbank, 1978; Buck & Brown, 1979), a process for which there is little evidence in nature (Crittenden, 1983).

Recently, we have attempted to determine the efficiency with which mats of *Stereocaulon paschale* and *Cladonia stellaris* capture nitrogen from incident rainfall in subarctic birch woodlands (P. D. Crittenden & R. P. Beckett, unpublished data). The experimental procedures have been detailed elsewhere (Crittenden,

Table 12.1. Net uptake from rainfall of inorganic nitrogen and potassium by reconstructed lichen mats.

	% Retention					
	NO$_3^-$		NH$_4^+$		K$^+$	
	range	mean	range	mean	range	mean
Stereocaulon paschale	86 to 100	95	40 to 99	86	-37 to +90	42
Cladonia stellaris	62 to 99	83	50 to 97	77	-978 to +65	-154

Data are based on analyses of bulked (i.e. unreplicated) collections from 10 replicate funnels of each type for 19 rainfall events between 27 May and 17 August 1984 (152 mm total rainfall). Total input (per 10 funnels): 460 μmol NO$_3^-$, 621 μmol NH$_4^+$, and 385 μmol K$^+$. Rainfall in event 19 was insufficient to exceed interception loss in both species (i.e. there was no lichen throughfall) resulting in 100% net uptake; this information is not included in the ranges and mean values. Net losses of potassium were greatest during the early rainfall events: mean net uptake for events 7 to 19 was 48% for *S. paschale* and -46% for *C. stellaris*. Potassium was determined by atomic absorption spectrophotometry; for other analytical methods see Table 12.3. Mean dry weights of lichen per funnel at the end of the experiment were 39·9 ± 0·9 and 14·8 ± 0·2 g for *S. paschale* and *C. stellaris* respectively.

1983). In brief, lichen thalli were arranged vertically on stainless steel grilles (3 × 3 mm mesh) so as to form single-layered lichen canopies of each species and at densities equivalent to those of undisturbed lichen mats. Only the well-pigmented upper 20-35 mm of podetia (*C. stellaris*) and pseudopodetia (*S. paschale*) were used in these reconstructions. Lichen-covered grilles were superimposed on funnels supported in trenches in the ground, with the mouth of the funnels (253 mm i.d.) and the lichen mats at ground level. During periods of rain throughfall from the lichen canopies, and unmodified rainfall from funnels covered by steel grilles only, were collected quantitatively and analyzed for ammonium, nitrate and potassium. The results for 19 consecutive rainfall events (Table 12.1) reveal that, on average, 80% or more of inorganic nitrogen delivered in rainfall was retained by these species, even during thunderstorms with comparatively high rainfall intensities. Results obtained from sequential sampling during individual rainfall events (Fig. 12.3) were in good accord with those obtained from bulked water samples and suggest that nitrogen escape in

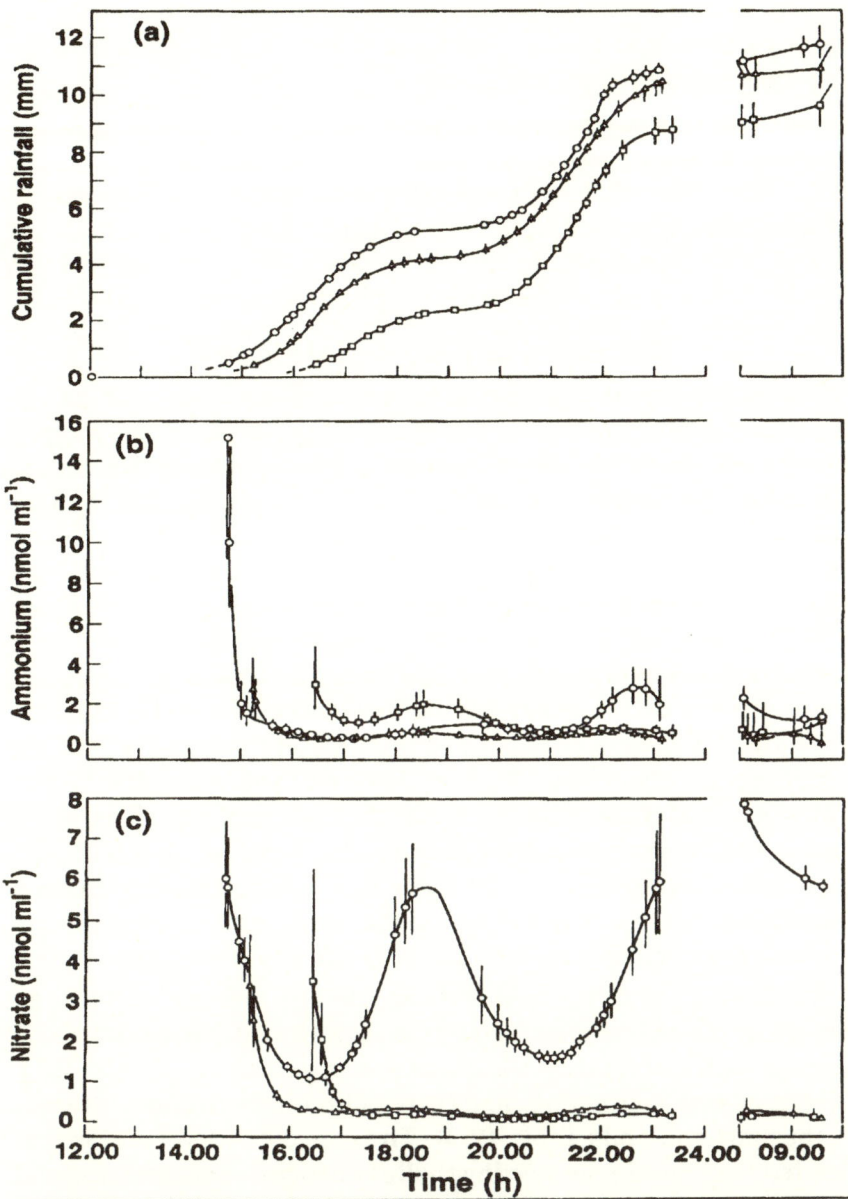

Fig. 12.3. Progress curves in rainfall event 14 (17-18 July, 1984) of cumulative rainfall (a) and the concentration of ammonium (b) and nitrate (c). Rainfall (circles), *Cladonia* throughfall (triangles) and *Stereocaulon* throughfall (squares) were collected sequentially and quantitatively in 20-30 ml aliquots from 5 replicate funnels of each type. Smooth lines were fitted using the computer program of Parsons & Hunt (1981) and fitted values (n = 5) are plotted together with 95% confidence limits (all values have been back-transformed from logarithms). Air temperature at ground level ranged between 9-14°C during the collection period. For analytical methods see Table 12.3.

throughfall was to a large extent due to low uptake efficiency during the early deliveries of rainfall.

Stereocaulon paschale and *C. stellaris* are clearly efficient at scavenging nitrogen from wet deposits. Occupying the inter-tree areas the lichen mats intercept rainfall as opposed to throughfall, although during wind-driven rain the quantities of rainfall received may be reduced in the lee of trees. Nitrogen inputs from dry deposition and mist (occult precipitation) were not quantified.

Nitrogen supply and demand

In the study reported above lichens suspended over rain gauges increased in biomass during the course of the experiment. Tagged thalli of *S. paschale* and *C. stellaris*, placed within the reconstructed lichen mats in the early part of the investigation, had increased in dry eight by $24.3 \pm 1.5\%$ (mean $\pm$ SEM; n = 10) and $26.4 \pm 1.1\%$ (n = 7) respectively by the end of the experiment (the respective mean relative growth rates were 0.0036 ± 0.0002 and 0.0039 ± 0.0001 d^{-1}). The observed growth and the absence of a major potassium efflux (Table 12.1) suggest that neither species was damaged by the experimental treatment. However, the concentration of thallus nitrogen was significantly lower ($P < 0.05$) in experimental material at the termination of the study than in undisturbed field material collected at the start and end of the work and in the following spring (Table 12.2). Net loss of soluble organic nitrogen during rainfall (Table 12.3) was too small to explain the differences and it was assumed that thallus nitrogen had been diluted by growth in the absence of an adequate nitrogen supply.

Table 12.3 gives observed and expected increments in nitrogen capital in the lichen mats during the experimental period derived from (i) the final total dry weight of lichen per funnel (Table 12.1), (ii) the growth increments, and (iii) thallus nitrogen concentration at the start and end (Table 12.2). The expected increment is that needed to maintain the starting thallus nitrogen concentration and is a measure of the growth demand for this element. The observed nitrogen gain in *S. paschale*, over and above that attributable to atmospheric deposits, provides a gravimetric estimate of nitrogen fixation. This value (669 mg N m^{-2}) compares favourably with rates of nitrogen fixation in this species derived from acetylene reduction assays by Crittenden & Kershaw (1979). Despite the large er-

Table 12.2. Total nitrogen conentration (% dry weight) in field and experimental material of *Stereocaulon paschale* and *Cladonia stellaris*.

	Samples from undisturbed lichen mats			Material from steel grilles at end of experiment
	May 1984	August 1984	June 1985	
S. paschale	1·08 ± 0·10	1·15 ± 0·04	1·11 ± 0·09	0·96 ± 0·06
C. stellaris	n.d.	0·40 ± 0·04	0·42 ± 0·05	0·35 ± 0·03

The top 25 mm of pseudopodetia and podetia were analyzed for total Kjeldahl nitrogen by the method of Bremner & Breitenbeck (1983). In the case of experimental material thalli of this length were selected; in the case of field material thalli were pruned to this length. Mean values (n=10) appear with 95% confidence intervals. n.d. = not determined.

Table 12.3. Increments, inputs and losses of nitrogen in reconstructed lichen mats during an 82 day growth period (mg N m^{-2} pure lichen cover).

	Stereocaulon paschale	*Cladonia stellaris*
Increment in total N capital[1]		
expected	1717 ± 380	252 ± 53
observed	758 ± 245	95 ± 51
Inorganic N in rainfall[2]		
deposited	31	31
retained	27	25
Net loss of soluble organic N[3]	18·7	11·4
Unaccountable N (error and/or dry deposition)	(81)	81[†]
Estimated N$_2$-fixation	669[††]	–

[1]Derived from mean values of 7 to 10 measurements (derived mean ± SEM).
[2,3]Water samples were filtered through prewashed cellulose acetate membrane filters (0·22 μm pore size) then ammonium was determined using a modification of the indophenol blue method given in Allen *et al.* (1974), nitrate by HPLC with uv detection, soluble organic N by a modification of Johnson's (1941) method as described by Crittenden (1983).
[†]Observed N increment - net N uptake from rainfall; it is assumed that the same value applies in the case of *S. paschale*.
[††]Observed N increment - (net N uptake from rainfall + unaccountable N).

rors associated with the estimated nitrogen demands an inesca-
pable fact emerges: nitrogen scavenged from rainfall accounts for
a surprisingly small fraction of the apparent nitrogen requirement
in these lichens (c. 2·6% in the case of *S. paschale*, over and above
that provided by nitrogen fixation, and c. 10% in *C. stellaris*). Ac-
cordingly, the possibility that other sources of nitrogen are available to undisturbed lichen mats *in situ* must be considered.

In transplanting lichen to the metal grilles two potentially important sources of nutrients had been removed: the soil and senescent basal parts of lichen thalli. On first consideration it seems
unlikely that significant quantities of nitrogen could diffuse or be
translocated upwards from the soil. These lichens do not make in-
timate contacts with the underlying soil and, after all, the most obvious explanation for the poor growth of mycorrhizal *Vaccinium
vitis-idaea* in organic surface soils gathered from beneath lichen
mats is that they are deficient in available forms of both inorganic
and organic nitrogen. However, in the selection of 'healthy' (well-
pigmented) apical parts of *C. stellaris* podetia and *S. paschale*
pseudopodetia for lichen mat reconstruction roughly equivalent
weights of discoloured basal thallus material were discarded. Nutrients in these older regions may represent an internal resource
that can be recycled to support new growth. In Fig. 12.4 the vertical distribution of thallus nitrogen in experimental and field material of both species is compared. Nitrogen concentration in *S.
paschale* progressively decreases with increasing distance from the
thallus tip. In all depth intervals nitrogen concentration is lower
in experimental material grown on steel grilles compared to previously undisturbed field material but the difference is least at the
apex (although still significant at $P<0.05$) where new growth is assumed to have occurred. These results are consistent with the
proposition that nitrogen is remobilized in older regions of the
thallus and moves to the growing apices. Data for *C. stellaris* is in-
complete but shows some consistency with that for *S. paschale*.
Thus the pronounced vertical gradients in nitrogen, phosphorus
and potassium reported by Pakarinen (1981a) in *C. arbuscula*
growing on bogs may owe more to tight internal nutrient recycling
than to decomposition involving the soil microflora.

There is no reason to believe that a reduced supply of nitrogen
(and perhaps other nutrients) retarded thallus growth in the rec-

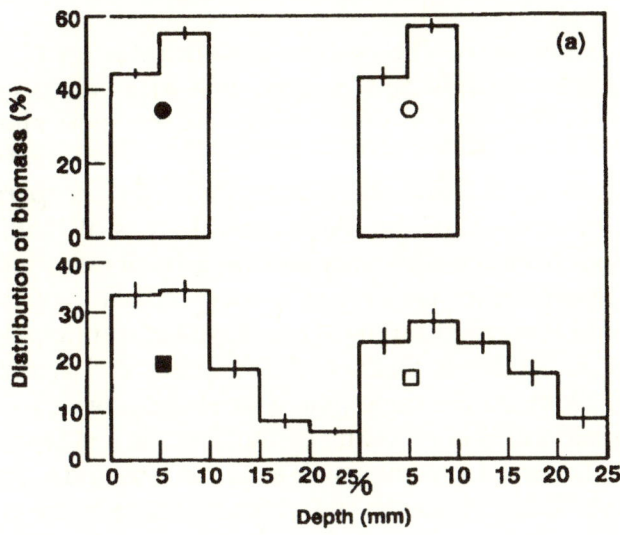

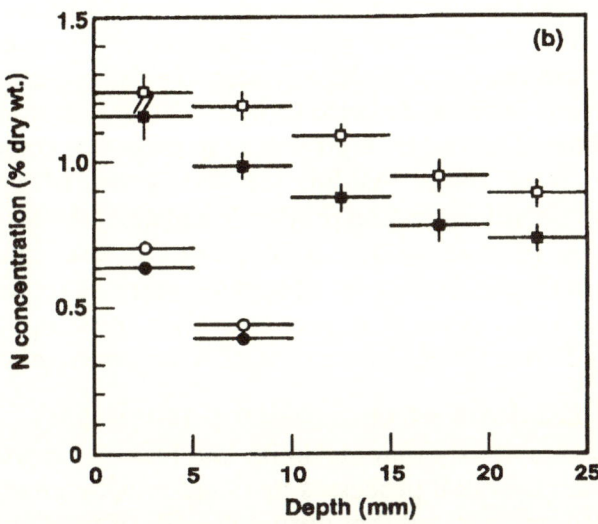

Fig. 12.4. Vertical distribution of biomass (a) and vertical variation in total nitrogen concentration (b) in *Stereocaulon paschale* (squares) and *Cladonia stellaris* (circles) taken from steel grilles at the end of the experiment (closed symbols) and from undisturbed lichen mats (open symbols). In the case of experimental material of *S. paschale*, pseudopodetia 25 mm in length were selected for analysis; in the case of field material pseudopodetia were pruned to this length. Because of the limited availability of material, the top 10 mm only of *C. stellaris* podetia were cut from samples not standardized for length. Total Kjeldahl nitrogen was determined by the method of Bremner & Breitenbeck (1983). Mean values (n = 10) are plotted together with 95% confidence limits.

onstructed lichen mats. The relative growth rates for *C. stellaris* and *S. paschale* reported above agree well with values given by Kärenlampi (1971) for the same species at the same site. Armstrong (1979, 1984) also found that the radial extension rate of peripheral lobes of the foliose lichen *Parmelia conspersa* was unaffected by removal of the older central areas of thalli. However, it is clear from the above results that growth rate (and it may follow, photosynthetic rate) is not a good indicator of the adequacy of mineral nutrient supply as may have been anticipated previously (cf. Carstairs & Oechel, 1978; Bailey & Larson, 1982; Armstrong, 1984). At the same time it would be unwise to assume that large scale internal recycling of nutrients is a universal phenomenon in lichens. In species such as *C. stellaris* and *S. paschale* that grow vertically upwards and characteristically form dense mats such a mechanism will allow much faster growth rates than could otherwise be sustained in a chronically oligotrophic environment. This could have several advantages. It would help to maintain a large mass of structurally intact basal litter giving depth to the mats and casting deep shade, a process by which these species are able to exert dominance *sensu* Grime (1979). It may also promote rapid lateral spread by regrowth of damaged or fragmented thalli and by branching or budding (cf. Yarranton, 1975). Rapid remobilization of resources and production of dead thallus material would be less of an advantage to foliose or crustose lichens that grow horizontally. In slow growing species nutrient requirement for new growth may perhaps be met entirely by new input as was suggested by Farrar (1976) in the case of *Hypogymnia physodes*.

Mutualism or controlled parasitism?

In a classic series of experiments Smith and his collaborators elucidated the fundamental processes involved in lichen carbon nutrition. They demonstrated that during periods of photosynthetic activity a large fraction of the total carbon assimilated by the photobiont moves in the form of single specific simple carbohydrates to the fungus where it accumulates in fungal products (Smith, Muscatine & Lewis, 1969; Smith, 1975, 1980). In so doing, they verified the hypothesis, first advanced by de Bary and Schwendener in the nineteenth century, that lichen fungi are sustained by assimilate provided by their algal symbionts (Quispel,

1943). However, as Smith has repeatedly pointed out (Smith, 1975, 1976, 1980; Smith & Douglas, 1987) there is only very limited evidence to support the long standing supposition that there is bilateral nutrient exchange, the fungus supplying the photobiont with mineral elements and/or growth factors. In view of this, Ahmadjian & Jacobs (1981, 1983) and Ahmadjian (1982) proposed that the symbiotic relationship in lichens is one of controlled parasitism rather than mutualism.

The data presented above for *C. stellaris* and *S. paschale* have been interpreted in terms of internal recycling and translocation of nitrogen from older parts of the thallus through mycobiont pathways to the apices. The question of whether nitrogen might pass from fungus to photobiont in these regions where growth and multiplication of algal cells are most active (Honegger, 1987) now deserves attention. An argument in favour of this proposal can be developed from a consideration of temporal relationships between nutrient deposition and growth in lichen mats (Fig. 12.5). With the onset of rainfall, mats are rewetted and respiratory and assimilatory processes recover rapidly (Ried, 1960; Smith & Molesworth, 1973; Farrar & Smith, 1976; Crittenden & Kershaw, 1979; Brown, MacFarlane & Kershaw, 1983; Groulx & Lechowicz, 1987; Coxson, 1988; Fritz-Sheridan & Coxson, 1988). Thallus water contents sufficient for growth may obtain for substantial periods after rain ceases. Indeed, pseudopodetia of *S. paschale* can measurably increase in dry weight during individual 'wet events' comparable to that depicted in Fig. 12.5 (K. A. Kershaw & P. D. Crittenden, unpublished). Nutrients are deposited in rainfall but the major inputs of most ions are often contained in the first deliveries of rain (Seymour & Stout, 1983; Crittenden, 1983; Pellett, Bustin & Harriss, 1984; P. D. Crittenden & R. P. Beckett, unpublished). Consequently, growth may occur during comparatively long periods when there is little nutrient input and when intercellular thallus water is depleted of key mineral ions. Under these conditions, rapid growth and multiplication of algae in thallus apices may be dependent on nitrogen and other nutrients provided by the fungus.

The photobiont in *C. stellaris* and *S. paschale* is probably a species of *Trebouxia* (Ahmadjian, 1980). In the symbiotic state *Trebouxia* reproduces by autosporulation and this occurs predomi-

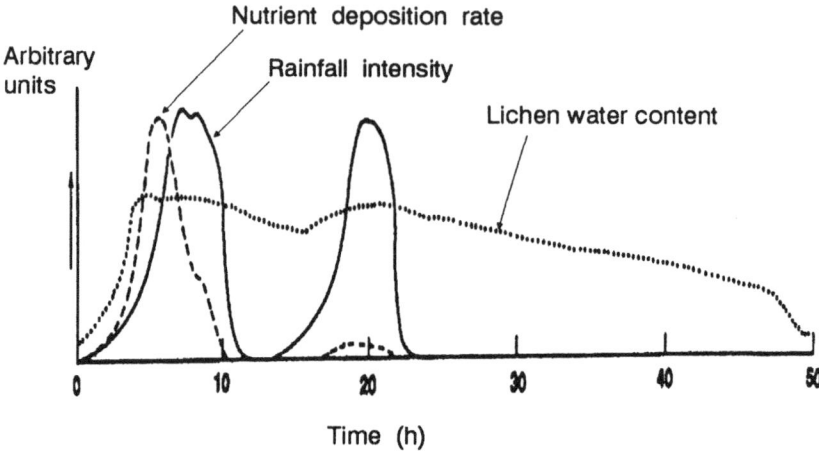

Fig. 12.5. Temporal variation in rainfall intensity, nutrient deposition rate and lichen water content during and following a hypothetical rainfall event. Time scale is indicated as an example. This scheme is based on the author's experience in northern boreal forests; data sets in partial support of it are given by Crittenden & Kershaw (1979) and Coxson (1987).

nantly in the growing apices of lichen thalli (Honegger, 1987). Young autospores are contacted by mycobiont hyphae that penetrate the old mother cell wall (Scott & Larson, 1984; Honegger, 1985). It is tempting to speculate that these contacts at an early growth stage of individual algal cells might establish symplastic pathways essential for mineral nutrient exchange.

Indirect evidence of a requirement for nitrogen in growing photobiont populations in lichen thalli has come from studies on nitrogen assimilating enzymes in cyanobacterial lichens of the genus *Peltigera*. Rowell, Rai & Stewart (1985) have shown that in *P. canina* the ammonium assimilating enzymes glutamate dehydrogenase (GDH) and glutamine synthetase (GS) have somewhat complementary distributions. GDH occurs throughout the thallus, located mainly in fungal hyphae lying close to the *Nostoc* photobiont, but activity is lower at the growing margins. GS, which is exclusive to the photobiont, is located chiefly in the young marginal regions although at levels of activity very much lower than those known in free-living *Nostoc*. Nitrogenase activity (mg thallus protein^{-1}) occurs in all parts of the thallus but declines progressively with increasing distance from the margin (note that on a unit thallus area basis nitrogenase activity increases with increasing distance from the growing margin (Millbank, 1972)). These results suggest that the capacity to assimilate photobiont generated ammonium exists in the fungus in all parts of the thallus but that there may be significant retention of fixed nitrogen by *Nostoc* in growing regions. Lallemant & Savoye (1985) noted similar differences between the distributions of GS and GDH in the mature thallus and actively growing lobules of *P. praetextata*. Studies on the tripartite lichen *P. aphthosa* have revealed that nitrogen can pass from fungus to alga: ^{15}N supplied as dinitrogen is fixed by *Nostoc* in surface cephalodia, rapidly assimilated by the cephalodial fungus (Rowell *et al.*, 1985), transferred to the main thallus and appears slowly in the *Coccomyxa* photobiont (Millbank & Kershaw, 1969; Kershaw & Millbank, 1970; Rai, Rowell & Stewart, 1981; Kershaw, 1985). The question of whether movement of label might be channelled towards new growth was not addressed.

The possible pathways by which mineral nutrients may reach the photobiont in a lichen thallus are summarized in Fig. 12.6 adapted from Harley & Smith's (1983) scheme for ectomycorrhizas. The model tentatively suggested here is similar to that proposed for sheathing mycorrhizas; acquisition of mineral nutrients by the photobiont is mainly from living fungal cells (B), diffusive movement through the external fungal tissue (A3) being of minor importance. If this is not the situation then it is to be assumed that the photobiont absorbs inorganic nutrients from surrounding intercellular water in competition with the fungus (cf. Smith 1975, 1976, 1980; Smith & Douglas 1987). Since the growing apices of *C. stellaris* and *S. paschale* form the uppermost surface of lichen mats they are well placed to intercept a large proportion of the incoming nutrients. Farrar & Smith (1976) and Smith (*op. cit.*) believe that leakage of metabolites from lichen symbionts during the rewetting phase at the onset of rainfall, and subsequent reabsorption, may provide an opportunity for exchange of nutrients in fa-

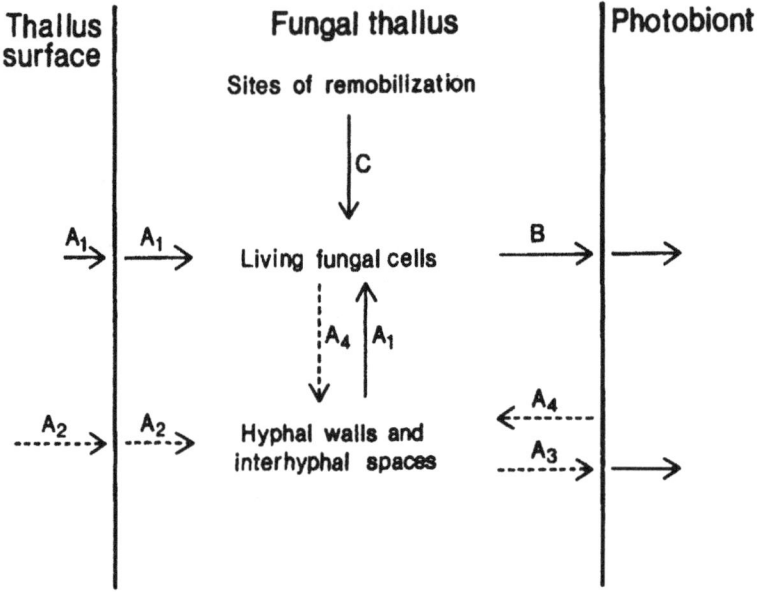

Fig. 12.6. Diagram showing potential pathways for mineral nutrient movement through fungal tissue in a lichen thallus (cf. sheath of ectomycorrhizas). Broken arrows show movement by diffusion or mass flow; solid arrows show metabolically dependent movement. Mineral nutrients move by diffusion or mass flow in surface water films. They are absorbed by metabolically dependent reactions A_1 into hyphae at the thallus surface and may either remain incorporated in the fungal biomass or be released to the photobiont by metabolic reaction B. Nutrients in older parts of the thallus are mobilized by reaction C and move to regions of growth (internal recycling). Nutrients may also pass non-metabolically from surface water films through the hyphal cell walls and interhyphal spaces of the cortex (*Stereocaulon*) or plectenchyma (*Cladonia stellaris* and other species of the subgenus *Cladina*; Ahti, 1984)(A_2) from where they may be accumulated into the living fungal cells (A_1) or into photobiont cells (A_3). During rewetting of dry lichen thalli at the onset of rainfall nutrients from both symbionts may be released passively into the free space (A_4). Adapted from Harley & Smith's (1983) scheme for ectomycorrhizas; cf. also Honegger's (1985) scheme for water transport in lichen thalli.

vour of the photobiont; again, such a process may be most evident in the thallus apices. Richardson & Nieboer (1980) argue that the growing regions in a lichen may develop a particularly favourable electrochemical gradient for ion uptake during the rewetting phase resulting in ion migration and accumulation in the apices. Resolving which of these pathways (Fig. 12.6) operate will necessitate a combination of astute observation and ingenious experimental approach.

Finally, recent evidence for pathway B has been provided by quite a different line of enquiry. Ahmadjian & Jacobs (1987) used axenic cultures of symbionts isolated from species of *Cladonia* to resynthesize lichens on mica strips. They maintained the cultures on mica in closed vessels at about 95% relative humidity for more than a year with no addition of nutrients. Not only did individual squamules survive this long period but new squamules formed. The authors concluded that nutrients for the new growth of both fungus and alga must have been derived largely from old mycelium that made up the original inoculum.

Concluding remarks

There are features common to the mineral nutrient relationships of ectomycorrhizas (as described by Harley & Smith, 1983) and to those proposed above for the lichens *C. stellaris* and *S. paschale*. These are (i) exploitation of periodic nutrient flushes, (ii) nutrient storage in fungal tissues, and (iii) remobilization of stored reserves to meet the growth requirements of the photobiont. In the case of lichens, inorganic nutrients are absorbed during rainfall which results in a flush of mineral ions in the form of wet deposits, and dry deposits brought into solution; nutrients in older parts of thalli are recycled and move to the growing apices (to this extent the thallus functions as a nutrient store) where there is nutrient exchange from fungus to alga promoting autosporulation. Ectomycorrhizas are believed to trap efficiently mineral nutrients delivered in seasonal flushes associated with periods of leaf senescence, fall and decomposition; nutrients in excess of the immediate demands of the host are stored in fungal sheaths and can subsequently be remobilized and passed to the photobiont during periods of nutrient shortage. Interestingly, polyphosphate storage granules characteristic of phosphorus accumulation in fungal sheaths of ectomycorrhizas have also been observed in large numbers in lichen fungi (Chilvers, Ling-Lee & Ashford, 1978).

Using ectomycorrhizas and coelenterates as examples, Lewis (1973) ascribed the relative success of mutualistic symbioses in nutrient poor environments to the possession therein of mechanisms for tight nutrient cycling; 'a tightness which is assisted by antibiotic production and the partial elimination of competitors'. Preliminary evidence reported here indicates that mat-forming lichens conform closely to this general model.

Acknowledgements. I am particularly indebted to the staff of the Kevo Subarctic Research Station (University of Turku) for the generous provision of facilities during the course of the fieldwork and to Dr R. P. Beckett (Department of Botany, University of Natal) for permitting me to present some of our previously unpublished data . Mamoru Matsuki kindly helped with collecting soils which were brought into the UK under MAFF Licence No. PHF 77/108 (86). The work was supported by a grant from the Erna and Victor Hasselblad Foundation.

References

Addison, P. A. & Puckett, K. J. (1980). Deposition of atmospheric pollutants as measured by lichen element content in the Athabasca oil sands area. *Canadian Journal of Botany*, **58**, 2323-2334.

Ahmadjian, V. (1980). Separation and artificial synthesis of lichens. In *Cellular Interactions in Symbiosis and Parasitism*, eds. C. B. Cook, P. W. Pappas & E. D. Rudolph, pp. 3-29. Ohio State University Press, Columbus, Ohio, USA.

Ahmadjian, V. (1982). Algal/fungal symbioses. In *Progress in Phycological Research*, vol 1, eds. F. E. Round & D. J. Chapman, pp. 179-233. Elsevier Biomedical Press, Amsterdam.

Ahmadjian, V. & Jacobs, J. B. (1981). Relationship between fungus and alga in the lichen *Cladonia cristatella* Tuck. *Nature*, **289**, 169-172.

Ahmadjian, V. & Jacobs, J. B. (1983). Algal-fungal relationships in lichens: recognition, synthesis, and development. In *Algal Symbiosis*, ed. L. J. Goff, pp. 147-172. Cambridge University Press, Cambridge, UK.

Ahmadjian, V. & Jacobs, J. B. (1987). Studies on the development of synthetic lichens. In *Progress and Problems in Lichenology in the Eighties*, ed. E. Peveling, pp. 47-58. J. Cramer, Berlin.

Ahti, T. (1977). Lichens of the boreal coniferous zone. In *Lichen Ecology*, ed. M. R. D. Seaward, pp. 145-181. Academic Press, London.

Ahti, T. (1984). The status of *Cladina* as a genus segregated from *Cladonia*. *Nova Hedwigia* Beiheft 79, 25-61.

Allen, S. E., Grimshaw, H. M., Parkinson, J. A. & Quarmby, C. (1974). *Chemical Analysis of Ecological Materials*. Blackwell Scientific Publications, Oxford, UK.

Armstrong, R. A. (1977). The response of lichen growth to additions of distilled water, rainwater and water from a rock surface. *New Phytologist*, **79**, 373-376.

Armstrong, R. A. (1979). Growth and regeneration of lichen thalli with the central portions artificially removed. *Environmental and Experimental Botany*, **19**, 175-178.

Armstrong, R. A. (1984). Growth of experimentally reconstructed thalli of the lichen *Parmelia conspersa*. *New Phytologist*, **98**, 497-502.

Auclair, A. N. D. (1983). The role of fire in lichen-dominated tundra and forest-tundra. In *The Role of Fire in Northern Circumpolar Ecosystems*, eds. R. W. Wein & D. A. MacLean, pp. 235-256. J. Wiley & Sons, Chichester.

Bailey, C. & Larson, D. W. (1982). Water quality and pH effects on *Umbilicaria mammulata* (Ach.) Tuck. *Bryologist*, **85**, 431-437.

Bakaeva, M. V. & Galanin, A. V. (1985). Ecological role of the lichen cover in the white-moss pine forests of the central Vychegda. *Soviet Journal of Ecology*, **16**, 80-84.

Bremner, J. M. & Breitenbeck, G. A. (1983). A simple method for determination of ammonium in semimicro-Kjeldahl analysis of soils and plant materials using a block digester. *Communications in Soil Science and Plant Analysis*, **14**, 905-913.

Brodo, I. M. (1973). Substrate Ecology. In *The Lichens*, eds. V. Ahmadjian & M. E. Hale, pp. 401-441. Academic Press, London.

Brown, D., MacFarlane, J. D. & Kershaw, K. A. (1983). Physiological-environmental interactions in lichens. XVI. A re-examination of resaturation respiration phenomena. *New Phytologist*, **93**, 237-246.

Brown, D. H. & Di Meo, J. A. (1972). Influence of local maritime conditions on the distribution of two epiphytic lichens. *Lichenologist*, **5**, 305-310.

Brown, R. T. & Mikola, P. (1974). The influence of fruticose soil lichens upon the mycorrhizae and seedling growth of forest trees. *Acta Forestalia Fennica*, **141**, 1-23.

Buck, G. W. & Brown, D. H. (1979). The effect of desiccation on cation location in lichens. *Annals of Botany*, **44**, 265-277.

Carstairs, A. G. & Oechel, W. C. (1978). Effects of several microclimatic factors and nutrients on net carbon dioxide exchange in *Cladonia alpestris* (L.) Rabh. in the Subarctic. *Arctic and Alpine Research*, **10**, 81-94.

Chilvers, G. A., Ling-Lee, M. & Ashford, A. E. (1978). Polyphosphate granules in the fungi of two lichens. *New Phytologist*, **81**, 571-574.

Crittenden, P. D. (1983). The role of lichens in the nitrogen economy of subarctic woodlands: nitrogen loss from the nitrogen-fixing lichen *Stereocaulon paschale* during rainfall. In *Nitrogen as an Ecological Factor*, eds. J. A. Lee, S. McNeill & I. H. Rorison, pp. 43-68. Blackwell Scientific Publications, Oxford, UK.

Crittenden, P. D. & Kershaw, K. A. (1978). Discovering the role of lichens in the nitrogen cycle in boreal-arctic ecosystems. *Bryologist*, **81**, 258-267.

Crittenden, P. D. & Kershaw, K. A. (1979). Studies on lichen-dominated systems. XXII. The environmental control of nitrogenase activity in *Stereocaulon paschale* in spruce-lichen woodland. *Canadian Journal of Botany*, **57**, 236-254.

Coxson, D. S. (1987). Photoinhibition of net photosynthesis in *Stereocaulon virgatum* and *S. tomentosum*, a tropical-temperate comparison. *Canadian Journal of Botany*, **65**, 1707-1715.

Coxson, D. S. (1988). Recovery of net photosynthesis and dark respiration on rehydration of the lichen, *Cladonia mitis*, and the influence of prior exposure to sulphur dioxide while desiccated. *New Phytologist*, **108**, 483-487.

Dawson, H. J., Hrutfiord, B. F. & Ugolini, F. C. (1984). Mobility of lichen compounds from *Cladonia mitis* in arctic soils. *Soil Science*, **138**, 40-45.

Ellis, K. M. & Smith, J. N. (1987). Dynamic model for radionuclide uptake in lichen. *Journal of Environmental Radioactivity*, **5**, 185-208.

Farrar, J. F. (1976). The uptake and metabolism of phosphate by the lichen *Hypogymnia physodes*. *New Phytologist*, **77**, 127-134.

Farrar, J. F. & Smith, D. C. (1976). Ecological physiology of the lichen *Hypogymnia physodes*. III. The importance of the rewetting phase. *New Phytologist*, **77**, 115-125.

Fisher, R. F. (1979). Possible allelopathic effects of reindeer-moss (*Cladonia*) on jack pine and white spruce. *Forest Science*, **25**, 256-260.

Fletcher, A. (1976). Nutritional aspects of marine and maritime lichen ecology. In *Lichenology: Progress and Problems*, eds. D. H. Brown, D. L. Hawksworth & R. H. Bailey, pp. 359-384. Academic Press, London

Foster, D. R. (1985). Vegetation development following fire in *Picea mariana* (black spruce) – *Pleurozium* forests of south-eastern Labrador, Canada. *Journal of Ecology*, **73**, 517-534.

Fritz-Sheridan, R. P. & Coxson, D. S. (1988). Nitrogen fixation on a tropical volcano, La Soufrière (Guadeloupe): the interaction of temperature, moisture, and light with net photosynthesis and nitrogenase activity in *Stereocaulon virgatum* and response to periods of insolation shock. *Lichenologist*, **20**, 63-81.

García-Junceda, E. & Xavier Filho, L. (1986). Solubilization of lichen phenolics from *Cladonia sprucei* by simulated rainfall. *Lichen Physiology and Biochemistry*, **1**, 61-69.

Garty, J., Galun, M. & Hochberg, Y. (1986). The accumulation of metals in *Caloplaca aurantia* growing on concrete roof tiles. *Lichenologist*, **18**, 257-263.

Gilbert, O. L. (1969). The effect of SO_2 on lichens and bryophytes around Newcastle upon Tyne. In *Air Pollution, Proceedings of the First European Congress on the Influence of Air Pollution on Plants and Animals, Wageningen 1968*, pp. 223-235. Centre for Agricultural Publishing and Documentation, Wageningen, The Netherlands.

Grime, J. P. (1979). *Plant Strategies and Vegetation Processes*. John Wiley & Sons, Chichester.

Groulx, M. & Lechowicz, M. J. (1987). Net photosynthetic recovery in subarctic lichens with contrasting water relations. *Oecologia*, **71**, 360-368.

Hällbom, L. & Bergman, B. (1983). Effects of inorganic nitrogen on C_2H_2 reduction and CO_2 exchange in the *Peltigera praetextata-Nostoc* and *Peltigera aphthosa-Coccomyxa-Nostoc* symbioses. *Planta*, **157**, 441-445.

Hämet-Ahti, L. (1963). Zonation of the mountain birch forests in northernmost Fennoscandia. *Annales Botanici Societatis Zoologicæ Botanicæ Fennicæ 'Vanamo'*, **34** (No. 4), 1-127.

Harley, J. L. & Smith, S. E. (1983). *Mycorrhizal Symbiosis*. Academic Press, London.

Honegger, R. (1985). Fine structure of different types of symbiotic relationships in lichens. In *Lichen Physiology and Cell Biology*, ed. D. H. Brown, pp. 287-302. Plenum Press, New York.

Honegger, R. (1987). Questions about pattern formation in the algal layer of lichens with stratified (heteromerous) thalli. In *Progress and Problems in Lichenology in the Eighties*, ed. E. Peveling, pp. 59-71. J. Cramer, Berlin.

Huss-Danell, K. (1977). Nitrogen fixation by *Stereocaulon paschale* under field conditions. *Canadian Journal of Botany*, **55**, 585-592.

Huss-Danell, K. (1978). Seasonal variation in the capacity for nitrogenase activity in the lichen *Stereocaulon paschale*. *New Phytologist*, **81**, 89-98.

Hutchison-Benson, E., Svoboda, J. & Taylor, H. W. (1984). The latitudinal inventory of ^{137}Cs in vegetation and topsoil in northern Canada, 1980. *Canadian Journal of Botany*, **63**, 784-791.

Johnson, M. J. (1941). Isolation and properties of a pure yeast polypeptidase. *Journal of Biological Chemistry*, **137**, 575-586.

Kallio, S. & Kallio, P. (1975). Nitrogen fixation in lichens at Kevo, north-Finland. In *Fennoscandian Tundra Ecosystems, Part 1. Plants and Micro-organisms (Ecological Studies, 16)*, ed. F. E. Wielgolaski, pp. 292-304. Springer-Verlag, Berlin.

Kärenlampi, L. (1971). Studies on the relative growth rate of some fruticose lichens. *Reports from the Kevo Subarctic Research Station*, **7**, 33-39.

Kershaw, K. A. (1977). Studies on lichen-dominated systems. XX. An examination of some aspects of the northern boreal lichen woodlands in Canada. *Canadian Journal of Botany*, **55**, 393-410.

Kershaw, K. A. (1985). *Physiological Ecology of Lichens*. Cambridge University Press, Cambridge, UK.

Kershaw, K. A. & Millbank, J. W. (1970). Nitrogen metabolism in lichens. II. The partition of cephalodial-fixed nitrogen between the mycobiont and phycobionts of *Peltigera aphthosa*. *New Phytologist*, **69**, 75-79.

Kershaw, K. A., Rouse, W. R. &. Bunting, B. T. (1975). *The Impact of Fire on Forest and Tundra Ecosystems*, INA Publication No. QS-8038-000-EE-A1. Ministry of Indian and Northern Affairs, Ottawa, Canada.

Lallemant, R. & Savoye, D. (1985). Lectins and morphogenesis: facts and outlooks. In *Lichen Physiology and Cell Biology*, ed. D. H. Brown, pp.335-350. Plenum Press, New York.

Lang, G. E., Reiners, W. A. & Heier, R. K. (1976). Potential alteration of precipitation chemistry by epiphytic lichens. *Oecologia*, **25**, 229-241.

Larson, D. W., Matthes-Sears, U. & Nash IIIrd, T. H. (1985). The ecology of *Ramalina menziesii*. I. Geographical variation in form. *Canadian Journal of Botany*, **63**, 2062-2068.

Lawrey, J. D. (1984). *Biology of Lichenized Fungi*. Praeger Publishers, New York.

Lawrey, J.D. (1986). Biological role of lichen substances. *Bryologist*, **89**, 111-122.

Lewis, D. H. (1973). The relevance of symbiosis to taxonomy and ecology, with particular reference to mutualistic symbioses and the exploitation of marginal habitats. In *Taxonomy and Ecology*, ed. V. H. Heywood, pp. 151-172. Academic Press: London.

Lewis Smith, R. I. (1978). Summer and winter concentrations of sodium, potassium and calcium in some maritime Antarctic cryptogams. *Journal of Ecology*, **66**, 891-909.

Malicki, J. (1965). The effect of lichen acids on the soil microorganisms. Part 1. The washing down of the acids into the soil. *Annales Universitatis Mariae Curie - Skł odowska, Sectio C*, **20**, 239-248.

Matthes-Sears, U., Nash IIIrd, T. H. & Larson, D. W. (1987). Salt loading does not control CO_2 exchange in *Ramalina menziesii* Tayl. *New Phytologist*, **106**, 59-69.

Millbank, J. W. (1972). Nitrogen metabolism in lichens. IV. The nitrogenase activity of the *Nostoc* phycobiont in *Peltigera canina*. *New Phytologist*, **71**, 1-10.

Millbank, J. W. (1978). The contribution of nitrogen fixing lichens to the nitrogen status of their environment. In *Environmental Role of Nitrogen-fixing Blue-green Algae and Asymbiotic Bacteria (Ecological Bulletins, 26)*, ed. U. Granhall, pp. 260-265. Swedish Natural Science Research Council, Stockholm.

Millbank, J. W. & Kershaw, K. A. (1969). Nitrogen metabolism in lichens. 1. Nitrogen fixation in the cephalodia of *Peltigera aphthosa*. *New Phytologist*, **68**, 721-729.

Moore, T. R. (1980). The nutrient status of subarctic woodland soils. *Arctic and Alpine Research*, **12**, 147-160.

Nieboer, E., Ahmed, H.M., Puckett, K. J. & Richardson, D. H. S. (1972). Heavy metal content of lichens in relation to distance from a nickel smelter in Sudbury, Ontario. *Lichenologist*, **5**, 292-304.

Nieboer, E., Padovan, D., Lavoie, P. & Richardson, D. H. S. (1984). Anion accumulation by lichens. II. Competition and toxicity studies involving arsenate, phosphate, sulphate and sulphite. *New Phytologist*, **96**, 83-93.

Nieboer, E. & Richardson, D. H. S. (1981). Lichens as monitors of atmospheric deposition. In *Atmospheric Pollutants in Natural Waters*, ed. S. J. Eisenreich, pp. 339-388. Ann Arbor Science Publishers Inc., Ann Arbor, Michigan, USA.

Nieboer, E., Richardson, D. H. S. & Tomassini, F. D. (1978). Mineral uptake and release by lichens: an overview. *Bryologist*, **81**, 226-246.

Oksanen, J. & Ahti, T. (1982). Lichen-rich pine forest vegetation in Finland. *Annales Botanici Fennici*, **19**, 275-301.

Pakarinen, P. (1981a). Nutrient and trace metal content and retention in reindeer lichen carpets of Finnish ombrotrophic bogs. *Annales Botanici Fennici*, **18**, 265-274.

Pakarinen, P. (1981b). Regional variation of sulphur concentrations in *Sphagnum* mosses and *Cladonia* lichens in Finnish bogs. *Annales Botanici Fennici*, **18**, 275-279.

Parsons, I.T. & Hunt, R. (1981). Plant growth analysis: a program for the fitting of lengthy series of data by the method of *B*-splines. *Annals of Botany*, **48**, 341-352.

Payette, S., Filion, L., Gauthier, L. & Boutin, Y. (1985). Secular climate change in old-growth tree-line vegetation of northern Quebec. *Nature*, **315**, 135-138.

Pellet, G. L., Bustin, R. & Harriss, R. C. (1984). Sequential sampling and variability of acid precipitation in Hampton, Virginia. *Water, Air, and Soil Pollution*, **21**, 33-49.

Pilegaard, K. (1978). Airborne metals and SO_2 monitored by epiphytic lichens in an industrial area. *Environmental Pollution*, **17**, 81-92.

Pilegaard, K. (1979). Heavy metals in bulk precipitation and transplanted *Hypogymnia physodes* and *Dicranoweisia cirrata* in the vicinity of a Danish steelworks. *Water, Air, and Soil Pollution*, **11**, 77-91.

Puckett, K.J. (1985). Temporal variation in lichen element levels. In *Lichen Physiology and Cell Biology*, ed. D. H. Brown, pp. 211-225. Plenum Press, New York.

Purvis, O. W. (1984). The occurrence of copper oxalate in lichens growing on copper sulphide-bearing rocks in Scandinavia. *Lichenologist*, **16**, 197-204.

Pyatt, F. B. (1973). Plant sulphur content as an air pollution gauge in the vicinity of a steelworks. *Environmental Pollution*, **5**, 103-115.

Quispel, A. (1943). The mutual relations between algae and fungi in lichens. *Recueil des travaux botaniques néerlandais*, **40**, 413-541.

Rai, A. N., Rowell, P. & Stewart, W. D. P. (1981). $^{15}N_2$ incorporation and metabolism in the lichen *Peltigera aphthosa* Willd. *Planta*, **152**, 544-552.

Read, D. J. (1983). The biology of mycorrhiza in the Ericales. *Canadian Journal of Botany*, **61**, 985-1004.

Read, D. J. (1984). The structure and function of the vegetative mycelium of mycorrhizal roots. In *The Ecology and Physiology of the Fungal Mycelium*, eds. D. H. Jennings & A. D. M. Rayner, pp. 215-240. Cambridge University Press, Cambridge, UK.

Reiners, W. A. & Olson, R. K. (1984). Effects of canopy components on throughfall chemistry: an experimental analysis. *Oecologia*, **63**, 320-330.

Richardson, D. H. S. & Nieboer, E. (1980). Surface binding and accumulation of metals in lichens. In *Cellular Interactions in Symbiosis and Parasitism*, eds. C. B. Cook, P. W. Pappas & E. D. Rudolph, pp. 75-94. Ohio State University Press, Columbus, Ohio, USA.

Ried, A. (1960). Stoffwechsel und Verbreitungsgrenzen von Flechten. II. Wasser- und Assimilationshaushalt, Entquellungs- und Submersionsresistenz von Krustenflechten benachbarter Standorte. *Flora, Jena*, **149**, 345-385.

Rowe, J. S. (1984). Lichen woodland in northern Canada. In *Northern Ecology and Resource Management*, eds. R. Olson, R. Hastings & F. Geddes, pp 225-237. University of Alberta Press, Edmonton, Alberta, Canada.

Rowe, J. S. & Acton, D. F. (1985). Taproots of jack pine (*Pinus banksiana* Lamb.) and soil tongues in Saskatchewan. *Canadian Journal of Forest Research*, **15**, 646-650.

Rowell, P., Rai, A. N. & Stewart, W. D. P. (1985). Studies on the nitrogen metabolism of the lichens *Peltigera aphthosa* and *Peltigera canina*. In *Lichen Physiology and Cell Biology*, ed. D. H. Brown, pp. 145-160. Plenum Press, New York.

Scott, M. G. & Larson, D. W. (1984). Comparative morphology and fine structure of a group of *Umbilicaria* lichens. *Canadian Journal of Botany*, **62**, 1947-1964.

Sepponen, P. (1985). The ecological classification of sorted forest soils of varying genesis in northern Finland. *Communicationes Instituti Forestalis Fenniae*, **129**, 1-77.

Seymour, M. D. & Stout, T. (1983). Observations on the chemical composition of rain using short sampling times during a single event. *Atmospheric Environment*, **17**, 1483-1487.

Sheard, J. W. (1986). Distribution of uranium series radionuclides in upland vegetation of northern Saskatchewan. I. Plant and soil concentrations. *Canadian Journal of Botany*, **64**, 2446-2452.

Smith, D., Muscatine, L. & Lewis, D. (1969). Carbohydrate movement from autotrophs to heterotrophs in parasitic and mutualistic symbiosis. *Biological Reviews*, **44**, 17-90.

Smith, D. C. (1962). The biology of lichen thalli. *Biological Reviews*, **37**, 537-570.

Smith, D. C. (1975). Symbiosis and the biology of lichenised fungi. In *Symbiosis*, Symposia of the Society for Experimental Biology, No. 29, pp. 373-405. Cambridge University Press, Cambridge, UK.

Smith, D. C. (1976). A comparison between the lichen symbiosis and other symbioses. In *Lichenology: Progress and Problems*, eds. D. H. Brown, D. L. Hawksworth & R. H. Bailey, pp. 497-513. Academic Press, London.

Smith, D. C. (1980). Mechanisms of nutrient movement between the lichen symbionts. In *Cellular Interactions in Symbiosis and Parasitism*, eds. C. B. Cook, P. W. Pappas & E. D. Rudolph, pp. 197-227. Ohio State University Press, Columbus, Ohio, USA.

Smith, D. C. & Douglas, A. E. (1987). *The Biology of Symbiosis*. Edward Arnold, London.

Smith, D. C. & Molesworth, S. (1973). Lichen Physiology. XIII. Effects of rewetting dry lichens. *New Phytologist*, **72**, 525-533.

Syers, J. K., Birnie, A. C. & Mitchell, B. D. (1967). The calcium oxalate content of some lichens growing on limestone. *Lichenologist*, **3**, 409-414.

Tomassini, F. D., Puckett, K. J., Nieboer, E., Richardson, D. H. S. & Grace, B. (1976). Determination of copper, iron, nickel, and sulphur by X-ray fluorescence in lichens from the Mackenzie Valley, Northwest Territories, and the Sudbury District, Ontario. *Canadian Journal of Botany*, **54**, 1591-1603.

Tuominen, Y. & Jaakkola, T.(1973). Absorption and accumulation of mineral elements and radioactive nuclides. In *The Lichens*, eds. V. Ahmadjian & M. E. Hale, pp. 185-223. Academic Press, London.

Ugolini, F. C., Reanier, R. E., Rau, G. H. & Hedges, J. I. (1981). Pedological, isotopic, and geochemical investigations of the soils at the boreal forest and alpine tundra transition in northern Alaska. *Soil Science*, **131**, 359-374.

Vainstein, E. A. & Ravinskaya, A. P. (1984). Biological degradation of lichen acids in the soil. *Botanicheskii zhurnal*, **69**, 1347-1351.

Yarranton, G. A. (1975). Population growth in *Cladonia stellaris* (Opiz.) Pouz. and Vezda. *New Phytologist*, **75**, 99-110.

Zakshek, E. M., Puckett, K. J. & Percy, K. E. (1986). Lichen sulphur and lead levels in relation to deposition patterns in Eastern Canada. *Water, Air, and Soil Pollution*, **30**, 161-169.

Chapter 13

Role of fungi in nitrogen, phosphorus and sulphur cycling in temperate forest ecosystems

J. Dighton & Lynne Boddy

Institute of Terrestrial Ecology, Merlewood Research Station, Grange-over-Sands, Cumbria LA11 6JU, UK & School of Pure and Applied Biology, University of Wales, College of Cardiff, Cardiff CF2 1XH, UK

Introduction

The continuing flow of energy through any ecosystem depends upon maintenance of nutrient supplies to the primary producers and since there is only a limited input of nutrients from outside, a balanced cycle within the ecosystem is essential. Three main types of cycle (Fig. 13.1) provide nutrients to the primary producer. The *external geological cycle* involves inputs to the ecosystem from the atmosphere (from natural sources and from pollutants), by weathering of soil parent material and from the addition of fertilizers, whereas losses result from leaching, burning and harvesting. The *biological cycle* involves exchange between soil, plants, herbivores and carnivores and the decomposer subsystem. *Internal cycling* occurs within these subsystems and within individual organisms, particularly trees. Nutrient cycling within forests has received detailed treatment by a number of authors, including Ovington (1962), Duvigneaud & Denaeyar de-Smet (1970), Borman & Likens (1979), Heal (1979), Miller (1979), Whittaker *et al.* (1979), Swank & Waid (1980), Clark & Rosswall (1981), Cole & Rapp (1981), Heal, Swift & Anderson (1981), yet the specific role of fungi has received much less attention (but see Lindberg, 1981).

The crucial role of fungi in the cycling of nutrients lies in the fact that fluxes of nitrogen, phosphorus and sulphur within the decomposer subsystem and other subsystems are mediated largely by

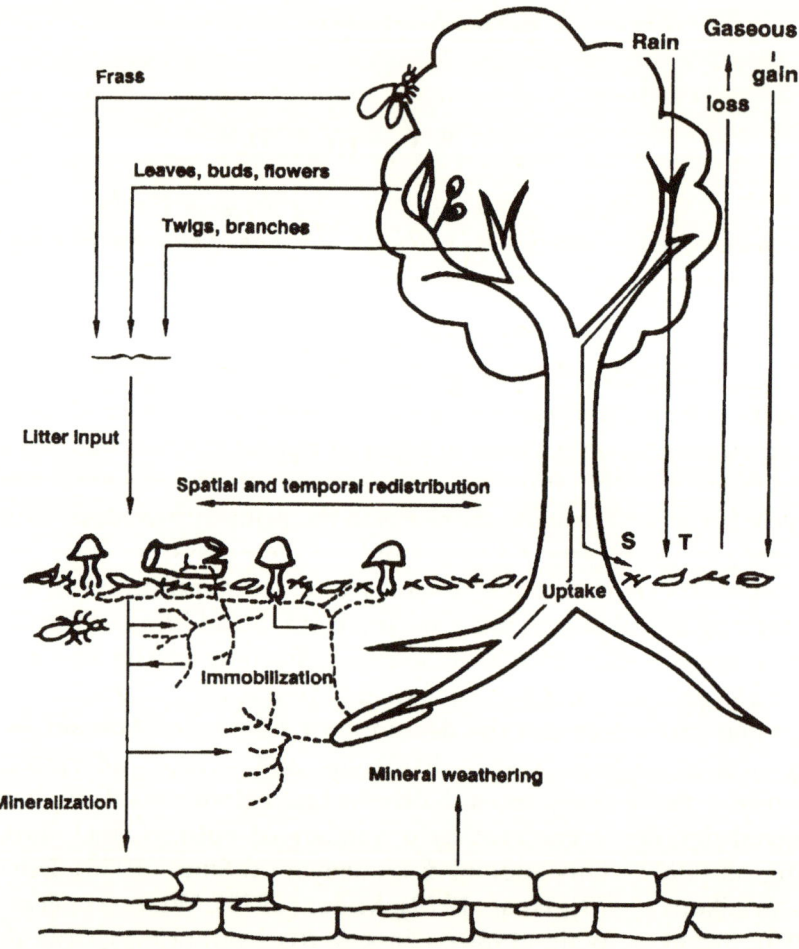

Fig. 13.1. Diagrammatic representation of nutrient inputs, immobilization and mineralization in a forest ecosystem. (S = stem flow; T = throughfall).

them. Fungi, as a result of saprotrophic nutrition, are the major decomposers of organic substrata in terrestrial ecosystems and consequently are responsible for the return of nutrients immobilized in dead plant, animal and microbial tissues to the soil pool. Fungi may also sequester nutrients which have 'leaked' from plants in, for example, the phylloplane and the rhizosphere, and obtain them directly by parasitic interactions. They also directly influence the nutrient relations of autotrophs by the mutualistic association of mycorrhizal fungi and plant roots, and of certain fungi and algae which form lichens. Nutrient fluxes between parasitic fungi and plants are explored by Walters (Chapter 7) and Paul (Chapter 8) and of lichens by Crittenden (Chapter 12), and will not be considered here. The role of mycorrhizas in nitrogen and phosphorous nutrition of plants is treated by Read, Leake & Langdale (Chapter 9), Hetrick (Chapter 10) and Gianinazzi-Pearson & Gianinazzi (Chapter 11), and will only be considered here in terms of their role in saprotrophic decomposition and overall ecosystem nutrient cycling. Emphasis in this chapter will be placed on the role of saprotrophic fungi in nitrogen and phosphorus cycling within temperate forests. Relatively little is known of the involvement of fungi in the sulphur cycle of forest ecosystems but some consideration is given to the possible impact of sulphur depositions arising from atmospheric pollution (see also Wainwright, Chapter 4).

Forest ecosystem development: changes associated with succession

Forests are static neither in space nor time; under suitable climatic conditions, bare ground will be successively covered by herbaceous vegetation and then by woody forms, ultimately forming climax forest communities. The latter are often of uneven age and may be patchy because of localized loss and re-establishment of the canopy. Thus, when considering the role of organisms in nitrogen and phosphorus cycling, it is essential to be aware of the discontinuous and dynamic nature of the ecosystem.

As the seral succession of plant species proceeds, important changes in many aspects of the structure and functioning of the ecosystem occur. Net primary production (NPP) progressively increases and then declines as the climax community becomes established. Associated with the initial increase in NPP is an increase in the total biomass of the system, with an increasing proportion of NPP going into the production of perennial above- and below-ground woody tissues. Associated with the shifts in plant species composition and production of perennial tissues are changes in the quality and quantity of nutrient resources. Thus the standing crop and litter input of herbaceous ground flora decreases, whilst the relative contribution of perennial components increases (Fig. 13.2). This provides increasing and often diversifying opportunities for the decomposer community resulting in a seral succession of the fungal community which will accompany that of the plant community.

Clearly then, dynamics of nitrogen, phosphorus and sulphur cycling processes in an ecosystem will differ according to the successional stage of the system, and changes in the supply of and demand for these elements by the forest (Miller *et al.*, 1979; Miller, 1981; Dighton, 1987; J. Dighton & A. F. Harrison, unpublished) may have significant effects on the physiology of the fungal components of the decomposer community.

Role of saprotrophic fungi in forest nutrient cycling

Input, standing crop and turnover of plant litter and hence nitrogen, phosphorus and sulphur: general considerations

Input and standing crop in temperate forests range, respectively, between 3,700-5,500 kg ha^{-1} yr^{-1} and 11,500-44,500 kg ha^{-1} which represents a large investment of nutrient capital (Table 13.1; Vogt, Grier & Vogt, 1986). Whilst input is largely determined by productivity of green plants, standing crop represents the ratio between input and turnover time, the latter depending on the decomposer organisms, resource quality and microclimate. Above-ground litter comprises various fractions which, in temperate woodlands, are approximately 55% leaves, 10% fruits, buds, flowers, etc., 20% twigs, 10% branches, 5% insect frass and miscellaneous (e.g. Bray & Gorham, 1964; Brown, 1974; Boddy & Swift, 1983). The amount of wood entering the decomposer system has, however, usually been underestimated and a more realistic figure is 40% (Swift, 1977). The input of below-ground

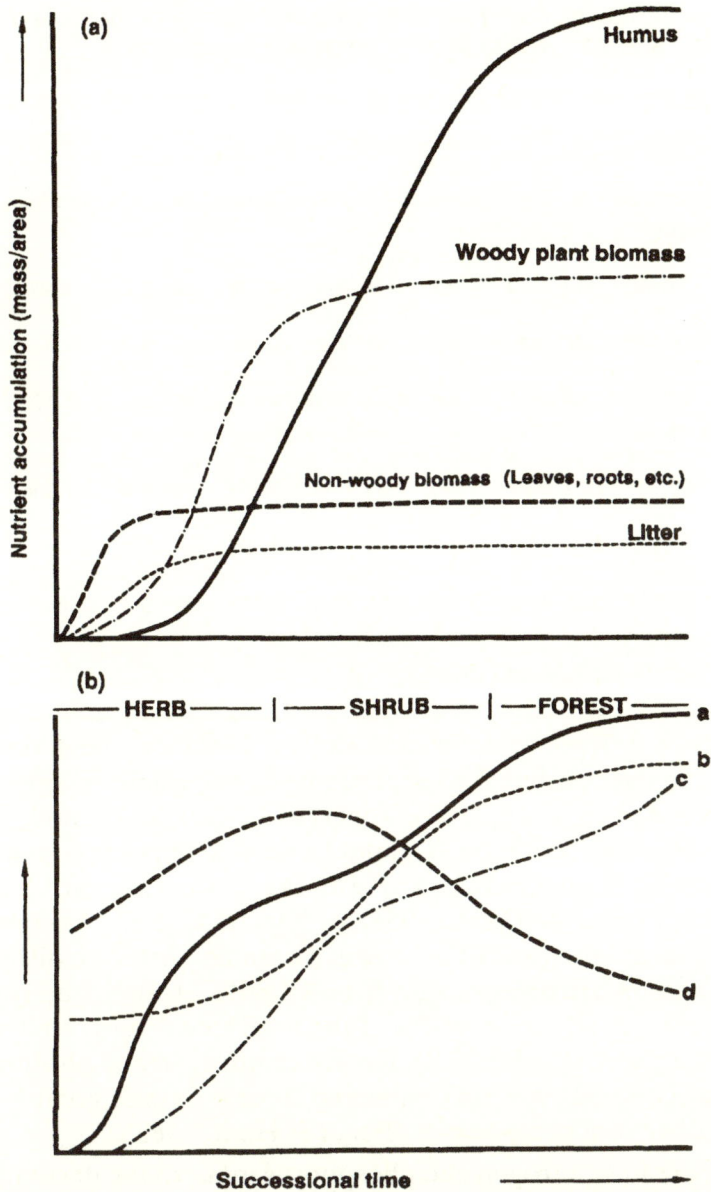

Fig. 13.2. (a) Generalized pattern of changes in nutrient pools during successional time in a forest ecosystem. (b) Hypothetical changes in resource quality and decomposition rates during primary succession in a forest ecosystem. (a = litter-fall mass, b = litter fall C:N, c = decomposition of forest floor organic matter (excluding litter), d = litter decomposition). (After Heal & Dighton, 1986).

Table 13.1. Above ground standing crop and input of nutrients and biomass in woody and non-woody litter in temperate woodlands.

	Forest floor (kg ha^{-1})			Mean residence time on forest floor (yr)			Non-woody litter-fall (kg ha^{-1} yr^{-1})			Woody litter-fall (kg ha^{-1} yr^{-1})		
	biomass	N	P	biomass	N	P	biomass	N	P	biomass	N	P
Warm temperate												
broadleaf deciduous	11,480	163	11·9	2·7	4·8	4·3	4,236	36	3·8	891	2·6	0·8
broadleaf evergreen	19,148	60	3·8	3·1	1·0	2·2	6,484	55	3·7	-	-	-
needleleaf evergreen	20,026	362	25·1	4·6	13·8	11·0	4,432	28	2·7	1,107	2·5	0·2
Cold temperate												
broadleaf deciduous	32,207	624	50·2	10·2	19·1	11·1	3,854	43	4·6	1,046	3·7	0·2
broadleaf evergreen	13,900	200	10·5	3·9	-	-	3,590	-	-	-	-	-
needleleaf evergreen	44,574	504	44·9	17·9	32·8	22·1	3,144	26	3·2	602	1·1	0·1

After Vogt, Grier & Vogt (1986).

---

biomass is difficult to determine and has frequently been omitted from calculations, but this input can be considerable and must not be ignored (Vogt *et al.*, 1986). Vogt *et al.* (1983), for example, have shown that fine root biomass input to the decomposer community is 200 to 300% greater than leaf litter fall in *Abies amabilis* forests. Similarly, Fogel & Hunt (1983) estimated that for Douglas fir (*Pseudotsuga menziesii*), the return of organic matter to soil by fine roots and mycorrhizas ranged from 78-84% of total tree return, which accounted for over 80% of the total tree return of nitrogen and phosphorus (Table 13.2). Return of nitrogen and phosphorus from mycorrhizas was also estimated to be 4-5 times greater than from other root components (Fogel & Hunt, 1983).

The greatest nitrogen and phosphorus input to the decomposer subsystem is from the non-woody components (Table 13.1) which may sometimes contribute over 35 times that of the woody components. Return of nitrogen and phosphorus immobilized in non-woody litter is also rapid compared to woody components: complete decomposition of most deciduous litter takes between 9

Table 13.2. Standing crops and flux of biomass and nutrients into the decomposer system of a Douglas fir stand.

Component	Standing crop (kg ha⁻¹)	Throughput (kg ha⁻¹ yr⁻¹)	Turnover time (yr)	Release and return of nutrients (kg ha⁻¹)[%] N	P
Tree					
foliage	14,732	2,410	6	15·3 [5]	4 [7]
wood	243,396	407	-	3 [1]	2 [3·5]
roots	64,303	≈10,000	1·6	103 [34]	37 [65]
Forest floor	19,034	3,032	6·3	38 [12·5]	2 [3·5]
Fungi					
mycorrhizas	12,794	8,262	1·5		
sporocarps	65	65	1·0	145·5 [47·5]	12 [21]
hyphae	7,035	6,991	1·1		
Stand Total	361,359	31,167		305	57

After Fogel & Hunt (1979, 1983).

---

months and 3 years, depending on species, with coniferous needle litter taking slightly longer, about 7 years, whereas complete decomposition of twigs, branches and small logs often takes 3 to 20 years and whole trees can take as long as 400 years (Rayner & Boddy, 1988). However, woody material, by virtue of this relatively slow decay rate, forms a large standing crop of biomass. For example, Harmon et al. (1986) estimated that the standing crop of large woody components on the floor of deciduous forests was 11 to 38 $\times$ 10^3 kg ha^{-1}, and in coniferous forests was 10 to 511 $\times$ 10^3 kg ha^{-1} which represents a large reserve of nitrogen and phosphorus. As such, it has been suggested that wood may act as a buffer when systems are perturbed (Ausmus, 1977; Swift, 1977).

Contribution of fungi to decomposition processes and hence to turnover of nitrogen and phosphorus

Decomposition of organic substrata results from a variety of interacting biotic and abiotic factors which can be considered as four distinct processes: non-enzymatic chemical reactions, leaching and/or volatilization, comminution and catabolism (Swift, Heal & Anderson, 1979; Boddy, 1986). Non-enzymatic chemical reactions, such as photosensitized oxidation by sunlight, weakening or direct cleavage of chain bonds by ultraviolet light are probably only significant in exposed situations and may, therefore, be unimportant in woodlands.

Leaching of inorganic ions and, to a lesser extent, soluble organic materials, occurs when water passes through and/or over dead organic substrata and living organisms. Leachates may be captured further down the canopy or soil profile or lost from the ecosystem. Some authors have not found leaching to result in significant weight loss of litter, although others have indicated that losses of 10 to 30% may occur (see Swift *et al.*, 1979; Boddy, 1986). Its significance relative to other components of the decomposition process probably depends on climate (particularly rainfall and temperature), soluble components of different substrata, size and extent of decay, and comminution. With regard to nutrients *per se*, losses can be quite large (Berg & Staaf, 1981). For example, in a mixed deciduous woodland in Belgium, 2 and 20% respectively of the total input of nitrogen and phosphorus at litter fall was attributable to leachate (Duvigneaud & Denaeyer de Smet, 1970). Gosz, Likens & Bormann (1973) found that up to 25% of total nitrogen was leached from *Fagus grandifolia* leaf litter in the field. Volatilization results in loss of oils, waxes, and resins, which will be of significance only in plant litters containing large quantities of these materials, e.g. creosote bush, eucalypts, etc. (Whitford *et al.*, 1981).

Comminution is the reduction in particle size of organic substrata which can be brought about by physical phenomena, such as wetting/drying and freezing/thawing cycles and wind abrasion, but in most ecosystems, including forests and woodlands, invertebrates are the main agents (Swift *et al.*, 1979). Amongst other effects, particle size influences susceptibility to leaching and alters the pattern of microbial colonization with small particles tending to select for unicellular as opposed to mycelial forms (Hanlon & Anderson, 1980; Swift & Boddy, 1984).

Catabolism is the component of decomposition processes which is entirely biologically mediated. Invertebrates catabolize both simple and complex organic molecules, either as a result of their own metabolism or in association with microorganisms, but *quan-*

titatively, their contribution is small, accounting for between about 1 to 15% of total heterotrophic metabolism (Edwards, Reichle & Crossley, 1970), although other invertebrate activities are directly or indirectly much more significant to nutrient cycling. Most of the heterotrophic metabolism in forests, therefore, results from microbial activity. For example, Reichle (1977) estimated that over 85% of the living heterotrophic biomass in a temperate *Liriodendron* forest was microbial. Frankland, (1982) estimated that approximately two-thirds of the total microbial biomass of the litter layer and soil of a temperate deciduous woodland was saprotrophic fungi (i.e. not mycorrhizal nor pathogenic) and amounted to approximately 75 kg ha^{-1}. This figure is, however, not necessarily related to activity, since turnover of bacteria is greater than for fungi. Furthermore, the ability to utilize different components of the litter varies between and within fungal and bacterial groups. With regard to the above- and below-ground woody litter, decomposition is brought about almost entirely by fungi, in particular by basidiomycetes and xylariacious ascomycetes, since they have the capacity to break down the lignocellulose complex, but microfungi are also important at early and late stages of decomposition (Rayner & Boddy, 1988).

Immobilization and mineralization of nitrogen and phosphorus by saprotrophic fungi

It is evident then that the heterotrophic microflora of the forest floor are ultimately responsible for releasing large quantities of nitrogen and phosphorus which would otherwise be retained in plant tissues. 'Ultimately' is, however, the operative word, since before these nutrient elements become mineralized (i.e. made available to the soil pool as soluble inorganic forms) they become immobilized in the fungal tissue of the saprotrophic community. The amounts of nutrients immobilized are dependent upon the group of fungi effecting decomposition of a specific substratum, the resource quality of that substratum and the season. Frankland (1982) estimated that some 30-40% of phosphorus and 18% of nitrogen from *Fraxinus* and *Betula* leaf litter was immobilized by *Mycena galopus* hyphae during incubation for six months. She also demonstrated that immobilization into wood decay fungi can be even greater. Stark (1972) showed that hyphae had 193 to 272%

greater nitrogen content and 104 to 223% greater phosphorus content than needle litter in a temperate *Pinus* forest, which indicated immobilization of these elements. Similarly, Bääth & Söderström (1979) showed that 10%, or 3·8 g m^{-2}, of Norway spruce (*Picea abies*) forest floor nitrogen and 20% of the phosphorus was immobilized into the fungal component.

Nitrogen, phosphorus and sulphur are released from fungal tissues on their death, following lysis or autolysis and during their life by excretion, secretion of extracellular enzymes, leaching and grazing by animals. The lifespan of an organism is, therefore, a major determinant of the duration of the immobilization phase. The lifespan of bacteria and ruderal fungi utilizing ephemeral organic compounds may be only a few days, hence they immobilize nutrients for relatively short periods of time. However, these nutrients may be rapidly sequestered and immobilized by other fungi, particularly if the latter caused the death and lysis of the former. At the other extreme, there are many fungi, almost exclusively basidiomycetes and xylariaceous ascomycetes, which utilize recalcitrant woody substrata and whose individual mycelial systems can persist for several years. Obvious examples include the perennial fruit bodies of certain of the Aphyllophorales, rhizomorphs of certain *Armillaria* spp. and mycelial cords of, for example, *Phanerochaete velutina*, *Phallus impudicus*, and *Tricholomopsis platyphylla* (see Thompson & Rayner, 1982; Dowson, Rayner & Boddy, 1988).

Not only are nitrogen and phosphorus immobilized in fungal structures, but they are also probably relocated within the ecosystem over considerable distances via fungi. Thus, systems of mycelial cords can ramify through leaf litter forming connections between woody and other resources, covering areas of over 2 m^2 (Thompson & Rayner, 1982; Dowson *et al.*, 1988), and individuals of the same genotype of *Armillaria bulbosa* and of *Phanerochaete velutina* have been found decaying dead trees over 50 m apart (Thompson & Boddy, 1983). Large fluxes and rapid rates of translocation (sometimes exceeding 25 cm h^{-1}) of carbon and mineral nutrients have been demonstrated from substrata through cord systems in the laboratory, in the comprehensive studies on the dry rot fungus *Serpula lacrimans* (which is not found in nature)(Watkinson, 1971; Brownlee & Jennings, 1982a & b), and in *Armillaria* sp. (Granlund, Jennings & Thompson, 1985), *Phallus impudicus, Mutinus caninus* and *Phanerochaete velutina* (Cairney, 1986; J. M. Wells, L. Boddy & J Dighton, unpublished). Rigorous field studies have, however, not been performed.

The implications are that nitrogen and phosphorus are translocated out of organic substrata by cord systems and into substrata which are being colonized. Indeed, such movement of nutrients is likely to occur in all fungi that are capable of mycelial extension out of the substrata in which they are growing. That nitrogen is imported into organic substrata is borne out by the fact that *absolute* concentrations (i.e. amount of nutrient expressed as a percentage of the original quantity or of the volume of the substratum, rather than as a percentage of weight of organic substratum) within leaf and wood litter often increase at early stages of decomposition, particularly in substrata with initially low nitrogen concentrations (Figs. 13.3 & 13.4; e.g. Aber & Melillo, 1980; Berg & Staff, 1981; Melillo, Aber & Muratore, 1982; Swift & Boddy, 1984). However, rigorous field studies are required to determine whether translocation of phosphorus and sulphur to new resources is significant.

Spore dispersal and subsequent colonization is another means of relocating nutrients. For example, spores of *Fomes applanatus* contain 3·05% nitrogen (by dry weight), and it has been estimated by Merrill & Cowling (1966) that all of the nitrogen in 33·7 kg of *Betula alleghaniensis* is required to supply the annual production of the spores of a typical sporophore. The nitrogen in 5·75 kg of wood would be required to produce the sporophore.

The resource as a determinant of rate of mineralization of nitrogen and phosphorus

It was implied above that the lifespan of fungi may be closely related to the degradability of the resource. Hence release of nutrients to the soil pool may largely be governed by the resource itself. The carbon : nitrogen and lignin : nitrogen ratios are known to be determinants of the resistance of substrata to decomposition, and the ultimate release of nutrients has also been explained in terms of these ratios (e.g. Swift *et al.*, 1979; Melillo *et al.*, 1982; Paustian & Schnurer, 1987). At early stages of decomposition the relative concentration (i.e. expressed as % dry weight) of nitrogen and phosphorus tends to increase as the process proceeds, or expressed another way the carbon : nutrient ratio decreases (Figs. 13.3 & 13.4). This occurs irrespective of whether nutrients are imported, because carbon is lost, as CO_2 in respiration, but nitrogen and phosphorus are incorporated into microbial biomass. It has been suggested (e.g. Swift *et al.*, 1979) that net mineralization does

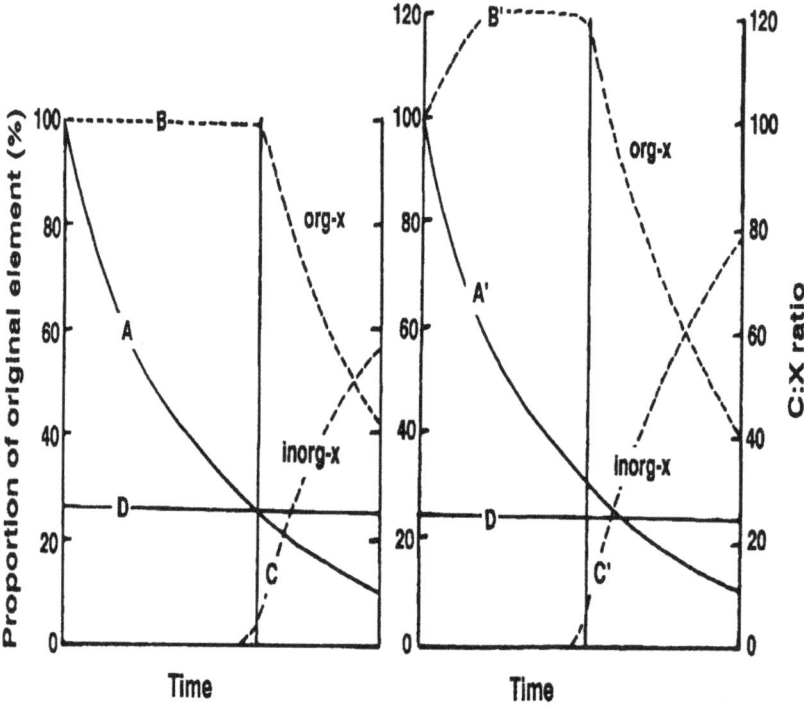

Fig. 13.3. A simple time-course model to illustrate the importance of C:nutrient ratios as indices of the equilibrium between immobilization and mineralization of an element X. Curve A = C:X ratio of an organic resource having an initial ratio of 800:1. Curve B = proportion of X in organic form (immobilised), (Curve B' = case where there is import of X from sources external to the resources being decomposed). Curve C = proportion of X in inorganic form (mineralized), (Curve C' = mineralization in response to alteration of C:X ratio by import of X as in case B'). D = C:X ratio of decomposer organism. After this point net mineralization occurs and curve A flattens off although carbon loss may continue on the same gradient. (From Swift, Heal & Anderson, 1979).

not occur until the carbon : nutrient ratio of the substratum reaches a threshold level which is approximately equivalent to the carbon : nutrient element ratio of the fungus (which is around 15 : 1 for phosphorus and 6 : 1 for nitrogen)(Fig. 13.3; Swift, 1977).

The initial carbon : nitrogen and carbon : phosphorus ratios vary considerably between different litter components. The carbon : nitrogen ratio of undecayed wood typically lies between 350 : 1 and 500 : 1, but can be as high as 1250 : 1, and the carbon : phosphorus ratio is >3500 : 1. In leaf litter these ratios range respec-

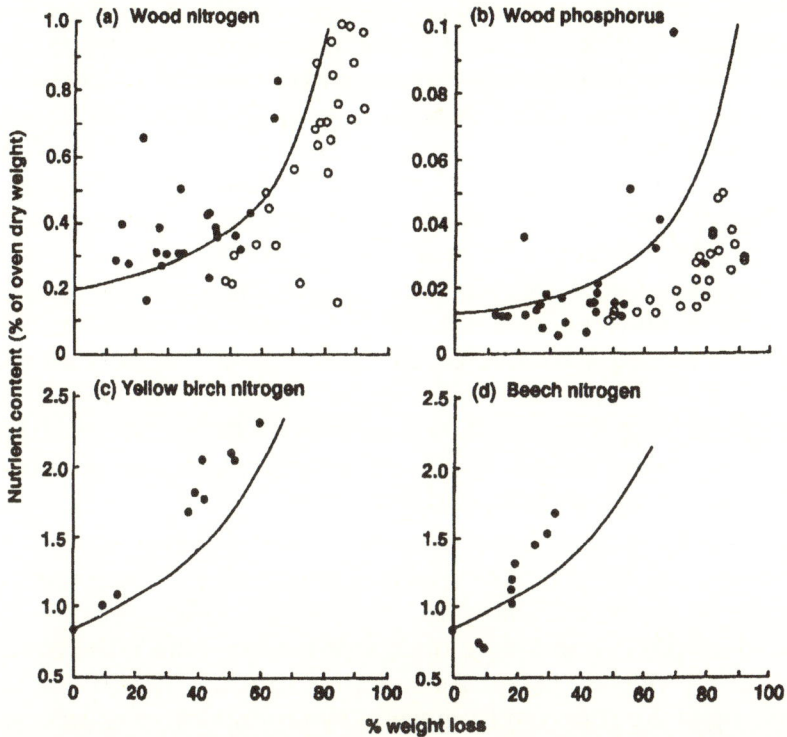

Fig. 13.4. The relationship between mineral nutrient concentration (as % of oven dry weight) and weight loss (% of original dry weight) of wood uninvaded (closed circles) and invaded (open circles) by invertebrates, and of leaf litter (panels c & d). The line on each graph indicates the predicted change in elemental concentration as decomposition proceeds, assuming no net gain or loss. Points above the line therefore represent increase in absolute concentration. [(a) and (b) from Swift & Boddy, 1984; data recalculated from Swift, 1977. Panels (c) and (d) redrawn from Aber & Melillo, 1980; original study by Gosz, Likens & Bormann, 1973].

tively between 25 : 1 to 100 : 1 and 450 : 1 to 1850 : 1 (Cowling, 1970; Swift *et al.*, 1979). Where the ratio is high, as in wood, the decomposer system becomes conservative and with increasing carbon : nutrient ratio there is an increasing trend for the fungi to immobilize nitrogen and phosphorus into their own biomass rather than to effect mineralization. Although the validity of the carbon : nitrogen ratio as a criterion for nitrogen supply is questionable (see Park, 1976; Dowding, 1981; Rayner & Boddy, 1988), it is likely that strong selective pressures act on wood decay fungi for 'economic' use of nitrogen (and phosphorus). As fungi may translocate nutrients it is likely that resources are recycled from 'redundant' hyphae to more active regions. Circumstantial evidence for this possibility is provided by the fact that several wood decay species can utilize their own dead mycelium, and that of other species, as sole source of nitrogen (Levi, Merrill & Cowling, 1968; Jennings, Chapter 1). Cord-forming fungi probably depend largely on 'exploitive' mycelium within the resource to supply the 'exploratory' systems with nitrogen and phosphorus, although it has been suggested that their requirements may be partly met by scavenging nutrients from the soil (Clipson, Cairney & Jennings, 1986). A consequence of efficient internal recycling, at least with cord-forming fungi, is that nitrogen and phosphorus will not necessarily be released to the soil pool when the carbon : nutrient ratio of the wood and the mycelium become similar, but rather will be conserved and translocated elsewhere.

Nitrogen and phosphorus availability may, however, be one factor governing the production of exit structures (i.e. outgrowth of mycelium, cords, rhizomorphs, sporocarps). Although conservation of nutrients is well developed in these fungi this is the part of the system which is 'leaky'. As mentioned earlier, large quantities of nitrogen are removed from wood by production of spores, and since few of these will establish new colonies, much of the nitrogen will be mobilised. Cord systems are also likely to 'leak'. Although no leakage into the soil was detected from major cords of *P. velutina* (L. Boddy & E. M. Owens, unpublished), considerable regression (i.e. dying back) of mycelium has been shown to occur during foraging outgrowth from resources (Dowson, *et al.*, 1986). In particular, mature cords develop following hyphal aggregation within mycelial fronts, with lysis of all hyphae and finely branched cords between the regions of greatest aggregation. Also, when exploratory systems encounter new resources there is often regression of non-connective cords and mycelium which emanate from the 'old' resource. Whilst some of the nutrients may be translocated back into the food base, the remainder will enter the soil pool. Quantitative studies are now necessary before the nutrient dynamics of such systems can be understood.

While carbon : nutrient ratio may partly regulate production of exit structures, at least some species, including cord-forming fungi, produce exit structures from relatively undecayed wood, providing the microclimate is suitable (e.g. Dowson *et al.*, 1986, 1988; Dowson, Boddy & Rayner, 1989). Since some lysis of these reproductive or vegetative structures occurs relatively rapidly, then some mineralization may occur when the carbon : nutrient ratio is still high.

Importance of interactions between soil fauna and fungi in mineralization

The soil fauna is a major determinant of the balance between mineralization and immobilization in organic substrata and microbial biomass, but rather than playing a direct role through catabolism or excretion it influences mineralization by interactions with the microflora (e.g. Anderson & Ineson, 1984). Thus grazing may result directly in release of nutrients as well as altering the relative contribution of different fungal species to the decomposition process (Parkinson, Visser & Whittaker, 1979; Newell, 1984a & b; Swift & Boddy, 1984). At present, however, the functional consequences of alterations in species composition are unclear.

The direct effects of grazing on mineralization and on stimulation/inhibition of fungal respiration and biomass production appear to be variable, probably at least partly because grazing interactions are subject to predator/prey regulation. For example, Visser, Whittaker & Parkinson (1981) found no evidence of enhanced nitrogen mineralization in systems grazed by collembola, and Seastedt & Crossley (1980) suggested that greater immobilization could result from the stimulation of fungal growth by microarthropod grazing. On the other hand there are several examples of increased mobilization of nutrients resulting from animal activity. In a microcosm experiment, Ineson, Leonard & Anderson (1982) found that fungal grazing by the collembolan *Folsomia candida*, on leaf litter, reduced fungal biomass and increased the mineralization of ammonium and nitrate nitrogen (Fig. 13.5). Similarly, consumption of litter (which contains fungi) by larger invertebrates also enhances mineralization. For example, Anderson, Ineson & Huish (1983) investigated the amount of leachable ammonium from microcosms containing oak

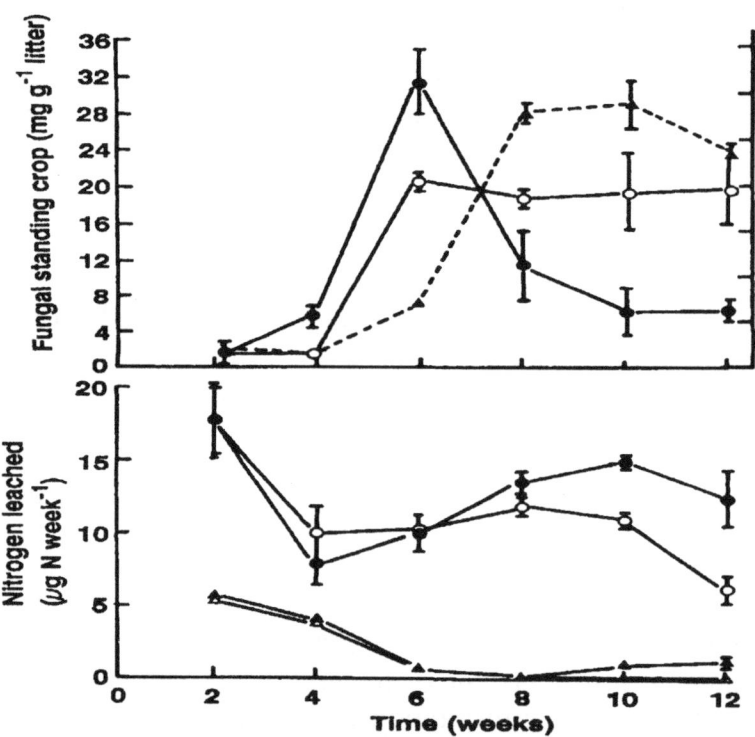

Fig. 13.5. Effect of collembolan feeding activities on fungal standing crop and nitrogen mobilization. (a) fungal standing crop on leaf litter with (closed circles) and without (open circles) collembola, and changes in collembolan populations (triangles); (b) mobilization of nitrogen as ammonium (circles) and nitrate (triangles) in presence (closed symbols) and absence (open symbols) of collembola. Values shown are means ± one SE; n = 4. From Ineson, Leonard & Anderson, 1982.

leaf litter colonised by one of six species of fungi, with and without the millipede *Glomeris marginata*. In all cases mineralization of nitrogen was enhanced in the presence of the millipede, but was strongly correlated with the different abilities of the fungi to effect mineralization. Likewise, enhanced mineralization of sulphur during grazing by the isopod *Oniscus asellus* has been demonstrated by Morgan & Mitchell (1987). Release of nitrogen from wood is strongly correlated with invasion by invertebrates, although appreciable phosphorus appears to occur before animal invasion (Fig. 13.4; Swift, 1977; Swift & Boddy, 1984).

Influence of mycorrhizas on nitrogen and phosphorus dynamics in the soil/plant system

Since much of the assimilable nitrogen and phosphorus in forest ecosystems is immobilized in living and dead plant biomass, the availability of nutrients for further plant growth depends upon recycling within the plant and mineralization by saprotrophic and pathogenic fungi. However, as implied earlier, nitrogen and phosphorus may be immobilized in fungal tissue for considerable periods before becoming available to plants. At early successional stages, nutrients may be relatively more available compared to mature woodlands, especially on poor soils, where the growth of trees is highly dependent on the release of nutrients from organic matter (Miller, 1979; Dighton, 1987; J. Dighton & A. F. Harrison, unpublished). In mature woodland tree roots tend to form associations with ectomycorrhizal fungi (Read, 1986). These not only take up inorganic nitrogen and phosphorus compounds, but can also utilize simple organic molecules (see Read, Leake & Langdale, Chapter 9). More importantly, some mycorrhizal species may be able to act as decomposers and thus short circuit the mineralization/nutrient uptake cycle (Went & Stark, 1968).

There is recent evidence that some ectomycorrhizal fungi have the ability to decompose organic forms of phosphorus and nitrogen by producing extracellular enzymes such as polyphenol oxidases (Giltrap, 1982), phosphatases (Alexander & Hardy, 1981; Ho & Zak, 1979; Dighton, 1983), cellulases (Linkins & Antibus, 1981) and proteinases (Abuzinadah & Read, 1986a & b; Abuzinadah, Finlay & Read, 1986; Dighton, Thomas & Latter, 1987). The ability of ectomycorrhizal fungi to produce degradative enzymes is probably related to the availability of nutrients in the soil. In general, coniferous forests planted on nutrient poor soils increasingly immobilise nutrients as the crop grows. There is a peak of immobilization around canopy closure where nitrogen and phosphorus are not released from organic complexes in the soil at a sufficient rate to supply the tree's demand (Miller, 1979; J. Dighton & A. F. Harrison, unpublished). With an increase in the organic : inorganic ratio of nutrients, the induction of decomposer activity in the mycorrhizal fungal flora would be of importance to the tree crop in the direct procurement of nutrients from organic sources.

It has been suggested by Dighton & Mason (1985) that change in the organic : inorganic ratio of soil nutrients may be one of the forces influencing mycorrhizal fungal succession in that the late stage fungi with 'K-' strategies have the enzymatic capabilities that the 'r-' fungi do not. Supporting evidence comes from the data of Abuzinàdah & Read (1986a) of growth of eight mycorrhizal fungi on protein, where it was the late stage fungi *Suillus bovinus*, *Rhizopogon roseolus* and *Amanita muscaria* which grew better than the early stage fungi *Laccaria laccata* and *Lactarius rufus*. Similarly, in a gnotobiotic experiment, the decomposition of organic substrates (hide powder and cellulose) was greater in the presence of *Suillus luteus* than *Hebeloma crustuliniforme* (Dighton *et al.*, 1987).

Counter to the opinion that mycorrhizal fungi may be involved in the direct release of nutrients from organic sources, Janos (1983) suggested that it is the juxtaposition of mycorrhizal and saprotrophic hyphae which enables mycorrhizas to remove mineralized nitrogen and phosphorus before they can enter the measurable soil nutrient pool. It is important, therefore, to consider the role of mycorrhizas in the decomposition of organic substrata and acquisition of nutrients from these and recalcitrant inorganic forms in relation to the total soil microbial population. These interactions were explored by Gadgil & Gadgil (1971, 1975) who demonstrated that the presence of mycorrhizal roots suppressed decomposition of needle litter in a *Pinus radiata* forest. However, Berg & Lindberg (1980) repeated these experiments but were unable to detect suppression of litter decomposition in the presence of mycorrhizal roots in a *Pinus sylvestris* forest.

Other interactions between mycorrhizas, saprotrophic organisms and soil fauna are also important. For example, where ectomycorrhizal fungi develop together with bacteria in controlled experiments, they synergistically enhance phosphate uptake by the host pines from recalcitrant inorganic and organic forms (Chakly & Berthelin, 1982; Laheurte & Berthelin, 1986). In an interaction study of the ability of mycorrhizal fungi to decompose organic resources in the presence of a saprotrophic fungus, Dighton *et al.* (1987) showed that the saprotroph inhibited the decomposition potential of the mycorrhizal fungus. Recently, Gransaull & Brown (1987) have shown the synergistic effect of adding both mycorrhizas and *Frankia*: these significantly increased the growth of *Pinus* and *Casuarina* species compared with plants with either mycorrhizas or nitrogen fixing actinomycetes alone. However, much work on such interactions is necessary to evaluate their importance in nutrient mineralization and nutrient sequestering by plants.

Effects of exogenous inputs of pollutants on nutrient cycling

Effects of exogenous nutrient inputs on the forest ecosystem fall into three main types: damage to plants; interactions with fungi; and direct nutrient effects. Pollutant gases may cause physical damage to the canopy foliage and premature leaf fall. This causes a temporal shift in the pulse of nutrient input into the soil system and can also affect the resource quality of the litter. These may in turn influence the rate of decomposition and microbial community structure and consequently nutrient cycling. For example, P. A. Wookey (personal communication) has shown that increasing the atmospheric SO_2 concentration in the range of 10 to 50 ppb significantly reduced microbial respiration in both mixed deciduous and pine leaf litter and altered the rate of mineralization of major nutrients. Sulphur dioxide has also been shown to affect differentially the growth of fungi, e.g. at concentrations of 15 to 50 ppb *Trichoderma* spp. increased, whereas both *Cladosporium cladosporioides* and *Coniothyrium olivaceum* decreased. In addition, atmospheric SO_2 can be used by fungi as a source of sulphur, with different species immobilizing sulphur at different rates (Craker & Manning, 1974; Wainwright, Chapter 4). However, these processes are only just beginning to receive attention. Atmospheric pollutants also affect plants, which act as nutrient sinks, by influencing the species composition and physiology of mycorrhizas (Ulrich, Mayer & Khanna, 1979; Hütterman, 1982; Becker, 1982; Stroo & Alexander, 1985; Reich et al., 1985; Jansen, Dighton & Bresser, 1988). For example, 'acid rain' caused soil chemical changes (Dighton & Skeffington, 1987) and acidified stemflow (Kumpfer & Heyser, 1986), which was correlated with a decreased frequency of corralloid mycorrhizas. Although the nutritional consequences of this imbalance have not been studied in detail, this community shift may represent a movement from 'K-' to 'r-' selected fungi and consequently a potential loss in the ability of the mycorrhizal system to sequester nutrients from organic resources in the soil. These effects may be due to mobilization of aluminium and Thompson & Medve (1984) demonstrated that aluminium concentrations of 146 μM could significantly reduce linear growth of agar cultures of the mycorrhizal fungi *Cenococcum*, *Pisolithus*, and *Thelephora*. Also, Oelbe-Farivar (1985) showed that aluminium toxicity reduced protein synthesis in mycorrhizal fungi.

Similarly, there is indirect evidence from the Netherlands that atmospheric nitrogen pollution can significantly affect the succession of mycorrhizal fungi under coniferous trees. By comparing recent fungal fruit body surveys with historical fungal foray records, Arnolds (1985, 1988) has demonstrated a trend of reduction in frequency of occurrence of late stage mycorrhizal fungal fruit bodies, which he attributes to an increase in nitrogen deposition. Also, numbers of *Russula* spp. fruit bodies in beech woods have been found to decline following the application of NPK fertilizer (Hall, 1978). Although care must be taken when interpreting such data (increasing nitrogen content of growth medium *in vitro* typically inhibits fruiting, but not production of mycelial biomass), the effects on plant nitrogen uptake of any such decline in mycorrhizas may be balanced by exogenous inputs of nitrogen which ameliorate conditions on poor soils. If, however, other major nutrients, such as phosphorus, do not increase concomitantly, imbalances could occur, resulting in a reduced yield of the tree crop.

Fungal conversion of inorganic N, P and S compounds

In addition to their saprotrophic abilities, some fungi are able to convert inorganic nitrogen, phosphorus and sulphur compounds from one form to another. Nitrogen exists in a number of inter-convertible inorganic forms:

$$N_2 \leftrightarrow N_2O \leftrightarrow (NO) \leftrightarrow NO_2^- \leftrightarrow NO_3^- \leftrightarrow NH_4^+$$

and there is evidence that certain fungi can perform nitrification (i.e. convert NH_4^+ to NO_3^-) e.g. *Penicillium citrinum, P. nigricans, Cladosporium herbarum, Verticillium lecanii* and several members of the genus *Aspergillus*, of which *A. flavus* has received most attention (e.g. Balasubramanya & Patel, 1980; Killham, 1986; Lang & Jagnow, 1986). Several different pathways of heterotrophic nitrification have been postulated and although it is uncertain which operates, it seems unlikely that energy is gained from the process (Killham, 1986). Thus, the benefits of nitrification are unclear. However, it has been suggested that nitrification may provide competitive advantages as certain of the products are known to be toxins, mutagens and microbial growth factors. Whilst it has been shown that the relevant fungi have the ability to nitrify following inoculation into autoclaved soil (Doxtader & Rovira, 1968), there does not appear to be any direct evidence that fungi play a significant role in nitrification in natural soils (Wainwright, 1988). Bacteria are assumed to be the major agents of nitrification (Keunen & Robertson, 1988), however, in acid coniferous forest soils the microbial biomass is often predominately fungal and fungi are believed to play a role in these situations (Killham, 1986).

Some fungi are also capable of nitrate reduction (i.e. convert NO_3^- to NH_4^+) (see also Jennings, Chapter 1; Tomsett, Chapter 2), e.g. *Aspergillus nidulans* and *Neurospora crassa* possess dissimilatory nitrate reduction pathways. Also, a number of species in the genera *Acremonium, Aspergillus* and *Fusarium* can perform various steps in the denitrification process (i.e. convert NO_3^- to other nitrogen oxides and dinitrogen) (Bleakley & Tiedje, 1982). These fungi cannot necessarily effect all steps of the process, for example, *Fusarium oxysporum* and *F. solani* can reduce nitrite to nitrous oxide at low oxygen tensions but are unable to reduce nitrate (Bollag & Tung, 1972). As with nitrification, fungi are only likely to make a significant contribution to denitrification in acid forest soils (Bleakley & Tiedje, 1982).

It has in the past been suggested that fungi can fix atmospheric dinitrogen, but Millbank (1969) reappraised the evidence and came to the conclusion that this was not so. More recently, work of Ginterova & Gallon (1979) indicated that the wood-rotting fungus *Pleurotus ostreatus* could reduce dinitrogen to ammonia. However, their results are open to several criticisms and the ability to fix dinitrogen remains unproven (Wainwright, 1988). Even if some species can fix dinitrogen it is likely to be quantitatively insignificant, but they may influence fixation in other ways. For example, Jenson & Holm (1975) have demonstrated that a number of soil fungi stimulate bacterial dinitrogen fixation.

Oxidation of inorganic sulphur compounds is analogous to nitrification of ammonia and nitrate, and a number of fungi are capable of oxidizing elemental sulphur and reduced sulphur compounds. The significance of these transformations to the organisms involved and to nutrient cycling remain unclear, but possibilities are considered by Wainwright (Chapter 4).

With regard to inorganic phosphorus, some fungi are able to release phosphate from insoluble inorganic phosphates, for example, species of *Aspergillus*, *Fusarium*, *Penicillium* and *Sclerotium* produce organic acids and/or chelating components which act as solubilising agents (Beever & Burns, 1980; Wainwright, 1981). Some mycorrhizal fungi are also known to have this ability (Gianinazzi-Pearson & Gianinazzi, Chapter 11). There is some evidence that some fungi, e.g. *Aspergillus niger*, *Alternaria* sp., *Chaetomium* sp. and *Penicillium notatum*, but not others, e.g. *Pythium debaryanum* and *Rhizoctonia solani*, can utilize phosphite as a source of phosphorus. The quantitative significance is again unclear but probably small, particularly as phosphites and hypophosphites are found only in small quantities in soils (Beever & Burns, 1980). Condensed phosphates are widespread in soils arising from physicochemical processes, living organisms and some fertilizers and detergents (Beever & Burns, 1980). Many fungi, e.g. *Aspergillus niger*, *Coprinus lagopus* and *Phytophthora parasitica*, are capable of utilizing pyrophosphate as sole sources of phosphorus.

Whether fungi can utilize condensed phosphates containing three or more phosphorus units is unclear (Beever & Burns, 1980).

Conclusions

Fungi are the major agents of organic matter decomposition in temperate forests and as such play a key role in releasing inorganic nutrients through mineralization. They do, however, immobilize nutrients, often for long periods, in their own biomass and the balance between immobilization and release is often very much dependent on interactions with other organisms, e.g. bacteria, fauna and plant roots. Fungi also participate in inorganic conversions and are affected by exogenous inputs. However, there is little quantitative information on the role of fungi in the dynamics of nutrient cycling. It is essential to the management of forest soil as resource for future crops that this situation is remedied.

Acknowledgment. We would like to thank John Wells for comments on a draft of this chapter.

References

Aber, J. D. & Melillo, J. M. (1980). Litter decomposition: measuring relative contributions of organic matter and nitrogen to forest soils. *Canadian Journal of Botany*, **58**, 416-421.

Abuzinadah, R. A. & Read, D. J. (1986a). The role of proteins in the nitrogen nutrition of ectomycorrhizal plants. I. Utilization of peptides and proteins by ectomycorrhizal fungi. *New Phytologist*, **103**, 481-493.

Abuzinadah, R. A. & Read, D. J. (1986b). The role of proteins in the nitrogen nutrition of ectomycorrhizal plants. III. Protein utilization by *Betula*, *Picea* and *Pinus* in mycorrhizal association with *Hebeloma crustuliniforme*. *New Phytologist*, **103**, 507-514.

Abuzinadah, R. A., Finlay, R. D. & Read, D. J. (1986). The role of proteins in the nitrogen nutrition of ectomycorrhizal plants. II. Utilization of protein by mycorrhizal plants *of Pinus contorta*. *New Phytologist*, **103**, 495-506.

Alexander, I. J. & Hardy, K. (1981). Surface phosphatase activity of Sitka spruce mycorrhizas from a serpentine soil. *Soil Biology and Biochemistry*, **13**, 301-303.

Anderson, J.M. & Ineson, P. (1984). Interactions between microorganisms and soil invertebrates in nutrient flux pathways of forest ecosystems. In *Invertebrate Microbial Interactions*, eds. J. M. Anderson, A. D. M. Rayner & D. W. H. Walton, pp. 59-88. Cambridge University Press, Cambridge, UK.

Anderson, J. M., Ineson, P. & Huish, S. A. (1983). Nitrogen and cation mobilization by soil fauna feeding on leaf litter and soil organic matter from deciduous woodlands. *Soil Biology and Biochemistry*, **15**, 463-467.

Arnolds, E. J. M. (ed.) (1985). Veranderingen in de paddestoelenflora. *Wetenschappelijke Mededeling van de Koninklijke Nederlandse Natuurhistorische Vereniging, 167*, Hoogwoud.

Arnolds, E. J. M. (1988). The changing macromycete flora in The Netherlands. *Transactions of the British Mycological Society*, **90**, 391-406.

Ausmus, B. (1977). Regulation of wood decomposition rates by arthropod and annelid populations. In *Soil Organisms as Components of Ecosystems*, Ecological Bulletin 25, eds. U. Lohm & T. Persson, pp. 180-192. Swedish Natural Science Research Council, Stockholm.

Bääth, E. & Söderström, B. (1979). Fungal biomass and fungal immobilization of plants nutrients in Swedish coniferous forest soils. *Revue d'Ecologie et de Biologie du Sol*, **16**, 477-489.

Balasubramanya, R. H. & Patel, R. B. (1980). Heterotrophic nitrification by microorganisms capable of degrading carboxin and oxycarboxin. *Indian Journal of Microbiology*, **20**, 294-297.

Becker, von A. (1982). Aussaatersuch mit Bucheckern im Gewächshaus. In *LOLF - Mitteilungen, Landesanstalt für Ökologie, Landschaftsentwicklung und Forstplanung Nordrhein - Westfalen*, pp. 37-42.

Beever, R. E. & Burns, D. J. W. (1980). Phosphorus uptake, storage and utilization by fungi. *Advances in Botanical Research*, **8**, 128-220.

Berg, B. & Lindberg, T. (1980). Is litter decomposition retarded in the presence of mycorrhizal roots in forest soil? *Swedish Coniferous Project Internal Report, 95*, Uppsala.

Berg, B. & Staff, H. (1981). Leaching, accumulation and release of nitrogen in decomposing forest litter. In *Terrestrial Nitrogen Cycles*, Ecological Bulletin 33, eds. F. E. Clark & T. Rosswall, pp. 163-178. Swedish Natural Science Research Council, Stockholm.

Bleakley, B. H. & Tiedje, J. M. (1982). Nitrous oxide production by organisms other than nitrifiers or denitrifiers. *Applied and Environmental Microbiology*, **44**, 1342-1348.

Boddy, L. (1986). Water and decomposition processes in terrestrial ecosystems. In *Water, Fungi and Plants*, eds. P. G. Ayres & L. Boddy, pp. 375-398, Cambridge University Press, Cambridge, UK.

Boddy, L. & Swift, M. J. (1983). Wood decomposition in an abandoned beech and oak coppiced woodland in S.E. England. I. Patterns of wood-litter fall. *Oikos*, **6**, 320-332.

Bollag, J. M. & Tung, G. (1972). Nitrous oxide release by soil fungi. *Soil Biology and Biochemistry*, **4**, 271-276.

Bormann, F. H. & Likens, G. E. (1979). *Pattern and Process in a Forested Ecosystem*. Springer, New York.

Bray, J. R. & Gorham, E. (1964). Litter production in forests of the world. *Advances in Ecological Research*, **2**, 101-157.

Brown, A. H. F. (1974). Nutrient cycles in oakwood ecosystems in N. W. England. In *The British Oak*, eds. M. G. Morris & F. H. Perring, pp. 141-161, BSBI, Classey, Faringdon.

Brownlee, C. & Jennings, D. H. (1982a). Long distance translocation in *Serpula lacrimans*: velocity estimates and the continuous monitoring of induced perturbation. *Transactions of the British Mycological Society*, **79**, 143-148.

Brownlee, C. & Jennings, D. H. (1982b). The pathway of translocation in *Serpula lacrimans*. *Transactions of the British Mycological Society*, **79**, 401-407.

Cairney, J. W. G. (1986). Basidiomycete linear mycelial structures as nutrient absorbing and translocating organs in soil. Ph.D. thesis, University of Liverpool.

Chakly, M. & Berthelin, J. (1982). Rôle d'une ectomycorhize *Pisolithus tinctorius - Pinus caribea* et d'une bactérie rhizosphérique sur la mobilisation du phosphore de phosphates minéraux et organiques insolubles. In *Les Mycorhizes: Biologie et Utilisation*, eds. S. Gianinazzi, V. Gianinazzi-Pearson & A. Trouvelot, pp. 215-220. INRA, Dijon.

Clark, F. E. & Rosswall, T. (eds.) (1981). *Terrestrial Nitrogen Cycles*, Ecological Bulletin 33. Swedish Natural Science Research Council, Stockholm.

Clipson, N. J. W., Cairney, J. W. G. & Jennings, D. H. (1986). The physiology of basidiomycete linear organs. I. Phosphate uptake by cords and mycelium in the laboratory and the field. *New Phytologist*, 104, 444-458.

Cole, D. W. & Rapp, M. (1981). Elemental cycling in forest ecosystems. In *Dynamic Properties of Forest Ecosystems*, ed. D. E. Reichle, pp. 341-409. Cambridge University Press, Cambridge, UK.

Cowling, E. B. (1970). Nitrogen in forest trees and its role in wood deterioration. *Acta Universitatius Upsalliensis*, 164.

Craker, L. E. & Manning, W. J. (1974). SO_2 uptake by soil fungi. *Environmental Pollution*, 6, 309-311.

Dighton, J. (1983). Phosphatase production by mycorrhizal fungi. *Plant and Soil*, 71, 455-462.

Dighton, J. (1987). Ecology and management of mycorrhizas in the U.K. In *Mycorrhizae in the Next Decade: Practical Applications and Research Priorities*, Proceedings of the 7th North American Conference on Mycorrhizae, eds. D. M. Sylvia, L. L. Hung & J. H. Graham, pp. 75-77. Institute of Food & Agricultural Science, University of Florida, Gainesville, Florida, USA.

Dighton, J. & Mason, P. A. (1985). Mycorrhizal dynamics during forest tree development. In *Developmental Biology of Higher Fungi*, eds. D. Moore, L. A. Casselton, D. A. Wood, & J. C. Frankland, pp. 117-139. Cambridge University Press, Cambridge, UK.

Dighton, J. & Skeffington, R. A. (1987). Effects of artificial acid precipitation on the mycorrhizas of Scots pine seedlings. *New Phytologist*, 107, 191-202.

Dighton, J., Thomas, E. D. & Latter, P. M. (1987). Interactions between tree roots, mycorrhizas, a saprotrophic fungus and the decomposition of organic substrates in a microcosm. *Biology and Fertility of Soils*, 4, 145-150.

Dowding, P. (1981). Nutrient uptake and allocation, during substrate exploitation by fungi. In *The Fungal Community, Its Organization and Role in the Ecosystem*, eds. D. T. Wicklow & G. C. Carroll, pp. 621-635. Marcel Dekker, New York & Basel.

Dowson, C. G., Boddy, L. & Rayner, A. D. M. (1989). Development and extension of mycelial cords in soil at different temperatures and moisture contents. *Mycological Research*, in press.

Dowson, C. G., Rayner, A. D. M. & Boddy, L. (1986). Outgrowth patterns of mycelial cord-forming basidiomycetes from and between woody resource units in soil. *Journal of General Microbiology*, 121, 203-211.

Dowson, C. G., Rayner, A. D. M. & Boddy, L. (1988). Outgrowth patterns of mycelial cord-forming basidiomycetes into woodland soils. II. Resource capture and persistence. *New Phytologist*, 109, 343-349.

Doxtader, K. G. & Rovira, A. D. (1986). Nitrification by *Aspergillus flavus* in sterilized soil. *Australian Journal of Soil Research*, **6**, 141-147.

Duvigneaud, P. & Denaeyer-de-Smet, S. (1970). Biological cycling of minerals in temperate deciduous forests. In *Analysis of Temperate Forest Ecosystems*, ed. D. E. Reichle, pp. 199-225. Springer-Verlag, New York.

Edwards, C. A., Reichle, D. E. & Crossley, D. A. (1970). The role of soil invertebrates in turnover of organic matter and nutrients. In *Ecological Studies. Analysis and Synthesis*, Vol. 1., ed. D. E. Reichle, pp. 147-172. Springer-Verlag, Berlin.

Fogel, R. & Hunt, G. (1979). Fungal and arboreal biomass in a western Oregon Douglas fir ecosystem: distribution patterns and turnover. *Canadian Journal of Forest Research*, **9**, 245-256.

Fogel, R. & Hunt, G. (1983). Contribution of mycorrhizae and soil fungi to nutrient cycling in a Douglas fir ecosystem. *Canadian Journal of Forest Research*, **13**, 219-232.

Frankland, J. C. (1982). Biomass and nutrient cycling by decomposer basidiomycetes, In *Decomposer Basidiomycetes: their Biology and Ecology*, eds. J. C. Frankland, J. N. Hedger & M. J. Swift, pp. 241-261. Cambridge University Press, Cambridge, UK.

Gadgil, R. L. & Gadgil, P. D. (1971). Mycorrhiza and litter decomposition. *Nature*, **233**, 133.

Gadgil, R. L. & Gadgil, P. D. (1975). Suppression of litter decomposition by mycorrhizal roots of *Pinus radiata*. *New Zealand Journal of Forest Science*, **5**, 33-41.

Giltrap, N. J. (1982). Production of polyphenol oxidases by ectomycorrhizal fungi with special reference to *Lactarius* spp. *Transactions of the British Mycological Society*, **78**, 75-81.

Ginterova, A. & Gallon, J. (1979). *Pleurotus ostreatus*: a nitrogen fixing fungus? *Biochemical Society Transactions*, **7**, 1293-1295.

Gosz, J. R., Likens, G. E. & Bormann, F. H. (1973). Nutrient release from decomposing leaf and branch litter in the Hubbard Brook Forest, New Hampshire. *Ecological Monographs*, **43**, 173-191.

Granlund, H. I., Jennings, D. H. & Thompson, W. (1985). Translocation of solutes along rhizomorphs of *Armillaria mellea*. *Transactions of the British Mycological Society*, **84**, 111-119.

Gransaull, R. E. & Brown, R. T. (1987). Relationships among mycorrhizae, nitrogen fixing actinomycetes and host plants. In *Mycorrhizae in the Next Decade, Practical Application and Research Priorities*, Proceedings of the 7th North American Conference on Mycorrhizae, eds. D. M. Sylvia, L. L. Hung & J. H. Graham, p. 21. Institute of Food & Agricultural Science, University of Florida, Gainesville, Florida, USA.

Hall, D. (1978). Experiments with NPK fertilizers in relation to the growth of toadstools in beechwoods. 1. Preliminary study. *Vasculum*, **63**, 25-41.

Hanlon, R. D. G. & Anderson, J. M. (1980). Influence of macroarthropod feeding activities on microflora in decomposing oak leaves. *Soil Biology and Biochemistry*, **12**, 255-261.

Harmon, M. E., Franklin, J. F., Swanson, F. J., Sollins, P., Gregory, S. V., Lattin, J. D., Anderson, N. H., Cline, S. P., Aumen, N. G., Seddell, J. R., Lienkaemper, G. W., Cromack Jr, K. & Cummins, K. W. (1986). Ecology of course woody debris in temperate ecosystems. *Advances in Ecological Research*, **15**, 133-302.

Heal, O. W. (1979). Decomposition and nutrient release in even-aged plantations, In *The Ecology of Even-Aged Plantations*, eds. E. D. Ford, D. C. Malcolm & J. Atterson, pp. 257-291. Institute of Terrestrial Ecology, Cambridge, UK.

Heal, O. W. & Dighton, J. (1986). Nutrient cycling and decomposition in natural terrestrial ecosystems. In *Microfloral and Faunal Interactions in Natural and Agroecosystems*, eds. M. J. Mitchell & J. P. Nakas, pp. 14-73. Martinus Nijhoff/Dr. W. Junk, Dordrecht.

Heal, O. W., Swift, M. J. & Anderson, J. M. (1981). Nitrogen cycling in United Kingdom forests: the relevance of basic ecological research. *Philosophical Transactions of the Royal Society of London, B,* **276**, 427-444.

Ho, I. & Zak, B. (1979). Acid phosphatase activity of six ectomycorrhizal fungi. *Canadian Journal of Botany,* **37**, 1203-1205.

Hütterman, A. (1982). Frühdiagnose von Immisionsschaden in Würzelbereich von Waldbäumen. In *LOLF - Mitteilungen, Landesanstalt für Ökologie, Landschaftsentwicklung und Forstplanung Nordrhein - Westfalen*, pp. 26-31.

Ineson, P., Leonard, M. A. & Anderson, J. M. (1982). Effect of collembolan grazing on nitrogen and cation leaching from decomposing leaf litter. *Soil Biology and Biochemistry,* **14**, 601-605.

Janos, D. P. (1983). Tropical mycorrhizas, nutrient cycles and plant growth. In *Tropical Rain Forest: Ecology and Management*, eds. S. L. Sutton, T. C. Whitmore & A. C. Chadwick, pp. 327-45. Blackwell, Oxford.

Jansen, A. E., Dighton, J. & Bresser, A. H. M. (eds.) (1988). *Ectomycorrhiza and Acid Rain*. CEC Air Pollution Research Report, **12**. Bilthoven, The Netherlands.

Jenson, V. & Holm, E. (1975). Associative growth of nitrogen fixing bacteria with other micro-organisms. In *Nitrogen Fixation by Free-living Micro-organisms*, ed. W. D. P. Stewart, pp. 101-119. Cambridge University Press, Cambridge, UK.

Keunen, J. G. & Robertson, L. A. (1988). Ecology of nitrification and denitrification. In *The Nitrogen and Sulphur Cycles*, eds. J. A. Cole & S. J. Ferguson, pp. 161-218. Cambridge University Press, Cambridge, UK.

Killham, K. (1986). Heterotrophic nitrification. In *Nitrification*, ed. J. I. Prosser, pp. 117-126. IRL Press,

Kumpfer, W. & Heyser, W. (1986). Effects of stemflow on the mycorrhiza of beech (*Fagus sylvaticus* L.). In *Mycorrhizae: Physiology and Genetics*, eds. V. Gianinazzi-Pearson & S. Gianinazzi, pp. 743-750. INRA, Paris.

Levi, M. P., Merrill, W. & Cowling, E. B. (1968). Role of nitrogen in wood deterioration. VI. Mycelial fractions and model nitrogen compounds as substrates for growth of *Polyporus versicolor* and other wood-destroying and wood inhabiting fungi. *Phytopathology,* **58**, 626-634.

Laheurte, F. & Berthelin, J. (1986). Interactions between endomycorrhizas and phosphate solubilizing bacteria: effects on nutrition and growth of maize. In *Mycorrhizae: Physiology and Genetics*, eds. V. Gianinazzi-Pearson & S. Gianinazzi, pp. 339-343. INRA, Paris.

Lang, E. & Jagnow, G. (1986). Fungi of a forest soil nitrifying at low pH values. *FEMS Microbiology Ecology,* **38**, 257-265.

Lindeberg, G. (1981). Role of litter-decomposing and ectomycorrhizal fungi in nitrogen cycling in the Scandinavian coniferous forest ecosystem. In *The Fungal Community: its Organisation and Role in the Ecosystem*, eds. D. T. Wicklow & G. C. Carroll, pp. 653-664. Marcel Dekker, New York.

Linkins, A. E. & Antibus, R. K. (1981). Mycorrhizae of *Salix rotundifolia* in coastal arctic tundra. In *Arctic and Alpine Mycology*, eds. G. A. Laursen & J. F. Ammirati, pp. 509-531. University of Washington Press, Washington.

Melillo, J. M., Aber, J. D. & Muratore, J. F. (1982). Nitrogen and lignin control of hardwood leaf litter decomposition dynamics. *Ecology*, 63, 621-626.

Merrill, W. & Cowling, E. B. (1966). The role of nitrogen in wood deterioration: amount and distribution of nitrogen in fungi. *Phytopathology*, 56, 1085-1090.

Millbank, J. W. (1969). Nitrogen fixation in moulds and yeasts - a reappraisal. *Archiv für Mikrobiologie*, 68, 32-39.

Miller, H. G. (1979). The nutrient budgets of even-aged forests, In *The Ecology of Even-Aged Forest Plantations*, eds. E. D. Ford, D. C. Malcolm & J. Atterson, pp. 221-256. Institute of Terrestrial Ecology, Cambridge.

Miller, H. G. (1981). Forest fertilization: some guiding concepts. *Forestry*, 54, 157-167.

Miller, H. G., Cooper, J. M., Miller, J. D. & Pauline, O. J. C. (1979). Nutrient cycles in pine and their adaptation to poor soils. *Canadian Journal of Forest Research*, 9, 19-26.

Morgan, C. R. & Mitchell, M. J. (1987). The effects of feeding by *Oniscus asellus* on leaf litter sulfur constituents. *Biology and Fertility of Soils*, 3, 107-111.

Newell, K. (1984a). Interaction between two decomposer basidiomycetes and a collembolan under Sitka spruce: distribution, abundance and selective grazing. *Soil Biology and Biochemistry*, 16, 227-233.

Newell, K. (1984b). Interactions between two decomposer basidiomycetes and a collembolan under Sitka spruce: grazing and its potential effects on fungal distribution and litter decomposition. *Soil Biology and Biochemistry*, 16, 235-239.

Oelbe-Farivar, M. (1985). Physiologische Reaktionen von Mykorrhizapilzen auf simulierte saure Bodenkbedingungen. Ph.D. Thesis, University of Göttingen.

Ovington, J. D. (1962). Quantitative ecology and the woodland ecosystem concept. *Advances in Ecological Research*, 1, 103-192.

Park, D. (1976). Carbon and nitrogen levels as factors influencing fungal decomposers, In *The Role of Terrestrial and Aquatic Organisms in Decomposition Processes*, eds. J. M. Anderson & A. Macfadyen, pp. 41-59, Blackwell Scientific Publications, Oxford.

Parkinson, D., Visser, S. & Whittaker, J. B. (1979). Effects of collembolan grazing on fungal colonization of leaf litter. *Soil Biology and Biochemistry*, 11, 529-555.

Paustian, K. & Schnürer, J. (1987). Fungal growth response to carbon to and nitrogen limitation: a theoretical model. *Soil Biology and Biochemistry*, 19, 613-620.

Rayner, A. D. M. & Boddy, L. (1988). *Fungal Decomposition of Wood*. John Wiley, Chichester.

Read, D. J. (1986). Non-nutritional effects of mycorrhizal infection. In *Mycorrhizae: Physiology and Genetics*, eds. V. Gianinazzi-Pearson & S. Gianinazzi, pp. 189-176. INRA, Paris.

Reich, P. B., Schoettle, A. W., Stroo, H. F., Troiano, J. G Amundson, R. G. (1985). Effects of O_3, SO_2 and acidic rain on mycorrhizal infection in northern red oak seedlings. *Canadian Journal of Botany*, 63, 2049-2055.

Reichle, D. E. (1977). The role of invertebrates in nutrient cycling. In *Soil Dynamics as Components of Ecosystems*, Ecological Bulletin 25, eds. U. Lohm & T. Persson, pp. 145-154. Swedish Natural Science Research Council, Stockholm.

Seastedt, T. R. & Crossley, D. A. Jnr. (1980). Effects of microarthropods on seasonal dynamics of nutrients in forest litter. *Soil Biology and Biochemistry*, 12, 237-342.

Stark, N. (1972). Nutrient cycling pathways and litter fungi. *Bioscience*, 22, 355-360.

Stroo, H. F. & Alexander, M. (1985). Effect of simulated acid rain on mycorrhizal infection of *Pinus strobus* L. *Water, Air, Soil Pollution*, 25, 107-114.

Swank, W. T. & Waide, J. B. (1980). Interpretation of nutrient cycling research in a management context: evaluating potential effects of alternative management strategies on site productivity. In *Forests: Fresh Perspectives from Ecosystem Analysis*, ed. R. Wareing, pp. 137-158. Oregon State University, Corvallis.

Swift, M.J. (1977). The role of fungi and animals in the immobilisation and release of nutrient elements from decomposing branch-wood, In *Soil Organisms as Components of Ecosystems*, Ecological Bulletin 25, eds. U. Lohm & T. Persson, pp. 193-202. Swedish Natural Science Research Council, Stockholm.

Swift, M. J. & Boddy, L. (1984). Animal-microbial interactions during wood decomposition. In *Invertebrate-Microbial Interactions*, eds. J. M. Anderson, A. D. M. Rayner & D. W. H. Walton, pp. 89-131. Cambridge University Press, Cambridge, UK.

Swift, M.J., Heal, O. W. & Anderson, J. M. (1979). *Decomposition in Terrestrial Ecosystems*. Blackwell Scientific, Oxford.

Thompson, G. W. & Medve, R. V. (1984). Effects of aluminium and manganese on the growth of ectomycorrhizal fungi. *Applied Environmental Microbiology*, 48, 556-560.

Thompson, W. & Boddy, L. (1983). Decomposition of suppressed oak trees in even-aged plantations. II. Colonisation of tree roots by cord and rhizomorph producing basidiomycetes. *New Phytologist*, 93, 277-291.

Thompson, W. & Rayner, A. D. M. (1982). Structure and development of mycelial cord systems of *Phanerochaete laevis* in soil. *Transactions of the British Mycological Society*, 78, 193-200.

Ullrich, B., Mayer, R. & Khanna, P. H. (1979). Deposition von Luftverunreinigungen und ihre Auswirkungen in Waldöksystemen im Solling. *Schriften aus der Forstlichen Fakultät der Universität Göttingen*, 58, J. D. Sauerlands Verlag, Frankfurt.

Visser, S., Whittaker, J. B. & Parkinson, D. (1981). Effects of collembolan grazing on nutrient release and respiration of a leaf litter inhabiting fungus. *Soil Biology and Biochemistry*, 13, 215-218.

Vogt, K. A., Grier, C. C. & Vogt, D. J. (1986). Production, turnover, and nutrient dynamics of above- and belowground detritus of world forests. *Advances in Ecological Research*, 15, 303-377.

Vogt, K. A., Moore, E. E., Vogt, D. J., Redlin, M. J. & Edmunds, R. L. (1983). Conifer fine root and mycorrhizal root biomass within the forest floors of Douglas fir stands of different ages and site productivities. *Canadian Journal of Forest Research*, 13, 429-437.

Wainwright, M. (1988). Metabolic diversity of fungi in relation to growth and mineral cycling in soil - a review. *Transactions of the British Mycological Society*, 90, 159-170.

Watkinson, S. C. (1971). Phosphorus translocation in stranded and unstranded mycelium of *Serpula lacrimans*. *Transactions of the British Mycological Society*, **57**, 535-
539.

Went, F. W. & Stark, N., (1968). The biological and mechanical role of soil fungi.
Proceedings of the National Academy of Sciences, U.S.A., **60**, 497-504.

Whitford, W. G., Meentemeyer, V., Seastedt, T. R., Cromack, K. Jr., Crossley, D. A.
Jr., Santos, P., Todd, R. L. & Waide, J. B. (1981). Exceptions to the AET model:
deserts and clear-cut forest. *Ecology*, **62**, 275-277.

Whittaker, R. H., Likens, G. E., Bormann, F. H., Eaton, J. S. & Siccama, T. G. (1979).
The Hubbard Brooke ecosystem study: forest nutrient cycling and element behaviour. *Ecology*, **60**, 203-220.

Index

A

For EU product safety concerns, contact us at Calle de José Abascal, 56–1°,
28003 Madrid, Spain or eugpsr@cambridge.org.

www.ingramcontent.com/pod-product-compliance
Ingram Content Group UK Ltd.
Pitfield, Milton Keynes, MK11 3LW, UK
UKHW010853090126
466816UK00011B/215